Zella Allen Dixson

The Comprehensive Subject Index to Universal Prose Fiction

Zella Allen Dixson

The Comprehensive Subject Index to Universal Prose Fiction

ISBN/EAN: 9783744694773

Printed in Europe, USA, Canada, Australia, Japan

Cover: Foto ©Thomas Meinert / pixelio.de

More available books at **www.hansebooks.com**

THE

COMPREHENSIVE

SUBJECT INDEX

TO

UNIVERSAL PROSE FICTION

𝕮𝖔𝖒𝖕𝖎𝖑𝖊𝖉 𝖆𝖓𝖉 𝕬𝖗𝖗𝖆𝖓𝖌𝖊𝖉

BY

ZELLA ALLEN DIXSON, A.M.

ASSOCIATE LIBRARIAN OF THE UNIVERSITY OF CHICAGO

NEW YORK

DODD, MEAD AND COMPANY

1897

𝕬𝖓𝖎𝖛𝖊𝖗𝖘𝖎𝖙𝖞 𝕻𝖗𝖊𝖘𝖘:

JOHN WILSON AND SON, CAMBRIDGE, U.S.A.

TO

MY STUDENTS,

WHEREVER THEY MAY BE SCATTERED ABROAD,

WHOSE NEEDS HAVE BEEN THE INSPIRATION FOR THE WORK,

This Volume

IS AFFECTIONATELY DEDICATED.

PREFACE.

————

THIS work is an arrangement into an alphabetical subject list of works of fiction which are founded on fact, historical, physical, psychological, or moral. Foreign fiction, having a common alphabet, is included, as well as English, with the addition to the lists of the 'English translations whenever they exist. No attempt has been made to give the different translations by their translators, as for the purposes of a subject index one translation is as good as another. It has not been the object of the author to include in this topical arrangement *all* works of fiction, but only novels with a purpose, those which are sent out into the world with a definite lesson to teach mankind. The author reserves the right, even in this narrow limit, to discard many purpose novels where the facts forming the foundation of the stories have been so misrepresented or misstated as to be an injury rather than a help to the reader. Such an index must necessarily include very largely the writers of our own century. The older works of fiction, with a few notable exceptions, deal with intense love affairs, and have no other object than the whiling away of an idle hour. During our own century the novel has come to have a very different sphere. It is no longer sufficient that the work of fiction be an avenue through which the reader may follow the working of an overwrought imagination. Better still, it has ceased to be a boast among educated and intellectual people that they are not novel readers. On the contrary, it is now an established fact that no one can claim to be well educated in the highest sense of the term who is unfamiliar with the best fiction of other literatures.

With an increasing use of fiction has come a demand for subject classification in fiction. Booksellers, librarians, and the reading public generally have felt the need of some arrangement of fiction lists other than by author and historical epochs. It is in the hope of in some degree meeting this demand that the author offers this contribution to bibliography. Most of the novel reading of to-day is disconnected and purposeless. It is believed by scholars of literature who have seen this Index that it will make it possible to use with a purpose even the leisure reading of the books we "take up for an hour" and will permit the accumulation of much useful and instructive information in the form of a sugar-coated pill.

The author has endeavored to give the Index the greatest possible simplicity and directness, in the hope that in such a form it may be able to go into every private home as a family reference librarian, giving to its inmates a wise and satisfactory answer to the much repeated question, "What can I read?" The work does not claim to be an *exhaustive* treatment of the subjects considered, but does give to its readers a *comprehensive* list sufficient for all working purposes, and far better for the average student than a larger and therefore less carefully selected list. The basis for the selection has not been the personal choice of the author, but is the result of practical experience among students using the lists. In every case the request was made of those to whom the references were given that they mark those found most useful to them in the lines marked out. The compilation of such selections carried on for over six years is the Index now offered, not for the use alone of my students and readers, but for all brain workers everywhere.

The author wishes in this connection to acknowledge publicly the assistance rendered to her in this work by the libraries of the British Museum, La Biblioteca Nacional at the City of Mexico, the Congressional Library, Boston Public, Columbia College, and the libraries of Chicago, both in their publications and the courtesy of personal attention.

It is earnestly hoped that inquirers after truth in all departments will find in this volume a great labor-saving device, as well as a wise counsellor and friend.

This is an age of specialization. Life has become so complex that one may no longer aim to know everything, if success is to be attained. " This one thing I do," must be written over the study door. By the help of this Index the student may, even in the hours of leisure and recreation, still follow out the lines of research upon which specialization is based.

ZELLA ALLEN DIXSON.

WISTERIA COTTAGE, GRANVILLE, OHIO,
July, 1897.

SUBJECT INDEX TO FICTION.

ABBEYS.

Combe.

Combe Abbey.   ...Selina Bunbury.

Dorchester.

Brian Fitz-Count : a story of Wallingford Castle and Dorchester Abbey.   ...Augustine D. Crake.

Innisfoyle.

Innisfoyle Abbey.   ...Dennis I. Moriarty.

Margam.

Margam Abbey   ...Anon.

Melrose.

Monastery.   ...Sir Walter Scott.

Nightmare.

Nightmare Abbey.   ...Thomas Love Peacock.

Northanger.

Northanger Abbey.   ...Jane Austen.
Pride and prejudice.   ...Jane Austen.

Reading.

Legend of Reading Abbey.   ...Charles Knight.

Sea-mew.

Sea-mew Abbey.   ...Florence A. James. (Pseud., Florence Warden.)

Tavystoke.

Friar Hildebrand's cross ; or, Monk of Tavystoke Abbaye. ...Margaret A. Paull.

Typhaines.

L'Abbaye de Typhaines.   ...Joseph Arthur Comte de Gobineau.
Typhaines : a tale of the twelfth century.   ...Joseph Arthur Comte de Gobineau.   English translation.

ABBEYS — *Continued.*

Westminster.

Westminster Abbey.   ...Joseph Addison.
Westminster Cloisters.   ...M. Bidder.

ACADIA.

Description, History, Manners, and Customs.

Constance of Acadia.   ...Edward Payson Tenney.
Lady of Fort Saint John.   ...Mary Hartwell Catherwood.
Lily and the cross : a tale of Acadia.   ...James De Mille.
Rivals of Acadia.   ...Harriet V. Cheney.
Stories of the land of Evangeline.   ...Grace D. MacLeod.
See also **CANADA.**

ACTING AND ACTORS.   See **THEATRICAL LIFE.**

ADVENTURES.   *A Selection.*

Adventures in India.   ...William Henry Giles Kingston.
Adventures in the far West.   ...William Henry Giles Kingston.
Adventures of a casket.   ...Justin J. E. Roy.   English translation.
Adventures of a younger son.   ...John Edward Trelawny.
Adventures of Ann : stories of colonial times.   ...Mary E. Wilkins.
Adventures of Arthur O'Leary.   ...Charles James Lever.   (Pseud ,
    Cornelius O'Dowd ; Harry Lorrequer.)
Adventures of Brigadier Gerard.   ...Arthur Conan Doyle.
Adventures of Captain Horn.   ...Frank Richard Stockton.
Adventures of Captain Mago ; or, Phœnician expedition B. C. 1000.
    ...Léon Cahun.   English translation.
Adventures of Harry Marline ; or, Notes from an American mid-
    shipman's lucky bag.   ...David D. Porter.
Adventures of Huckleberry Finn.   ...Samuel Langhorne Clemens.
    (Pseud., Mark Twain.)
Adventures of Rob Roy.   ...James Grant.
Adventures of Sherlock Holmes.   ...Arthur Conan Doyle.
Adventures of three Englishmen and three Russians in South
    Africa.   ...Jules Verne.   English translation.
Antony Waymouth ; or, Gentlemen adventurers.   ...William Henry
    Giles Kingston.
As we sweep through the deep : a story of the stirring times of old.
    ...William Gordon Stables.
Les aventures de Capitaine Magon ; ou, Une exploration phé-
    nicienne mille ans avant l'ère chrétienne.   ...Léon Cahun.
Les aventures de Saturrin Fichet.   ...M. F. Soulié.
Les aventures de 3 Russes et de 3 Anglais.   ...Jules Verne.
Les aventures d'une cassette.   ...Justin J. E. Roy.
La bannière bleue, aventures d'un Musulman, d'un Chrétien, et

ADVENTURES — *Continued.*

Gun-runner.   ...Bertram Mitford.

Hair-breadth escapes of three boys in South Africa.   ...Henry C. Adams.

Heart of the world.   ...H. Rider Haggard.

Hector Servadac.   ...Jules Verne.

Hector Servadac.   ...Jules Verne.   English translation.

Hendricks the hunter: a tale of Zululand.   ...William Henry Giles Kingston.

History of Don Quixote de la Mancha.   ...Miguel de Cervantes Saavedra.   English translation.

El ingenioso hidalgo Don Quixote de la Mancha.   ...Miguel de Cervantes Saavedra.

In the Eastern seas.   ...William Henry Giles Kingston.

In the heart of the Rockies: a story of adventure in Colorado. ...George Alfred Henty.

Iron pirate: a plain tale of strange happenings on the sea.   ...Max Pemberton.

Jack Hinton, the guardsman.   ...Charles James Lever.   (Pseud., Cornelius O'Dowd; Harry Lorrequer.)

Jacob Faithful.   ...Frederick Marryat.

Jacobite exile: adventures in the service of Charles XII. of Sweden. ...George Alfred Henty.

James Braithwaite, the supercargo. ...William Henry Giles Kingston.

John Dean: historic adventures by land and sea.   ...William Henry Giles Kingston.

Kéraban-le-têtu.   ...Jules Verne.

Keraban the inflexible.   ...Jules Verne.   English translation.

King's own.   ...Frederick Marryat.

Knight of Gwynne.   Charles James Lever.   (Pseud., Cornelius O'Dowd; Harry Lorrequer.)

Life, adventures, and piracies of the famous Captain Singleton. ...Daniel Defoe.

Life and adventures of Major Roger Sherman Potter.   ...Francis C. Adams.   (Pseud., Pheleg Van Trusedale.)

Life and strange adventures of Robinson Crusoe, of York, mariner. ...Daniel Defoe.

List, ye landmen.   ...William Clark Russell.

Lost canyon of the Toltecs: an account of strange adventures in Central America.   ...Charles Sumner Seeley.

Manco: the Peruvian chief.   ...William Henry Giles Kingston.

Mark Seaworth: a tale of the Indian Ocean.   ...William Henry Giles Kingston.

Masterman Ready; or, Wreck of the Pacific.   ...Frederick Marryat.

Maurice; or, Red jar.   ...Frances Villiers, Countess of Jersey.

ADVENTURES — *Continued.*

Maurice Tiernay, the soldier of fortune.   ...Charles James Lever. (Pseud., Cornelius O'Dowd ; Harry Lorrequer.)

Memoirs of Sherlock Holmes.   ...Arthur Conan Doyle.

Micah Clarke : his statement as made to his three grandchildren. ...Arthur Conan Doyle.

Mission ; or, Scenes in Africa.   ...Frederick Marryat.

Mr. Midshipman Easy.   ...Frederick Marryat.

Modern buccaneer.   ...Thomas A. Brown. (Pseud., Rolf Boldrewood.)

My Boer chum.   ...J. Gordon.

My Comrades.   ...Howard Hinton.

My first voyage to Southern seas.   ...William Henry Giles Kingston.

Narrative of Prince Charlie's escape.   ...John Macdonald.

Narrative of the travels and adventures of Monsieur Violet, in California, Sonora, and Western Texas.   ...Frederick Marryat.

Nevermore.   ...Thomas A. Brown.   (Pseud., Rolf Boldrewood.)

New voyage around the world.   ...Daniel Defoe.

Orchid seekers.   ...Ashmore Russan and Frederick Boyle.

Oscar in Africa.   ...Charles A. Fosdick. (Pseud., Harry Castlemon.)

Our home in the silver west : a story of struggle and adventure   ...William Gordon Stables.

Out of bounds.   ...A. Garry.

Pacha of many tales.   ...Frederick Marryat.

Paddy Finn ; or, Adventures of a midshipman afloat and ashore. ...William Henry Giles Kingston.

Pampas : a story of adventure in the Argentine Republic.   ...Robert Hope Moncrieff. (Pseud., Ascott R. Hope.)

Peter Simple.   ...Frederick Marryat.

Pirate and Moonshine.   ...Frederick Marryat.

Prisoner of Zenda, being the history of three months in the life of an English gentleman.   ...Anthony Hope Hawkins. (Pseud., Anthony Hope.)

Privateer's-man one hundred years ago.   ...Frederick Marryat.

Queer race : the story of a strange people.   ...William Westall.

Rattlin the reefer.   ...Frederick Marryat.

Red cockade.   ...Stanley J. Weyman.

Robbery under arms.   ...Thomas A. Brown.   (Pseud., Rolf Boldrewood.)

Ronald Morton ; or, Fireships.   ...William Henry Giles Kingston.

Sidney-side Saxon.   ...Thomas A. Brown. (Pseud., Rolf Boldrewood.)

Silas Verney : being the story of his adventures in the days of King Charles II.   ...Edgar Pickering.

Simple adventures of a memsahib.   ...Sarah Jeanette Duncan.

Snarlyow ; or, Dog fiend.   ...Frederick Marryat.

Sport in Ashanti ; or, Melinda the Caboceer.   ...J. A. Skertchly.

ADVENTURES — *Continued.*

Story of a trooper. ...Francis C. Adams. (Pseud., Pheleg Van Trusedale.)

Story of the filibusters. ...James Jeffrey Roche.

Strange adventures of a house boat. ...William Black.

Strange adventures of a phaeton. ...William Black.

True-blue ; or, Life and adventure of a British seaman of the old school. ...William Henry Giles Kingston.

Twice lost : a story of shipwreck and of adventure in the wilds of Australia. ...William Henry Giles Kingston.

Two supercargoes ; or, Adventures in savage Africa. ...William Henry Giles Kingston.

Up and down the Nile ; or, Young adventurers in Africa. ...William Taylor Adams. (Pseud., Oliver Optic.)

Up the Tapajos ; or, Adventures in Brazil. ...Edward S. Ellis.

Westward Ho ! or, Voyages and adventures of Sir Amyas Leigh. ...Charles Kingsley.

Wild tales of the North. ...Anne Bowman.

With Maceo in Cuba : adventures of a Minnesota boy. ...Franc R. E. Woodward.

Wonderful adventures by land and sea of the seven queer travellers who met at an inn. ...Josiah Barnes.

Wood of the brambles. ...Frank Mathew.

Young Franc-Tireurs and their adventures in the Franco-Prussian war. ...George Alfred Henty.

See also **DETECTIVE STORIES ; IMAGINARY CITIES, LANDS, AND INSTITUTIONS ; SOLDIERS** and **SEA STORIES.**

AFRICA.

Description, History, Manners, and Customs.

Adventures of Herbert Massey in Eastern Africa. ...Verney Lovett Cameron.

Adventures of three Englishmen and three Russians in South Africa. ...Jules Verne. English translation.

Aventures de 3 Russes et de 3 Anglais. ...Jules Verne.

Black ivory : a tale of adventure among the slaves of East Africa. ...Robert Michael Ballantyne.

Blue lights ; or, Hot work in the Soudan. ...Robert Michael Ballantyne.

By sheer pluck : a tale of the Ashanti war. ...George Alfred Henty.

Callista : a sketch of the third century. ...John Henry Newman.

Captain Enderis, First West African Regiment. ...Arthur P. Crouch.

Cinq semaines en ballon. ...Jules Verne.

Daireen. ...F. Frankfort Moore.

Drowned gold. ...Robert H. Newell. (Pseud., O. C. Kerr.)

AFRICA — *Continued.*
Dream life and real life : a little African story.   ...Olive Schreiner. (Pseud., Ralph Iron.)
Five weeks in a balloon.   ...Jules Verne.   English translation.
Flaming sword, being an account of the extraordinary adventures and discoveries of Dr. Percival.   ...Dr. Percival.
For honour, not honours : being the story of Gordon of Khartoum. ...George Gordon Stabbles.
Fossicker.   ...Ernest Glanville.
Gentleman digger.   ...Anna de Bremont.
Ghamba.   ...W. C. Scully.
Gorilla hunters.   ...Robert Michael Ballantyne.
Gregorio.   ...Percy Hemingway.
Gun-runner.   ...Bertram Mitford.
Hair-breadth escapes of three boys in South Africa.   ...Henry C. Adams.
Hendricks the hunter : a tale of Zululand.   ...William Henry Giles Kingston.
King Remba's point.   ...John Lander.
King's Assegai.   ...Bertram Mitford.
Long odds.   ...H. Rider Haggard.
Lost heiress.   ...Ernest Glanville.
Luck of Gerard Ridgley.   ...Bertram Mitford.
Mary Musgrave.   ...Anon.
Mission ; or, Scenes in Africa.   ...Frederick Marryat.
Mixed humanity.   ...J. R. Cowper.
Modern Telemachus.   ...Charlotte Mary Yonge.
My Kalulu, prince, king, and slave.   ...Henry Moreland Stanley.
Mystery of Sasassa Valley.   ...Arthur Conan Doyle.
My Boer chum.   ...J. Gordon.
Off to the wilds : the adventures of two brothers.   ...George Manville Fenn.
Oscar in Africa.   ...Charles A. Fosdick. (Pseud., Harry Castlemon.)
Le pays des diamants.   ...Jules Verne.
Pirate city.   ...Robert Michael Ballantyne.
Playing with fire : a story of the Soudan war.   ...James Grant.
Quickening of Caliban.   ...J. Compton Rickett.
Red Sultan.   ...John Maclaren Corban.
Romance of N'Shabe.   ...A. A. Anderson and A. Wall.
Rose of Paradise : adventures in connection with the pirate Edward England in 1720.   ...Howard Pyle.
Scapegoat.   ...Thomas Henry Hall-Caine.
Settler and the savage.   ...Robert Michael Ballantyne.
Six months at the Cape.   ...Robert Michael Ballantyne.
Story of an African farm.   ...Olive Schreiner. (Pseud., Ralph Iron.)

AFRICA — *Continued.*

Trooper Peter Halket of Mashonaland. ...Olive Schreiner. (Pseud., Ralph Iron.)

Two supercargoes; or, Adventures in savage Africa. ...William Henry Giles Kingston.

Virgil. ...Charles Montague.

Ula in veldt and laager. ...Charles H. Eden.

Vanished diamonds. ...Jules Verne. English translation.

War of the axe; or, Adventures in South Africa. ...John Percy Groves.

Young Carthaginian. ...George Alfred Henty.

Young colonists. ...George Alfred Henty.

AGNOSTICISM.

Agnostic island. ...F. J. Gould.

Calmire. ...Anon.

Gerard. ...Mary E. Braddon.

New continent. ...Mrs. Worthey.

ALABAMA.

At love's extremes. ...Maurice Thompson.

Flush times of Alabama and Mississippi. ...Joseph G. Baldwin.

Richard Hurdis : a tale of Alabama. ...William Gilmore Simms.

ALASKA.

Description, History, Manners, and Customs.

Boy explorers; or, Adventure of two boys in Alaska. ...Harry Prentice.

Fur-seal's tooth : a story of Alaskan adventure. ...Kirk Munroe.

Kin-da-shon's wife : an Alaskan story. ...Mrs. Eugene S. Willard.

Lorita : an Alaskan maiden. ...Susan G. Clark.

Red Mountains of Alaska. ...Willis Boyd Allen.

ALEXANDRIA, EGYPT.

Description, History, Manners, and Customs.

Beyond recall. ...Adeline Sergeant.

Chapter of adventures; or, Through the bombardment of Alexandria. ...George Alfred Henty.

Hypatia ; or, New friends with old faces. ...Charles Kingsley.

Mr. Potter of Texas. ...Archibald Clavering Gunter.

Per aspera. ...Georg Moritz Ebers.

Per aspera : a thorny path. ...Georg Moritz Ebers. English translation.

Der Scheik von Alessandria und seine Sklaven. ...Wilhelm Hauff.

Serapis. ...Georg Moritz Ebers.

Serapis. ...Georg Moritz Ebers. English translation.

Sheik of Alexandria and his slaves. ...Wilhelm Hauff. English translation.

ALFRED THE GREAT. *Reigned 871–901.*
Chronicle of Ethelfled. ...Anne Manning.
Danes in England. ...Alfred H. Engelbach.
Dragon and the raven; or, Days of King Alfred. ...George Alfred
Henty.
Glastonbury. ...A. Payne.
Sea kings of England. ...Edwin Atherstone.
See also **ENGLISH HISTORY.**

ALLEGORIES.
Fables. ...Æsop. English translation.
Jungle Book. ...Rudyard Kipling.
Memorable voyages of Rebel and Victory. ...A. B. King.
Out of the past. ...E. Anson More.
Peter Schlemihl's wunderbare Geschichte. ...Adelbert von Chamisso.
Pilgrim's progress. ...John Bunyan.
Shĕ : a history of adventure. ...H. Rider Haggard.
Der weiss König. ...Maximilian I. Arranged by Maria Treitzsaurwein.
Wonderful history of Peter Schlemihl. ...Adelbert von Chamisso.
English translation.

ALTRUISM.
Altruria. ...Titus K. Smith.
Experiment in Altruism. ...Margaret Pollock Sherwood. (**Pseud.,**
Elizabeth Hastings.)
Traveller from Altruria. ...William Dean Howells.
See also **FORECASTS.**

AMERICA. *A Selection.*
Description, History, Manners, and Customs.
Central America.
Captain Brand of the Centipede. ...Henry A. Wise. (Pseud.,
Harry Gringo.)
Damsel of Darien. ...William Gilmore Simms.
Darien; or, Merchant prince. ...E. B. G. Warburton.
In tropic seas : a tale of the Spanish main. ...William Westall.
Last days of Tul. ...Metta V. Victor. (Pseud., Seeley Regester.)
Lost canyon of the Toltecs : an account of strange adventures in
Central America. ...Charles Sumner Seeley.
Montezuma's gold mines. ...Frederick Albion Ober.
Phantom city : a volcanic romance. ...William Westall.
Sir Henry Morgan. ...E. Howard.
Under Drake's flag : a tale of the Spanish main. ...George Alfred
Henty.
Westward Ho ! or, Voyages and adventures of Sir Amyas Leigh.
...Charles Kingsley.
Years ago : domestic life in the eighteenth century. ...Hannah Lynch.

AMERICA — *Continued.*

North America.

Abdalla the Moor and the Spanish knight.   ...Robert Montgomery Bird.

Adventures of a young naturalist.   ...Lucien Biart.   English translation.

Agnes Surriage.   ...Edwin L. Bynner.

Algonquin maiden.   ...G. M. Adam and A. E. Wertherald.

Les anciens Canadiens.   ...Philippe Aubert de Gaspé.

Arthur Mervyn; or, Memoirs of the year 1793.   ...Charles Brockden Brown.

Aventures d'un jeune naturaliste.   ...Lucien Biart.

Aztec treasure house.   ...Thomas Allibone Janvier.

Bastonnais.   ...John Lesperance.

Bénito Vazquez.   ...Lucien Biart.

Bernard Lile.   ...Jeremiah Clemens.

Black hollow; or, Dragoon's bride: a tale of the Ramapo in 1779. ...N. C. Iron.

Braddock: a story of the French and Indian wars. ...John R. Musick.

Brandon: a tale of the American colonies.   ...Osgood Tiffany.

Bride of Fort Edward.   ...Delia S. Bacon.

British partisan.   ...M. E. Davis.

Brother soldiers.   ...Mary S. Robinson.

Burton; or, Sieges.   ...Joseph Holt Ingraham.

Buttonwoods; or, Refugees of the revolution.   ...Anon.

By right of conquest; or, With Cortez in Mexico.   ...George Alfred Henty.

Calavar; or, Knight of the conquest.   ...Robert Montgomery Bird.

Cameron Hall.   ...Mary A. Cruse.

Camp Charlotte: a tale of 1774.   ...Anon.

Camp-fires of the revolution.   ...Henry C. Watson.

Canadians of old.   ...Philippe Aubert de Gaspé.   English translation.

Le capitaine Paul.   ...Alexandre Dumas.

Captain Molly.   ...Ellen T. H. Putnam.   (Pseud., Thrace Talmon.)

Captain Paul.   ...Alexandre Dumas.   English translation.

Century too soon: a story of Bacon's rebellion.   ...John R. Musick.

Chainbearer.   ...James Fenimore Cooper.

Charles Morton.   ...Mary S. B. D. Shindler.

Chauncey Judd; or, Stolen boy: a tale of the revolution.   ...Israel P. Warren.

"Le Chien d'or": a legend of Quebec.   ...William Kirby.

Chinampa; or, Island home.   ...Anon.

Claudius the cowboy of Ramapo valley.   ...P. Demarest Johnson.

Clayton's rangers.   ...E. H. Williamson.

Humbled pride: a story of the Mexican War. ...John R. Musick.

Hurricane Harry. ...William Henry Giles Kingston.

Independence: a story of the American revolution. ...John R. Musick.

Inez: a tale of the Alamo. ...Augusta J. Wilson.

Infidel; or, Fall of Mexico. ...Robert Montgomery Bird.

L'Intendant Bigot, 1758. ...Joseph Marmette.

In the valley. ...Harold Frederick.

Iron furnace. ...J. H. Anghey.

Isidra. ...Willis Steell.

Israel Potter: his fifty years of exile. ...Herman Melville.

Izram. ...Charlotte Elizabeth Tonna. (Pseud., Charlotte Elizabeth.)

Jacques et Marie. ...N. Bourassa.

John Gray: a Kentucky tale of the olden time. ...James Lane Allen.

John March, Southerner. ...George Washington Cable.

Johnson Manor: a tale of olden time in New York. ...James Kent.

La jongleuse. ...H. Raymond Casgrain.

Karmel, the scout. ...Sylvanus Cobb, Jr.

Kate Aylesford. ...Charles J. Peterson.

Katherine Walton. ...William Gilmore Simms.

Kin-da-shon's wife. ...Mrs. Eugene S. Willard.

King's warrant. ...A. H. Engelback.

Lady of Fort Saint John. ...Marý Hartwell Catherwood.

Last of the foresters. ...John Esten Cooke.

Last of the Mohicans. ...James Fenimore Cooper.

Leather stockings and silk; or, Hunter John Myers and his times: a story of the Valley of Virginia. ...John Esten Cooke.

Lebensbilder aus der westlichen Hemisphäre. ...Charles Sealfield (formerly Karl Postl).

Legends of the West. ...James Hall.

Life in the new world; or, Sketches of American society. ...Charles Sealfield (formerly Karl Postl). English translation.

Lily and the cross. ...James De Mille.

Lionel Lincoln; or, Leaguer of Boston. ...James Fenimore Cooper.

Lone star of Texas. ...J. W. Dallam.

Lord of himself. ...Francis H. Underwood.

Lucia Dare. ...Sarah A. Dorsey. (Pseud., Filia.)

Les Machabées de la Nouvelle France. ...Joseph Marmette.

Maid of Esopus. ...N. C. Iron.

Malmiztic the Toltec, and the cavaliers of the cross. ...W. W. Fosdick.

Man without a country. ...Edward Everett Hale.

Marcus Blair. ...Caleb E. Wright.

AMERICA — *Continued.*

Margaret Moncrieff, the first love of Aaron Burr.   ...Charles Burdett.

Marrying by lot: a tale of the primitive Moravians.   ...Charlotte B. Mortimer.

Mehetabel.   ...Mrs. H. C. Gardner.

Mellichampe : a legend of the Santee.   ...William Gilmore Simms.

Mercedes of Castile ; or, Voyage to Cathay.   ...James Fenimore Cooper.

Meredith ; or, Mystery of the Meschianza.   ...Anon.

Merton.   ...Eliza Ann Dupuy.

Merry-Mount : a romance of the Massachusetts colony.   ...John Lothrop Motley.

Mexican ; or, Love and land.   ...J. M. Dagnall.

Mexican prince ; or, Story of Montezuma.   ...Anon.

Might not right ; or, Stories of the discovery and conquest of America.   ...Anon.

Miss MacRéa ; Roman historique.   ...Michael René Hilliard d'Auberteuil.

Miss Ravenal's conversion from secession to loyalty.   ...John W. De Forest.

Mohun ; or, Last days of Lee and his paladins.   ...John Esten Cooke.

Monarchist.   ...John B. Jones.

Monody on Major Andrè.   ...Anna Seward.

Montezuma, the last of the Aztecs.   ...E. Maturin.

Montezuma, the serf.   ...Joseph Holt Ingraham.

More than she could bear.   ...H. Bendbow.

Mustang Gray.   ...Jeremiah Clemens.

My comrades.   ...Howard Hinton.

Nameless nobleman.   ...Jane Goodwin Austin.

Nazarene.   ...George Lippard.

Near to nature's heart.   ..Edward Payson Roe.

Nick of the woods ; or, Jibbenainosay : a tale of Kentucky.   ...Robert Montgomery Bird.

El nigromántico Mejicano.   ...I. M. Pusalgas.

Norsemen in the West ; or, America before Columbus.   ...Robert Michael Ballantyne.

Oath of Marion.   ...Charles J. Peterson.

Old bell of independence ; or, Philadelphia in 1776.   ...Henry C. Watson.

Old Dominion ; or, Southampton massacre.   ...George Payne Rainsford James.

Old Fort Duquesne ; a tale of the early toils, struggles, and adventures of the first settlers at the forks of the Ohio in 1754.   ...Anon.

AMERICA — *Continued.*

AMERICA — *Continued.*

Yorktown : an historical romance.   ...Anon.
Young rebels; a story of the battle of Lexington.   ...Robert Hope
Moncrieff. (Pseud., Ascott R. Hope.)
See also individual states and the different wars of the United States.

South America.

Un Acontecimiento en Tucuman.   ...J. O. Bustamente.
Admirable Lady Biddy Fane.   ...Frank Barrett.
Afloat in the forest.   ...Mayne Reid.
L'Araucan.   ...Gustave Aimard.
Dolores : ein Charaktergemälde aus Süd-Amerika.   ...Harro Paul
Harring.
Dolores : a historical novel of South America.   ...Harro Paul Har-
ring. English translation.
Les Espagnols au Pérou; ou, La mort de Rolla.   ...August Frie-
drich Ferdinand von Kotzebue.
Forest exiles ; or, Perils amid the wilds of the Amazon.   ...Mayne
Reid.
Frank Redcliffe : a story of travel and adventure in the forests of
Venezuela.   ...Achilles Daunt.
Golden magnet ; the Land of the Incas.   ...George Manville Fenn.
History of Oroonoko ; or, Royal slave.   ...Aphra Behn.
Hoffnungen in Peru.   ...Ernest, Baron von Bibra.
Inca queen ; or, Lost in Peru.   ...John Evelyn.
Les Incas ; ou, La destruction de l'empire du Pérou.   ...Jean Fran-
çois Marmontel.
Incas ; or, Destruction of the empire of Peru.   ...Jean François
Marmontel. English translation.
In New Granada.   ...William Henry Giles Kingston.
Jilma ó continuacion de Los Pizarros.   ...F. Perez.
Ein Juwel : südamerikanischer Roman.   ...Ernest, Baron von Bibra.
Letters written by a Peruvian princess.   ...F. H. de Graffigny.
English translation.
Lettres d'une Peruvienne.   ...F. H. de Graffigny.
Lost Inca.   ...P. Ozollo.
Lost in the wilds.   ...Edward S. Ellis.
Lucia Miranda.   ...Celestina Fúnes.
Lucia Miranda.   .. Eduardo Mansilla de Garcia.
Lutchmee and Dilloo.   ...Edward Jenkins.
Manco the Peruvian chief.   ...William Henry Giles Kingston.
Man of mark.   ...Anthony Hope Hawkins. (Pseud., Anthony Hope.)
Maria : a South American romance.   ...Jorge Isaacs. English trans-
lation.
María : Novela Americana.   ...Jorge Isaacs.

AMERICA — *Continued.*

Une mission au Paraguay. ...Élie Berthet.

Near the Lagunas; or, Scenes in the states of La Plata. ...Anon.

On the banks of the Amazon. ...William Henry Giles Kingston.

Out on the Pampas; or, Young settlers. ...George Alfred Henty.

Pampas; or, Story of adventure in the Argentine Republic. ...Robert Hope Moncrieff. (Pseud., Ascott R. Hope.)

Pizarro; or, Death of Rolla. ...August Friedrich Ferdinand von Kotzebue. English translation.

Los Pizarros. ...F. Perez.

Ponce de Leon; or, Rise of the Argentine Republic. ...Anon.

Der Rodeo in Chile. ...Ernest, Baron von Bibra.

Der Schatz des Inka. ...Franz Hoffmann.

Secret of the Andes: a romance. ...Friedrich Hassaurek.

La Serera, legenda historica. ...Enrique E. Rivaroli.

Die südamerikanische Reise des Doctor H. ...Ernest, Baron von Bibra.

Einige Tage in Rio Janeiro. ...Ernest, Baron von Bibra.

Théodore; ou, Les Peruviens. ...G. L. A. Pigault Lebrun.

Treasures of the Inca. ...Franz Hoffmann. English translation.

Under the Southern cross. ...Deborah Alcock.

Unter der Penchuenchen. ...Friedrich Gerstäcker.

Up the Tapojoos; or, Adventures in Brazil. ...Edward S. Ellis.

Voice of Urbano: romance of adventure on the Amazons. ...James William Wells.

Young Llanero. ...William Henry Giles Kingston.

ANARCHISM, FENIANISM, and NIHILISM.

Career of a Nihilist. ...Sergius Stepniak, Pseud. Translated from the Russian.

Condemned as a Nihilist: a story of escape from Siberia. ...George Alfred Henty.

Crime of the 'Liza Jane. ...Fergus W. Hume.

Dynamitards. ...Reginald Tayler.

Fair Saxon. ...Justin McCarthy.

Female Nihilist. ...E. Lavigna.

Fitzgerald the Fenian. ...J. D. Maginn.

Green book; or, Freedom under the snow. ...Mor Jökai. English translation.

Hartmann the Anarchist. ...E. Douglas Fawcett.

Light and shade. ...Charlotte O'Brien.

Marzio's crucifix. ...Francis Marion Crawford.

Narka, the Nihilist. ...Katharine O'Meara.

Nest of nobles. ...Ivan Sergyevich Turgenieff. Translated from the Russian.

ANARCHISM, FENIANISM, and NIHILISM — *Continued.*

Nihilist princess.   ...Louise Gageneur.
Our radicals.   ...Frederick G. Burnaby.
Out of the jaws of death.   ...Frank Barrett.
Paris.   ...Émile Zola.   English translation.
Prince of Balkistan.   ...Allen Upward.
Red route.   ...William Sime.
Ridgeway.   ...Scian Dubl, Pseud.
Rudine.   ...Ivan Sergyevich Turgenieff.   Translated from the Russian.
Storm light; or, Nihilist's doom.   ...J. E. Muddock.
Le ventre de Paris.   ...Émile Zola.
Vital question.   ...Nikolai Tchernishevsky.   Translated from the Russian.
When we were boys.   ...William O'Brien.
See also **LABOR AND CAPITAL, CONFLICT OF.**

ANCIENT HISTORY.   *A Selection.*

Twentieth Century Before Christ.

Hebrew heroes.   ...Charlotte Tucker.   (Pseud., A. L. O. E.)
Patriarchal times.   ...C. M. O'Keefe.
Sarchedon: a legend of the great Queen.   ...George John Whyte Melville.

Eighteenth Century Before Christ.

José et Benjamin.   ...Adolf Franz Delitzsch.
Joseph and Benjamin.   ...Adolf Franz Delitzsch.   English translation.

Fifteenth Century Before Christ.

Ephraim and Helah: a story of the Exodus.   ...Edwin Hodder.
Joshua: a story of biblical times.   ...Georg Moritz Ebers.   English translation.
Josua.   ...Georg Moritz Ebers.
Pillar of fire; or, Israel in bondage.   ...Joseph Holt Ingraham.
Rameses.   ...Edward Upham.
Rescued from Egypt.   ...Charlotte Tucker.   (Pseud., A. L. O. E.)
Spell of Ashtaroth.   ...Duffield Osborne.

Fourteenth Century Before Christ.

Uarda: Roman aus dem alten Aegypten.   ...Georg Moritz Ebers.
Uarda: a romance of ancient Egypt.   ...Georg Moritz Ebers.   English translation.

Thirteenth Century Before Christ.

Triumph over Midian.   ...Charlotte Tucker.   (Pseud., A. L. O. E.)

ANCIENT HISTORY — *Continued.*

Twelfth Century Before Christ.

Samson.   ...S. W. Odell.

Eleventh Century Before Christ.

Shepherd of Bethlehem.   ...Charlotte Tucker.   (Pseud., A. L. O. E.)

Tenth Century Before Christ.

Adventures of Captain Mago ; or, Phœnician expedition B. C. 1000.
...Léon Cahun.   English translation.
Les aventures de Capitaine Magon ; ou, Une exploration Phéni-
cienne mille ans avant l'ère Chrétienne.   ...Léon Cahun.
Leah.   ...Mrs. Alexander S. Orr.

Eighth Century Before Christ.

Numa Pompilius.   ...Jean Pierre Claris de Florian.

Seventh Century Before Christ.

Exiles in Babylon.   ...Charlotte Tucker.   (Pseud., A. L. O. E.)

Sixth Century Before Christ.

Belteshazzar : a romance of Babylon.   ...Edward R. Roe.
Egyptian princess.   ...Georg Moritz Ebers.   English translation.
Eine aegyptische Königstochter.   ...Georg Moritz Ebers.

Fifth Century Before Christ.

Fountain of Arethusa.   ...Robert Landor.
Pausanias, the Spartan.   ...Edward George Earle Lytton Bulwer-
Lytton.
Philothea.   ...Lydia Maria Child.
Prince of Argolis.   ...J. Moyr Smith.

Fourth Century Before Christ.

Apelles and his contemporaries.   ...Henry Greenough.
Heroes of Ancient Greece.   ...Ellen Palmer.
Voyage du jeune Anacharsis en Grèce, vers le milieu de quatrième
siècle avant Jésus Christ.   ...Jean Jacques Barthélemy.
Young Anacharsis.   ...Jean Jacques Barthélemy.   English trans-
lation.

Second Century Before Christ.

Charicles ; or, Illustrations of the private life of the ancient Greeks.
...Wilhelm Adolf Becker.   English translation.
Charikles.   ...Wilhelm Becker.
Emperor.   ...Georg Moritz Ebers.   English translation.

ANCIENT HISTORY — *Continued.*

Helon's pilgrimage to Jerusalem : a picture of Judaism in the century which preceded the advent of Our Saviour.   ...Friedrich Abraham Strauss.   English translation.

Helon's Wallfahrt nach Jerusalem hundert neun Jahr vor der Geburt unsers Herrn.   ...Friedrich Abraham Strauss.

Der Kaiser.   ...Georg Moritz Ebers.

Die Schwestern.   ...Georg Moritz Ebers.

Sisters.   ...Georg Moritz Ebers.   English translation.   .

First Century Before Christ.

Fawn of Sertorius.   ...Robert Landor.

Gallus : historischer Roman.   ...Wilhelm Adolf Becker.

Gallus ; or, Roman scenes in the time of Augustus.   ...Wilhelm Adolf Becker.   English translation.

Idumean.   ...J. M. Leavitt.

Zipporah.   ...Mrs. M. E. Bewsher.

First Century Anno Domini.

Agathocles.   ...Caroline Pichler.

Ahasvérus.   ...Edgar Quinet.

Am Kreuz : ein Passions roman aus Oberammergau.   ...Wilhelmine von Hillern.

Antipas, son of Chuza, and others whom Jesus loved.   ...Louisa Seymour Houghton.

Arius, the Libyan.   ...Nathan C. Kouns.

Aurelia ; or, Jews of Capernagate.   ...Abel Quinton.

Barabbas.   ...Marie Corelli.

Ben Hur : a tale of the Christ.   ...Lew Wallace.

Beric, the Breton.   ...George Alfred Henty.

Burning of Rome ; or, Story of the days of Nero.   ...Alfred John Church.

Daybreak in Britain.   ...Charlotte Tucker.   (Pseud., A. L. O. E.)

Dion and Sybilis.   ...M. G. Keon.

Doom of the holy city.   ...Lydia Hoyt Farmer.

Dorcas, the daughter of Faustina.   ...Nathan C. Kouns.

Early Christianity.   ...Charlotte Tucker.   (Pseud., A. L. O. E.)

Early dawn ; or, Sketches of Christian life in England in the olden times.   ...Elizabeth Charles.

Edol, the Druid.   ...William Henry Giles Kingston.

Emmanuel : story of the Messiah.   ...William Forbes Cooley.

Gaudentius.   ...Gerald S. Davies.

Gladiators : a tale of Rome and Judæa.   ...George John Whyte Melville.

Glaucia.   ...Emma Leslie.

ANCIENT HISTORY — *Continued.*
Golden age. ...Kenneth Grahame.
Helena's household : a tale of Rome in the first century. ...James De Mille.
Julia of Baioe. ...J. W. Brown.
Julian. ...William Ware.
Lapsed but not lost. ...Elizabeth Charles.
Light from the catacombs. ...E. L. M.
Naomi ; or, Last days of Jerusalem. ...Mrs. J. B. Peploe (Webb).
Neither Rome nor Judæa. ...E. Hoven.
Onesimus : memoirs of a disciple of Saint Paul. ...Edwin A. Abbott.
On the cross. ...Wilhelmina von Hillern. English translation.
Philip ; or, What may have been : a story of the first century. ...E. M. Cuttin.
Philochristus. ...Anon.
Pomponia ; or, Gospel in Cæsar's household. ...Mrs. J. B. Peploe (Webb).
Prince of the house of David ; or, Three years in the holy city : scenes in the life of Jesus. ...Joseph Holt Ingraham.
Quiet King. ...Caroline Atwater Mason.
Quo Vadis. ...Henryk Sienkiewicz.
Quo Vadis : a narrative of the time of Nero. ...Henryk Sienkiewicz. English translation.
Roman traitor. ...Henry W. Herbert. (Pseud., Frank Forester.)
Saint Paul in Greece. ...Gerald S. Davies.
Salathiel, the immortal. ...George Croly.
Story of the other wise man. ...Henry Van Dyke.
Triumphs of the cross. ...John Mason Neale. (Pseud., Aurelius Gratianus.)
Valeria. ...Ballydear and Bowden.
Valerius : a Roman story. ...John Gibson Lockhart.
Victory of the vanquished. ...Elizabeth Charles.
Zillah. ...Horace Smith. (Pseud., Paul Chatfield.)

Second Century Anno Domini.

Flavia. ...Emma Leslie.
Letters from Rome. ...Eustace Wace.
Roman exile. ...Guglielmo Gajani.
Three Bernices. ...Amanda M. Bright.
Valerius : a Roman story. ...John Gibson Lockhart.

Third Century Anno Domini.

Aurelian. ...William Ware.
Callistra : a sketch of the third century. ...John Henry Newman.

ANCIENT HISTORY — *Continued.*

Diotima : eine culturhistorische Novelle aus der Zeit der diocleti-anischen Verfolgung.   ...Wilhelm Tangermann.

Epicurean.   ...Thomas Moore.

Farm of Aptonga : a story for children of the times of Saint Cyprian.   ...John Mason Neale.   (Pseud., Aurelius Gratianus.)

Letters from Palmyra.   ...William Ware.

Martyrs of Carthage.   ...Mrs. J. B. Peploe (Webb).

Money God ; or, Empire and the papacy.   ...Abel Quinton.

Probus ; or, Rome in the third century.   ...William Ware.

Theban Legion : a tale of the times of Diocletian.   ...William M. Blackburn.

Fourth Century Anno Domini.

Alexandrians.   ...Anon.

Claudius.   ...Mrs. R. K. Causton.

Constantine.   ...Edmund Spencer.

Daughters of Pola.   ...Anon.

Dorcas, the daughter of Faustina.   ...Nathan C. Kouns.

Egyptian wanderers.   ...John Mason Neale.   (Pseud., Aurelius Gratianus.)

Evanus : a tale of the days of Constantine the Great.   ...Augustine D. Crake.

Homo sum.   ...Georg Moritz Ebers.

Homo sum.   ...Georg Moritz Ebers.   English translation.

Julian's dream.   ...Gerald S. Davies.

Last Athenian.   ...Abraham Victor Rydberg.   English translation.

Norma.   ...Ellen Palmer.

Out of the mouth of the lions.   ...Anon.

Parthenia ; or, Last days of Paganism.   ...Eliza Buckminster Lee.

Pearl of Antioch : a picture of the East at the end of the fourth century.   ...Marc A. Bayle.

Serapis : historischer Roman.   ...Georg Moritz Ebers.

Serapis.   ...Georg Moritz Ebers.   English translation.

Den siste Athenaren.   ...Abraham Victor Rydberg.

Fifth Century Anno Domini.

Alypius of Tagaste : a tale of the early Church.   ...Mrs. J. B Peploe (Webb).

Antonina ; or, Fall of Rome.   ...Wilkie Collins.

Attila.   ...George Payne Rainsford James.

Conquering and to conquer : a story of Rome in the days of Saint Jerome.   ...Elizabeth Charles.

Eudoxia, die Kaiserinn.   ...Ida Hahn-Hahn.

Fabiola ; or, Church of the Catacombs.   ...Nicholas Patrick Stephen Wiseman.

ANCIENT HISTORY — *Continued.*

Hypatia. ...Charles Kingsley.

Julemerk. ...Mrs. J. B. Peploe (Webb).

Kathleen : a tale of the fifth century. ...E. A.

Maid and Cleon. ...Elizabeth Charles.

Pearl of Antioch: a picture of the East at end of fourth century. ...Marc A. Bayle.

Quadratus : a tale of the world in the Church. ...Emma Leslie.

For the chronological continuation, see **EUROPE, HISTORY OF MEDIEVAL AND MODERN.**

ANNE BOLEYN. *Lived 1507–1536.*

Anne Boleyn; or, Suppression of the religious houses. ...Anon.

Anne Boleyn. ...Mrs. K. Thompson.

Bride of Bucklersbury : a tale of the Grocers' Company. ...Elizabeth M. Stewart.

Star of the court ; or, Maid of honour and the Queen of England, Anne Boleyn. ...Selina Bunbury.

Windsor Castle. ...William Harrison Ainsworth.

See also **HENRY VIII. OF ENGLAND.**

ANNE OF AUSTRIA. See **LOUIS XIV.**

ANNE OF ENGLAND. *Reigned 1702–1714.*

Cornet of the horse: a tale of Marlborough's wars. ...George Alfred Henty.

Devereux. ...Edward George Earle Lytton Bulwer-Lytton.

Esther Vanhomrich. ...Margaret L. Woods.

History of Henry Esmond. ...William Makepeace Thackeray.

Irish pearl : a tale of the times of Queen Anne. ...Anon.

John Law, the projector. ...William Harrison Ainsworth.

Maiden's lodge : a tale of the time of Queen Anne. ...Emily Sarah Holt.

Queen's jewel. ...M. P. Blyth.

Saint James's; or, Court of Queen Anne. ...William Harrison Ainsworth.

ARABIA.

Description, History, Manners, and Customs.

Amurath, Prince of Persia : an Arabian tale. ...Anon.

Arabian nights entertainment. ...Anon.

Beyond recall. ...Adeline Sergeant.

Frank Hilton ; or, Queen's own. ...James Grant.

From bondage to freedom. ...Anon.

Harun al Rascid. ...Anne Manning.

History of the Caliph Vathek. ...William Beckford.

ARABIA — *Continued.*

Homo sum. ...Georg Moritz Ebers.
Homo sum. ...Georg Moritz Ebers. English translation.
Khaled : a tale of Arabia. ...Francis Marion Crawford.
Lance of Kanana : a story of Arabia. ...Henry W. French.
Shaving a Shagpat: an Arabian entertainment and Farina. ...George Meredith.
Vathek : an Arabian tale. ...William Beckford.

ARCHÆOLOGY.

Aztec treasure house : a romance of contemporaneous antiquity. ...Thomas A. Janvier.
Last days of Tul. ...Metta V. Victor. (Pseud., Seeley Regester.)
Lost canyon of the Toltecs: an account of strange adventures in Central America. ...Charles Sumner Seeley.
Man who married the moon ; and other Pueblo Indian folk-stories. ...Charles F. Lummis.
Montezuma's daughter. ...H. Rider Haggard.
Story of Ab. ...Stanley Waterloo.
See also **FOLK-LORE.**

ARKANSAS.

Description, History, Manners, and Customs.

Colonel Thorpe's scenes in Arkansas. ...Anon.
Life and adventure of an Arkansaw doctor. ...M. Lafayette Byrn.
Trappers in Arkansas. ...Gustave Aimard. English translation.
Les trappeurs de l'Arkansas. ...Gustave Aimard.
We all. ...Alice French. (Pseud., Octave Thanet.)

ARMADA. See **SPANISH HISTORY.**

ARMINIUS. See **HERMANN.**

ARMY AND NAVY LIFE. *A Selection.*

Adventures of Brigadier Gerard. ...Arthur Conan Doyle.
Army society : life in a garrison town. ...Henrietta E. V. Stannard. (Pseud., John Strange Winter.)
Away westward. ...Frederick A. Whittaker.
Between the lines: a story of the war. ...Charles King.
Boy life in the United States Navy. ...H. H. Clark.
Bubble reputation. ...Katharine King.
Cadet button. ...Frederick A. Whittaker.
Cadet days. ...Charles King.
Cadet life at West Point. ...George C. Stronge.
Le Capitaine Paul. ...Alexandre Dumas.
Captain Paul. ...Alexandre Dumas. English translation.
Cavalry life ; or, Sketches and stories in barracks and out. ...Henrietta E. V. Stannard. (Pseud., John Strange Winter.)

ARMY AND NAVY LIFE — *Continued.*

Mignon's husband. ...Henrietta E. V. Stannard. (Pseud., John Strange Winter.)

Military mosaics : a set of tales and sketches on soldierly themes. ...John A. O'Shea.

Miss Bagg's secretary : a West Point romance. ...Clara Louise Burnham.

Night watch. ...Anon.

On the offensive : an army story. ...George Israel Putnam.

Our sailors ; or, Anecdotes of the engagements and gallant deeds of the British Navy during the reign of Queen Victoria. ...William Henry Giles Kingston.

Ouzel galley. ...William Henry Giles Kingston.

Paddy Finn ; or, Adventures of a midshipman afloat and ashore. ...William Henry Giles Kingston.

Plain tales from the hills. ...Rudyard Kipling.

Pluck. ...Henrietta E. V. Stannard. (Pseud., John Strange Winter.)

Privateer : a tale of the nineteenth century. ...Cecil P. Stone.

Queen's cadets. ...James Grant.

Second to none : a military romance. ...James Grant.

Servitude et grandeur militaires. ...Alfred Victor, Comte de Vigny.

Shoulder straps. ...Henry Morford.

Son of Mars. ...Arthur Griffiths.

South Sea whaler. ...William Henry Giles Kingston.

Starlight ranch, and other stories of army life on the frontier. ...Charles King.

Sustained honor : a story of the war of 1812. ...John R. Musick.

Tales of military life. ...William Maginn.

Three lieutenants ; or, Naval life in the nineteenth century. ...William Henry Giles Kingston.

Three midshipmen. ...William Henry Giles Kingston.

Tom Clifton ; or, Western boys in Grant's and Sherman's army. ...W. L. Goss.

Two admirals. ...James Fenimore Cooper.

Two soldiers and Dunraven Ranch. ...Charles King.

True blue ; or, Life and adventure of a British seaman of the old school. ...William Henry Giles Kingston.

Under Drake's flag : a tale of the Spanish main. ...George Alfred Henty.

Veterans of Chelsea Hospital. ...George R. Gleig.

Will Watch : from the autobiography of a British officer. ...William J. Neale.

Won at West Point. ...Fush, Pseud.

Young scout : the story of a West Point lieutenant. ...Edward S. Ellis.

ARMY AND NAVY LIFE—*Continued.*
Young Tom Bowling   ...John Conroy Hutcheson.
See also **SEA STORIES** and **SOLDIERS**.

ARNOLD, BENEDICT. *Lived 1740–1801.*
Arnold; or, British spy.   ...Joseph Holt Ingraham.
Gideon Godbold.   ...N. C. Iron.
Great treason.   ...Mary A. M. Hoppus.
Near to nature's heart.   ...Edward Payson Roe.
Rejected wife.   ...Ann S. W. Stephens.
See also **UNITED STATES HISTORY, Eighteenth Century.**

ART.
Alida Craig.   ...Pauline King.
Artist lovers.   ...Caroline Pichler.
Artists' wives.   ...Alphonse Daudet.   English translation.
Aspasia : a romance of art and love in ancient Hellas.   ...Robert
    Hamerling.
Atelier du Lys.   ...Margaret Roberts.
Breuchel brothers.   ...Alexander Ungern-Sternberg.   English trans-
    lation.
Die Brüder; oder, Das Geheimniss.   ...Alexander Ungern-Sternberg.
Coast of Bohemia.   ...William Dean Howells.
Les femmes d'artistes.   ...Alphonse Daudet.
Le fils du Titien.   ...Louis Charles Alfred de Musset.
Guenn : a wave on the Breton coast.   ...Blanche Willis Howard.
Higher than the church.   ...Wilhelmina von Hillern.   English trans-
    lation.
Hoc men schilder wordt.   ...Hendrik Conscience.
Höher als die Kirche.   ...Wilhelmina von Hillern.
How one becomes a painter.   ...Hendrik Conscience.   English
    translation.
Insignificant woman.   ...Bertha Behrens.   (Pseud., W. Heimburg.)
    English translation.
Margaret von Ehrenberg, the artist's wife.   ...William and Mary
    Howitt.
Master of Tanagra.   ...Ernst von Wildenbruch.   English translation.
Der Meister von Tanagra : eine Künstlergeschichte aus Alt-Hellas.
    ...Ernst von Wildenbruch.
Miss Angel.   ...Anna Isabella Thackeray (Mrs. Richmond Ritchie).
Premier and the Painter.   ...Isaac Zangwill and Louise Cowen.
Real thing.   ...Henry James.
Rowena in Boston.   ...Maria Louisa Poole.
Story of a modern woman.   ...Ella Hepworth Dixon.
Stumbler in wide shoes.   ...Anon.
Eine Unbedeutende Frau.   ...Bertha Behrens. (Pseud., W. Heimburg.)
William Hogarth : Roman.   ...Albert E. Brachvogel.

ARTHUR, KING. See **KNIGHTS AND KING ARTHUR.**

ASKEW, ANNE. *Lived 1521–1546.*

Lincolnshire tragedy. ...Anne Manning.

Passages in the life of the fair gospeller, Mistress Anne Askew. ...Anne Manning.

See also **REFORMATION** and **PROTESTANTISM.**

ASTRONOMY.

Astrologer's daughter. ...Rose E. Temple.

Dog of stars. ...Standish O'Grady.

House: an episode in the lives of Reuben Baker, astronomer, and his wife Alice. ...Eugene Field.

Johann Kepler. ...Julie Burow.

Journey in other worlds. ...John Jacob Astor.

ATHENS, GREECE.

Description, History, Manners, and Customs.

Aspasia: a romance of art and love in ancient Hellas. ...Robert Hamerling.

Callias: a tale of the fall of Athens. ...Alfred John Church.

Charmione: a tale of the great Athenian revolution. ...Edward A. Leatham.

Fair Athens. ...Elizabeth M. Edmonds.

Few days in Athens. ...Francis Wright D'Arusmont.

Glaucia. ...Emma Leslie.

Heroes of ancient Greece: a story of the days of Socrates, the Athenian. ...Ellen Palmer.

Last Athenian. ...Abraham Viktor Rydberg. English translation.

Pericles: a tale of Athens. ...Caroline Frances Cornwallis.

Den siste Athenaren. ...Abraham Viktor Rydberg.

Thoth: a romance. ...Joseph Shield Nicholson.

Three Greek children. ...Alfred John Church.

See also **GREECE.**

AUSTRALIA.

Description, History, Manners, and Customs.

Alfreda Holme: a story of social life in Australia. ...Elizabeth B. Bayly.

Australian heroine. ...Rose Murray Prior Praed (Mrs. Campbell Praed).

Australian millionaire. ...Mrs. A. Blitz.

Bail up. ...Hume Nisbet.

Boss of Taroomba. ...Ernest W. Hornung.

Colonial reformer. ...Thomas A. Brown. (Pseud., Rolf Boldrewood.)

Final reckoning: a tale of bush life in Australia. ...George Alfred Henty.

AUSTRALIA — *Continued.*

Girl at Birrell's. ...Thomas Heney.

Grif: a story of Australian life. ...Benjamin Leopold Farjeon.

Harry Heathcote of Gongoil. ...Anthony Trollope.

Head station: a novel of Australian life. ...Rose Murray Prior Praed (Mrs. Campbell Praed).

His first kangaroo. ...Arthur Ferres.

His natural life. ...Marcus Clarke.

Kidnapped squatter, and other Australian tales. ...Andrew Robertson.

Knight of the white feather. ...Jessie Fraser. (Pseud., Tasma.)

Maori and settler: a story of the New Zealand war. ...George Alfred Henty.

Miner's right: a tale of the Australian gold fields. ...Thomas A. Brown. (Pseud., Rolf Boldrewood.)

Modern buccaneer. ...Thomas A. Brown. (Pseud., Rolf Boldrewood.)

Moondyne: a story of the under-world. ...John Boyle O'Reilly.

Mrs. Tregaskiss. ...Rose Murray Prior Praed (Mrs. Campbell Praed).

Nevermore. ...Thomas A. Brown. (Pseud., Rolf Boldrewood.)

Nuggets in the Devil's punch-bowl, and other Australian stories. ...Andrew Robertson.

Outlaw and lawmaker. ...Rose Murray Prior Praed (Mrs. Campbell Praed).

Pilgrims. ...W. Carlton Dawe.

Recollections of Geoffry Hamlyn. ...Henry Kingsley.

Rogue's march. ...Ernest W. Hornung.

Spinx of Eaglehawk: a tale of old Bendigo. ...Thomas A. Brown. (Pseud., Rolf Boldrewood.)

Squatter's dream: a story of Australian life. ...Thomas A. Brown. (Pseud., Rolf Boldrewood.)

Sydney-side Saxon. ...Thomas A. Brown. (Pseud., Rolf Boldrewood.)

Tales of Australian early days. ...Price Warung.

Three Miss Kings: an Australian story. ...Ada Cambridge.

Thunderbolt: an Australian story. ...J. Middleton Macdonald.

Uncle Piper of Piper's hill. ...Jessie Fraser. (Pseud., Tasma.)

Winning a wife in Australia. ...A. Donnison.

AUSTRIA.

Description, Manners, and Customs.

Aus Herrn Walther's jungen Tagen. ...Victor Wodiczka.

Daughters of Pola. ...Anon.

Appelein von Geilingen. ...Franz Trautmann.

Gestalten und Bilder aus dem Tiroler Volksleben. ...Max Stichlberger.

AUSTRIA — *Continued.*

In the land of marvels : folk-tales from Austria and Bohemia.
...Friedrich Theodor Vernaleken. English translation.
Die Jakobiner in Oesterreich. ...Eduard Ruesser.
Kooroona : a tale of South Australia. ...Mrs. Mannington Caffyn.
(Pseud., Iota.)
Die letzten Tage von Alt-Oesterreich. ...Eduard Ruesser.
Majesty of man. ... "Alien," Pseud.
Der Meisterschuss. ...Franz Isidor Proschko.
Österreichische Kinder- und Hausmärchen. ...Friedrich Theodor
   Vernaleken.
Peter in der Luft. ...Franz Isidor Proschko.
Der schwarze Junker. ...Victor Wodiczka.
Der schwarze Mann. ...Franz Isidor Proschko.
Silent sea. ...Mrs. Alick Macleod.
Der Sohn des Regiments. ...Julius von Wickede.
Der Teufel am Traunsee. ...Franz Isidor Proschko.
Die Türken vor Wiens. ...Carl Mueller.
Der Untergang des Protestantism in Oberösterreich. ...Franz Lubo-
   jatzky.
Wien vor vierhundert Jahren. ...Eduard Breier.
Wrong man. ...Dorothea Gerard.
See also **AUSTRIAN HISTORY.**

AUSTRIAN HISTORY.

Fifteenth Century.

Das Buch von den Wienern. ...Michael Beheim.
Die Revolution der Wiener im 15 Jahrhundert. ...Eduard Breier.

Seventeenth Century.

Rüdiger von Starhemberg. ...Franz X. Huber.

Nineteenth Century.

Andreas Hofer. ...Clara M. Mundt. (Pseud., Louise Mühlbach.)
Andreas Hofer. ...Clara M. Mundt. (Pseud., Louise Mühlbach.)
   English translation.
At odds. ...Jemima Montgomery, Baroness von Tautphœus.
Auf dem Wiener Kongress. ...Julius Bacher.
Aus dem Ghetto. ...Leopold Kompert.
Aus der schönen, wilden Lieutenants Zeit. ...Carl Torresani.
Babel. ...Alfred Meissner.
Der Congress zu Wien. ...Eduard Breier.
Countess Irene. ...J. Fogerty.
1809, Historischer Roman. ...Eduard Breier.
1856–1889. ...Bruno Rodwald.

AUSTRIAN HISTORY — *Continued.*
Erlach count. ...Lola Kirschner. (Pseud., Ossip Schubin.)
For the right. ...Karl Emil Franzos. English translation.
Die Geheimnisse des Waldschlosses. ...Reinhard Edmund Hahn.
Imperia. ...Mary A. I. Seymour. (Pseud., Octavia Hensel.)
Jews of Barnow. ...Karl Emil Franzos. English translation.
Die Juden von Barnow. ...Karl Emil Franzos.
Ein Kampf um's Recht. ...Karl Emil Franzos.
Lotta Schmidt and other stories. ...Anthony Trollope.
Der neue Hiob. ...Leopold Sachermasoch.
New Job. ...Leopold Sachermasoch. English translation.
Orthodox. ...Dorothea Gerard.
Perlen aus der Krone des letzten deutschen Kaisers. ...Franz Isi-
dor Proschko.
Peter Mayr. ...Peter K. Rosegger.
Romance of Vienna. ...Frances Trollope.
Scenes from the Ghetto. ...Leopold Kompert. English translation.
Year nine : a tale of the Tyrol. ...Anne Manning.

AUTHORSHIP.
Alide : an episode of Goethe's life. ...Emma Lazarus.
Charlotte Ackermann : a theatrical romance. ...Otto Mueller. Eng-
lish translation.
Charlotte Ackermann : ein Hamburger Roman aus dem vorigen
Jahrhundert. ...Otto Mueller.
Clever wife. ...W. Pett Ridge.
Frau von Staël : biographischer Roman. ...Amalie C. E. M. Boelte.
George Mandeville's husband. ...C. E. Raimond, Pseud.
Goethe and Schiller. ...Clara M. Mundt. (Pseud., Louise Mühl-
bach.) English translation.
Göthe und Schiller. ...Clara M. Mundt. (Pseud., Louise Mühlbach.)
Heinrich Heine's erste Liebe. ...Katharine Diez.
Immeritus Redivivus. ...Anne Manning.
Lessing. Alexander Ungern-Sternberg.
Literary courtship under the auspices of Pike's Peak. ...Anna
Fuller.
Love affairs of a bibliomaniac. ...Eugene Field.
Madame de Staël : an historical novel. ...Amalie C. E. M. Boelte.
English translation.
Molière. ...Alexander Ungern-Sternberg.
Nobody's fault. ...Netta Syrett.
Poet hero. ...Minny Bothmer.
Rome. ...Émile Zola.
Rome. ...Émile Zola. English translation.
Schiller, Kulturhistorischer Roman. ...J. Scherr.

AUTHORSHIP — *Continued.*

School for husbands. ...Rosina W. L. Bulwer-Lytton.

Shelley, biographische novelle. ...Wilhelm Ritter von Hamm.

Story of a modern woman. ...Ella Hepworth Dixon.

Strawberry hill. ...Robert Folk Williams.

Trente ans de Paris. ...Alphonse Daudet.

Thirty years of Paris and of my literary life. ...Alphonse Daudet. English translation.

Venetia. ...Benjamin Disraeli.

What the dragon-flies told the children. ...Amy Brooks.

When a man 's single. ...James Matthew Barrie.

BABYLON.

Description, History, Manners, and Customs.

Belteshazzar ; a romance of Babylon. ...Edward R. Roe.

Exiles in Babylon. ...Charlotte Tucker. (Pseud., A. L. O. E.)

Handwriting on the wall. ...Edwin Atherstone.

Master of the Magicians. ...Elizabeth Stuart Phelps (Mrs. Herbert D. Ward) and Herbert D. Ward.

Sarchedon : a legend of the great queen. ...George John Whyte Melville.

Tales from Blackwood. Volume 2, second series.

BACH, WILHELM FRIEDEMANN. ' *Lived 1710–1784.*

Bach ; or, Fortunes of an Idealist. ...Albert E. Brachvogel. English translation.

Friedmann Bach. Roman. ...Albert E. Brachvogel.

Tone masters. ...Charles Barnard. Volume 3.

BACON'S REBELLION (Virginia). See **REBELLIONS,** *Bacon's.*

BALBOA, VASCO NUNEZ. *Lived 1475–1518.*

Damsel of Darien. ...William Gilmore Simms.

BARBAROSSA. See **FREDERICK I. OF GERMANY.**

BARON'S WAR.

Boy's adventure in the Baron's war. ...John George Edgar.

How I won my spurs. ...John George Edgar.

Last of the barons. ...Edward George Earle Lytton Bulwer-Lytton.

Siege of Kenilworth. ...L. S. Stanhope.

BATTLES. *A Selection.*

Agincourt. 1415.

Agincourt. ...George Payne Rainsford James.

Sword of De Bardwell. ...Lucy Ellen Guernsey.

BATTLES — *Continued.*

Aughrim. 1691.

Denounced. ...John Banim. (Pseud., The O'Hara family.)
Redmond Count of O'Hanlon, the Irish Rapparee. ...William
Carleton.

Bosworth. 1485.

Bosworth field. ...John Beaumont.
Bosworth field. ...Paul Leicester.

Bothwell Bridge. 1679.

Old mortality. ...Sir Walter Scott.

Bouvines. 1214.

Philip Augustus; or, Brothers in arms. ...George Payne Rainsford
James.

Boyne. 1690.

Boyne-water. ...John Banim. (Pseud., The O'Hara family.)
Old house by the Boyne. ...Mary A. Sadlier (Mrs. J. Sadlier).

Brandywine. 1777.

Blanche of Brandywine. ...George Lippard.

Bunker Hill. 1775.

Grandmother's story of Bunker Hill battle. ...Oliver Wendell
Holmes.

Champ de Mars. 1790.

La Comtesse de Charny. ...Alexandre Dumas.
Countess of Charny. ...Alexandre Dumas. English translation.

Cressey. 1346.

Cressey and Poictiers. ...John George Edgar.

Edgehill. 1642.

Arrah Neil. ...George Payne Rainsford James.

Jena. 1806.

Der Spion. ...Franz T. Wangenheim.

Killiecrankie. 1689.

Last of the cavaliers. ...Rose Piddington.
Scottish cavaliers. ...James Grant.

Langside. 1568.

Abbot. ...Sir Walter Scott.

BATTLES — *Continued.*

Leipsic. 1813.

Der Spion.   ...Franz T. Wangenheim.
Die Volkerschlacht bei Leipzig.   ...Joseph von Hinsberg.

Lepanto. 1571.

Word, only a word.   ...Georg Moritz Ebers.   English translation.
Ein Wort.   ...Georg Moritz Ebers.

Lexington. 1775.

Young rebels : a story of the battle of Lexington.   ...Robert Hope
Moncrieff.   (Pseud., Ascott R. Hope.)

Melegnano. 1515.

Die Schlacht von Marignano.   ...Carl A. F. von Witzleben.

Mühlberg. 1547.

Christian prince, Wolfgang, Prince of Anhalt.   ...Franz Hoffmann.
English translation.
Fürst Wolfgang : historische Erzählung.   ...Franz Hoffmann.

New Orleans. 1812.

Signal boys.   ...George C. Eggleston.
Subaltern in America.   ...George R. Gleig.

Pavia. 1525.

Saint Leon.   ...Caleb Williams Godwin.

Pinkie. 1547.

Mary of Lorraine.   ...James Grant.

Rosbach. 1757.

Rückwirkungen; oder wer regiert denn ?   ...Johann H. D. Zschokke.

Trenton. 1777.

Kate Aylesford.   ...Charles J. Peterson.
Water-waif : a story of the revolution.   ...Emma S. Bladen.

Wagram. 1809.

Der Spion.   ...Franz T. Wangenheim.

Waterloo. 1815.

Catherine.   ...Frances M. Peard.
La chartreuse de Parme.   ...Henri Beyle.
Elba und Waterloo : Forsetzung von 1813.   ...Anon.
Geordie Stuart.   ...M. B. Manwell.

BATTLES — *Continued.*

Great shadow and beyond the city. ...Arthur Conan Doyle.

Die grossen Tage des Junius 1815. ...Martin H. A. Schmidt.

Harry, the drummer. ...Agnes T. Deane.

Les Misérables. ...Victor Hugo.

Les Miserables. ...Victor Hugo. English translation.

One of the 28th : a tale of Waterloo. ...George Alfred Henty.

Quatre Bras. ...Arthur T. Pask.

Return to England. ...Anon.

Stories of Waterloo. ...William H. Maxwell.

Vanity Fair. ...William Makepeace Thackeray.

Waterloo. ...Auguste M. de Barthélemy and Joseph Méry.

Waterloo. ...Émile Erckmann and Alexandre Chatrian. English translation.

Waterloo: suite du Conscrit de 1813. ...Émile Erckmann and Alexandre Chatrian.

See also history of countries for shorter descriptions of battles.

BEETHOVEN, LUDWIG VAN. *Lived 1770–1827.*

Beethoven. Historischer Roman. ...Anon.

Rumor. ...Elizabeth S. Sheppard. (Pseud., Beatrice Reynolds.)

Eine stille Liebe zu Beethoven. ...Fanny del Rio.

Tone masters. ...Charles Barnard. Volume 3.

Unrequited love: an episode in the life of Beethoven. ...Fanny del Rio. English translation.

See also **MUSIC.**

BERLIN, GERMANY.

Description, History, Manners, and Customs.

Berlin and Sans-Souci ; or, Frederick the Great and his friends. ...Clara M. Mundt. (Pseud., Louise Mühlbach.) English translation.

Buchholz family. ...Julius Ernest Wilhelm Stinde. English translation.

Burgomaster of Berlin. ...Georg Wilhelm Heinrich Haering. English translation.

Die Familie Buchholz. ...Julius Ernest Wilhelm Stinde.

Die Franzosen in Berlin: 1806–8. ...Friedericke H. Unger.

Frau Wilhelmine. ...Julius Ernest Wilhelm Stinde.

Frau Wilhelmine : Sketches of Berlin life. ...Julius Ernest Wilhelm Stinde. English translation.

Friedrich der Grosse und sein Hof. ...Clara M. Mundt. (Pseud., Louise Mühlbach.)

Frederick the Great and his court. ...Clara M. Mundt. (Pseud., Louise Mühlbach.) English translation.

BERLIN, GERMANY — *Continued.*
Gold und Blut. ...Johann Ferdinand Martin Oskar Meding. (Pseud.,
    Gregor Samarow.)
Johann Gotzkowsky, der Kaufman von Berlin.   ...Clara M. Mundt.
    (Pseud., Louise Mühlbach.)
Kathinka ; Roman aus dem Berliner Leben.   ...O. Heller.
Merchant of Berlin.   ...Clara M. Mundt.   (Pseud., Louise Mühl-
    bach.) English translation.
Der Roland von Berlin.   ...Georg Wilhelm Heinrich Haering.
See also **GERMAN HISTORY** and **GERMANY.**

BERN, DIETRICH OF. See **DIETRICH OF BERN.**

BERNHARD, DUKE OF WEIMAR. *Lived 1604-1639.*
Bernhard Herzog zu Sachsen Weimar.   ...F. G. Schlenkert.
Herzog Bernhard.   Eine Geschichte vom Oberrhein aus den Jahren
    1638-9.   ...Hans Blum.
Herzog Bernhard.   ...Heinrich Laube.

BIBLE.
                Special History and Characters.
Ahasvérus.   ...Edgar Quinet.
Barabbas.   ...Marie Corelli.
Belteshazzar : a romance of Babylon.   ...Edward R. Roe.
Broken walls of Jerusalem and the rebuilding of them.   ...Susan
    Warner.
Come forth.   ...Elizabeth Stuart Phelps Ward (Mrs. Herbert D.
    Ward) and Herbert D. Ward.
Delilah.   ...S. W. Odell.
Ephraim and Helah : a story of the Exodus.   ...Edwin Hodder.
Exiles in Babylon.   ...Charlotte Tucker.   (Pseud., A. L. O. E.)
Flaming sword, being an account of the extraordinary adventures
    and discoveries of Dr. Percival in the wilds of Africa.   ...Anon.
    (Garden of Eden.)
Hammer : a story of Maccabean times.   ...Alfred John Church and
    Richmond Seeley.
Hebrew heroes.   ...Charlotte Tucker.   (Pseud., A. L. O. E.)
Helon's pilgrimage to Jerusalem : a picture of Judaism in the cen-
    tury which preceded the advent of our Saviour.   ...Friedrich
    Abraham Strauss.   English translation.
Helon's Wallfahrt nach Jerusalem, hundert neun Jahr vor der
    Geburt unsers Herrn.   ...Friedrich Abraham Strauss.
Idumean.   ...J. M. Leavitt.
José und Benjamin.   ...Adolf Franz Delitzsch.
Joseph and Benjamin.   ...Adolf Franz Delitzsch.   English trans-
    lation.

BIBLE — *Continued.*
Joshua : a story of biblical times. ...Georg Moritz Ebers. English
translation.
Josua. ...Georg Moritz Ebers.
Julian. ...William Ware.
King of Tyre : a tale of the times of Ezra and Nehemiah. ...James
M. Ludlow.
Leah : a tale of ancient Palestine. ...Mrs. Alexander S. Orr. (Ahab.)
Manuscript man. ...Elizabeth H. Walshe.
Master of the magicians. ...Elizabeth Stuart Phelps Ward (Mrs.
Herbert D. Ward) and Herbert D. Ward.
Onesimus : memoirs of a disciple of St. Paul. ...Edwin A. Abbott.
Patriarchal times. ...C. M. O'Keefe.
Pillar of fire ; or, Israel in bondage. ...Joseph Holt Ingraham.
(Moses.)
Price of peace : a story of the times of Ahab, king of Israel.
...A. W. Ackerman.
Prince of Peace. ...A. W. Ackerman.
Prince of the house of David ; or, Three years in the holy city :
scenes in the life of Jesus. ...Joseph Holt Ingraham.
Quiet king. ...Caroline Atwater Mason.
Rameses. ...Edward Upham.
Samson. ...S. W. Odell.
Sarchedon : a legend of the great queen. ...George John Whyte
Melville. (Samiramsi.)
Shepherd of Bethlehem. ...Charlotte Tucker. (Pseud., A. L. O. E.)
(David.)
Spell of Ashtaroth. ...Duffield Osborne. (Joshua and Achan.)
Story of Sodom. ...W. C. Kitchin.
Story of the other wise man. ...Henry Van Dyke.
Throne of David. ...Joseph Holt Ingraham. (Absalom's rebellion.)
Triumph over Midian. ...Charlotte Tucker. (Pseud., A. L. O. E.)
Zipporah. ...Mrs. M. E. Brewsher.
See also **CHRIST, LIVES OF.**

BICYCLING.
Arctic night. ...Roger Pocock.
On the down grade. ...Winifred Graham.
Two on a tandem. ...Charles James.
Vashti, old and new. ...Marvel Kayve.
Wheels : a bicycling romance. ...A Wheeler, Pseud.
Wheels of chance. ...H. G. Wells.

BISMARK-SCHOENHAUSEN, OTTO E. LEOPOLD. *Lived 1815–*
Prince Bismark, friend or foe ? ...Minny Bothmer.
Schach–Bismark. ...J. G. Findel.
See also **GERMAN HISTORY, Nineteenth Century.**

BLÜCHER, GEBHARD LEBERECHT VON.   *Lived 1742–1819.*

Der Feldmarschall Blücher und der Pfarrer Kretzschmer. ...Philipp
Friedrich Wilhelm Örtel.   (Pseud., W. O. von Horn.)

Napoleon in Deutschland. ...Clara M. Mundt.   (Pseud., Louise
Mühlbach.)

Napoleon in Germany. ...Clara M. Mundt.   (Pseud., Louise Mühlbach.)   English translation.

See also **GERMAN HISTORY, Nineteenth Century.**

BOHEMIA.

> *Description, History, Manners, and Customs.*

Babica. ...B. Němec.

Bohemians in the 15th century. ...Henri Guenot.   English translation.

Les Bohémiens au XVᵉ siècle. ...Henri Guenot.

Böhmische Juden. ...Leopold Kompert.

Böhmen von 1414 bis 1424. ...G. C. R. Herloss-sohn.

Conrad. ...Emma Leslie.

Crushed yet conquering : a story of Constance and Bohemia.
...Deborah Alcock.

Gisela, ein Roman aus der Zeit des Conciliums von Constanz.
...M. Lehmann.

Grandmother. ...B. Němec.   English translation.

In the land of marvels : folk-tales from Austria and Bohemia.
...Friedrich Theodor Vernaleken.   English translation.

Jean Ziska. ...George Sand, psued. (Amantine L. A. D. Dudevant.)

Johann Ziska. ...Franz T. Wangenheim.

Myths and folk-tales of the Russians, western Slaves, and Magyars.
...Jeremiah Curtin.

Österreichische Kinder- und Hausmärchen. ...Friedrich Theodor
Vernaleken.

Ottokar von Falkenburg. ...L. Lehnert.

Prokop Veliký. ...Věnceslav Lipovský.

Der Rabbi von Liegnitz. ...A. Sammter.

Upálem Jana Husa Čili. ...Felix Deriége.

"Vorhang-Purim." ...Math. Kisch.

Wenzel's inheritance. ...Annie Lucas.

Die Wrschowitze. ...E. C. V. Dietrich.

BOSTON, MASS.

> *Description, History, Manners, and Customs.*

American politician. ...Francis Marion Crawford.

Barclays of Boston. ...Elizabeth B. Otis (Mrs. Harrison Gray Otis).

Curse of the Old South Church of Boston. ...James J. Kane.

Lamplighter. ...Maria Susanna Cummins.

BOSTON, MASS. — *Continued.*
Lionel Lincoln; or, Leaguer of Boston. ...James Fenimore Cooper.
Looking backward. 2000–1887. Edward Bellamy.
Maggie: a girl of the streets. ...Stephen Crane.
Miss Curtis. ...Kate Gannett Wells.
Miss Eyre from Boston and others. ...Ellen L. Moulton.
Mrs. Keats Bradford. ...Maria Louise Pool.
Naomi; or, Boston two hundred years ago. ...E. B. Lee.
Old Boston. ...Augusta De Grasse Stevens.
Pirate Gold. ...Frederic J. Stimson. (Pseud., J. S. of Dale.)
Rebels; or, Boston before the revolution. ...Lydia Maria Child.
Rowena in Boston. ...Maria Louise Pool.
Shawmut. ...C. K. True.
Story of the siege of Boston. ...Horace E. Scudder. (In his Stories
    and romances.)
Two gentlemen of Boston. ...Anon.
White chief among the red men. ...J. T. Adams.
See also **MASSACHUSETTS.**

BOYNE, IRELAND.
Baldearg Donnell. ...A. S. G. Canning.
Boyne-water. ...John Banim. (Pseud., The O'Hara family.)
Denounced. ...John Banim. (Pseud., The O'Hara family.)
Derry: a tale of the revolution. ...Charlotte Elizabeth Tonna.
    (Pseud., Charlotte Elizabeth.)
Florence O'Neill, the rose of Saint Germains. ...Agnes M. Stewart.
Gap of Barnermore, the Irish highlands. ...Isaac Butts.
Last baron of Crana. ...John Banim. (Pseud., The O'Hara family.)
Leixip Castle. ...M. L. O'Byrne.
Old house by the Boyne. ...Mary A. Sadlier (Mrs. J. Sadlier).
Orange and green: a tale of the Boyne and Limerick. ...George
    Alfred Henty.
Peter of the castle. ...John Banim. (Pseud., The O'Hara family.)
Under which king? ...William Johnston.

BRABANT, GENEVIÈVE OF. See **GENEVIÈVE DE BRABANT.**

BUCCANEERS.
Admirable Lady Biddy Fane. ...Frank Barrett.
Buccaneer chief: a romance of the Spanish main. ...Gustave
    Aimard. English translation.
Buccaneers. ...Randolph Jones.
Buccaneers. ...S. B. H. Judah.
Captain Brand of the Centipede. ...Henry A. Wise. (Pseud.,
    Harry Gringo.)
Darien; or, Merchant prince. ...E. B. G. Warburton.

BUCCANEERS — Continued.

Demigod.   ...Edward P. Jackson.
Eppelein von Geilingen.   ...Franz Trautmann.
La grande flibuste.   ...Gustave Aimard.
Iron pirate : a plain tale of strange happenings on the sea.   ...Max
    Pemberton.
Lafitte, the pirate of the Mexican Gulf.   ...Joseph Holt Ingraham.
Life, adventures, and piracies of the famous Captain Singleton.
    ...Daniel Defoe.
King of the mountains.   ...Edmond About.   English translation.
Love afloat : a story of the American navy.   ...Frances H. Sheppard.
Master Ardick, buccaneer.   ...F. II. Costello.
Modern buccaneer.   ...Thomas A. Brown.   (Pseud., Rolf Boldrewood.)
Modern Telemachus.   ...Charlotte Mary Yonge.
New voyage around the world.   ...Daniel Defoe.
Pirate and moonshine.   ...Frederick Marryat.
Pirate city.   ...Robert Michael Ballantyne.
Le roi des montagnes.   ...Edmond About.
Rover's secret.   ...William J. C. Lancaster.   (Pseud.,   Harry
    Collingwood.)
Seven brothers of Wyoming.   ...Anon.
Sir Henry Morgan.   ...E. Howard.
Treasure Island.   ...Robert Louis Stevenson.
Westward Ho ! or, Voyages and adventures of Sir Amyas Leigh.
    ...Charles Kingsley.
Wild Western scenes.   ...J. B. Jones.

BUDA, HUNGARY.

Siege of Buda.   ...Anon.
Zord idő.   Zsigmund Kemény.
See also HUNGARY.

BUNYAN, JOHN.   Lived 1628–1688.

Mary Bunyan, the dreamer's blind daughter.   ...Sallie Rochester
    Ford.

BURR, AARON.   Lived 1756–1836.

Conspirators.   ...Eliza Ann Dupuy.
Rivals.   ...Jeremiah Clemens.
See also UNITED STATES HISTORY.

CADETS.

Cadet button.   ...Frederick M. Whittaker.
Cadet days : a story of West Point.   ...Charles King.
Cadet life at West Point.   ...Hugh T. Reed.
Cadet life at West Point.   ...George C. Strong.
Cuts : a story of West Point.   ...G. I. Cervus.

CADETS — *Continued.*

Edgar Fairfax : a story of West Point. ...Florence Nightingale Craddock.

From cadet to captain. ...John Percy Groves.

Miss Bagg's secretary : a West Point romance. ...Clara Louise Burnham.

Queen's cadet. ...James Grant.

Won at West Point. ...Fush, Pseud.

Young scout : the story of a West Point lieutenant. ...Edward S. Ellis.

See also **ARMY AND NAVY LIFE**.

CALIFORNIA. *A Selection.*

Description, History, Manners, and Customs.

Abandoned claim. ...Flora Haines Loughead.

American coin. ...Anon.

Amulet : a tale of Spanish California. ...Anon.

Argonauts of North Liberty. ...Francis Bret Harte.

Before the gringo came. ...Gertrude F. Atherton.

Boy emigrants. ...Noah Brooks.

Braxton's Bar ; a tale of pioneer life in California. ...R. M. Daggett.

Bridge of the gods : a romance of Indian Oregon. ...F. H. Balch.

By shore and sedge. ...Francis Bret Harte.

California Crusoe. ...Anon.

California sketches. ...O. P. Fitzgerald.

California sketches. ...Leonard Kip.

Captain Bayley's heir : a tale of the gold fields of California. George Alfred Henty.

Los Cerritos : a romance of the modern times. ...Gertrude F. Atherton.

Le cher chœur de pistes. ...Gustave Aimard.

Counter-currents. ...Sophie Winthrop Weitzel.

Digging for gold. ...Robert Michael Ballantyne.

L'École des Robinsons. ...Jules Verne.

Erema. ...Richard Doddridge Blackmore.

Feud of Oakfield Creek : a novel of California life. ...Josiah Royce.

First family of Tasajara. ...Francis Bret Harte.

La fièvre d'or. 1860. ...Gustave Aimard.

'49 ; or, Gold seeker of the Sierras. ...Cincinnatus Heine Miller. (Pseud., Joaquin Miller.)

Free prisoners : a story of California. ...Jane W. Bruner.

Gem of the mines : a narrative of California life. ...Mrs. J. B. Peploe (Webb).

Gerald French's friends. ...George H. Jessop.

CALIFORNIA — *Continued.*

Godfrey Morgan: a California mystery.   ...Jules Verne.   English translation.

Gold seekers : a tale of California.   ...Gustave Aimard.   English translation.

Golden days of '49: a tale of the California diggings. ...Kirk Munroe.

Golden dream.   ...Robert Michael Ballantyne.

In the valley of Havilah.   ...Frederick T. Clark.

Judge Lynch : a romance of California vineyards.   ...George H. Jessop.

Log school-house on the Columbia.   ...Hezekiah Butterworth.

Luck of Roaring Camp and other stories.   ...Francis Bret Harte.

Manuelita.   ...Marian Calvert Wilson.

Maruja.   ...Francis Bret Harte.

Monica, the Mesa maiden.   ...Evelyn Raymond.

Mysteries and miseries of San Francisco, California.   ...Anon.

Musgrove Ranch : a tale of Southern California.   ...T. M. Browne.

New and the old ; or, California and India in romantic aspects. ...John W. Palmer.   (Pseud., John Coventry.)

Off to California.   ...James F. Cobb.

Oregon and Eldorado ; or, Romance of the rivers.   ...Thomas Bulfinch.

Picture of pioneer times in California.   ...William Grey.

Ramona.   ...Helen Hunt Jackson.

Red Mountains of Alaska.   ...Willis B. Allen.

Rifle, rod, and gun in California: a sporting romance.   ...Theodore Stronge Van Dyke.

Romance dust from the historic placer.   ...William S. Mayo.

San Rosario Ranch.   ...Maud Howe.

Society in search of truth ; or, Stock-gambling in San Francisco. ...J. F. Clark.

Stories of the foot hills.   ...Margaret Collier Graham.

Stories of the Sierras and other sketches.   ...Francis Bret Harte.

Susy: a story of the plains.   ...Francis Bret Harte.

Summer in a canyon.   ...Kate Douglas Wiggin (Mrs. George Riggs).

Tisáyac of the Yosemite.   ...Mary B. M. Toland.

Trail hunter: a tale of the far West.   ...Gustave Aimard. English translation.

Twice bought.   Oregon gold-fields.   ...Robert Michael Ballantyne.

Village drama.   ...Vesta S. Simmons.

Le whip-poor-will; ou, Les pionniers de l'Orégon.   ...Amédée Bouis.

Zanita : a tale of the Yosemite.   ...Maria Thérèsa Longworth Yelverton.

CANADA.

Description, History, Manners, and Customs.

Algonquin maiden. ...G. M. Adam and A. E. Wertherald.
Bastonnais. ...John Lesperance.
Boys of 1745, at the capture of Louisburg. ...James Otis Kaler. (Pseud., James Otis.)
Captain Nelson: a romance of colonial days. ...Samuel A. Drake.
Chase of Saint Castin, and other stories of the French in the New World. ...Mary Hartwell Catherwood.
Le chien d'or : a legend of Quebec. ...William Kirby.
Constance of Acadia. ...Edward Payson Tenney.
Englishman's haven. ...W. J. Gordon.
François de Bienville. ...Joseph Marmette.
L'Intendant bigot. Roman Canadien. ...Joseph Marmette.
Jacques et Marie. ...N. Bourassa.
La jongleuse. ...H. Raymond Casgrain.
King's warrant: a story of old New France. ...Alfred H. Engelback.
Lady of Fort Saint John. ...Mary Hartwell Catherwood.
Lady and the cross. ...James De Mille.
Lily and the cross : a tale of Acadia. ...James De Mille.
Les Machabées de la Nouvelle France. ...Joseph Marmette.
Marjorie's Canadian winter. ...Agnes M. Machar.
Old judge ; or, Life in a colony. ...Thomas C. Haliburton. (Pseud., Sam Slick.)
Prisoner of the Border. ...Hamilton Myers.
Refugees : a tale of two continents. ...Arthur Conan Doyle.
Ridgeway. ...Scian Dubh.
Rivals of Acadia. ...Harriet V. Cheney.
Romance of Dollard. ...Mary Hartwell Catherwood.
Seats of the mighty. ...Gilbert Parker.
Settlers in Canada. ...Frederick Marryat.
Sinners twain. ...J. Mackie.
Stories of the land of Evangeline. ...Grace D. McLeod.
Stories of the New France. ...Agnes M. Machar.
Story of Tonty. ...Mary Hartwell Catherwood.
Their wedding journey. ...William Dean Howells.
Thirty-nine men and one woman. H. Émile Chevalier. English translation.
Trente-neuf hommes pour une femme. ...H. Émile Chevalier.
U. E. : a tale of Upper Canada. ...Anon.
White islander. ...Mary Hartwell Catherwood.
With Wolfe in Canada. ...George Alfred Henty.

CASTLES.

General and Imaginary.

Castle and town.   ...Frances M. Peard.
Castle builders.   ...Charlotte Mary Yonge.
Castle Comfort.   ...Helen Hays (Mrs. W. J. Hays).
Castle Foam.   ...Henry W. French.
Castle in Spain.   ...James De Mille.
Castle in the air.   ...Hugh B. Ewing.
Castle Nowhere.   ...Constance Fenimore Woolson.
Castles in the air.   ...Catherine G. F. Gore.
Castles in the air.   ...Louis R. Upton.
City and castle.   ...Anna Lucas.

Special.

Athlin.

Castles of Athlin and Dunbayne.   ...Anne Radcliffe.

Blair.

Blair Castle.   ...Flora L. Shaw.

Bolsover.

Bolsover Castle: a tale from Protestant history of the sixteenth century.   ...Anon.

Camber.

Chronicles of Camber Castle.   ...Anon.

Carisbrooke.

King's namesake: a tale of Carisbrooke Castle.   ...Catherine M. Phillimore.

Corfe.

Brave Dame Mary; or, Siege of Corfe Castle.   ...Louisa Hawtrey.
Eldrick, the Saxon.   ...A. S. Bride.

Coulying.

Coulying Castle; or, Knight of the olden days.   ...Agnes Giberne.

Daly.

Castle Daly: the story of an Irish home thirty years ago.   ...Annie Maria Keary.

Delany.

Two knights of Delany Castle.   ...Mary M. Sherwood.

Dublin.

Robber chieftain.   ...Anon.

CASTLES — *Continued.*

Dunbayne.

Castles of Athlin and Dunbayne.   ...Anne Radcliffe.

Ehrenstein.

Castle of Ehrenstein.   ...George Payne Rainsford James.

Glenmoyle.

Biblicals.   ...Anon.

Heidelberg.

Klytia. Historische Roman aus dem 16-Jahrh.   ...Adolf Hausrath.
Clytia: an historical novel.   ...Adolph Hausrath.   English trans-
    lation.

Herleck.

Gladys of Herleck.   ...Anon.

Kilsorrell.

Kilsorrell Castle : an Irish story.   ...Albert S. G. Canning.

Leixlip.

Leixlip Castle : a romance of the penal days of 1690.   ...M. L.
    O'Byrne.

Norwich.

Siege of Norwich Castle : a story of the last struggle against the
    Conqueror.   ...M. M. Blake.

Wallingford.

Brian Fitz-count : a story of Wallingford Castle and Dorchester
    Abbey.   ...Augustine D. Crake.

Windsor.

Windsor Castle.   ...William Harrison Ainsworth.

CATHERINE DE' MEDICI. See **CHARLES IX. OF FRANCE.**

CATHOLICISM.

Abbot.   ...Sir Walter Scott.
Bertha : a historical romance of the time of Henry IV. of Germany.
    ...Joseph E. C. Bischoff.   English translation.
Bertha ; or, Pope and the Emperor.   ...William B. MacCabe.
Castle of the Three Mysteries.   ...Anon.
Christlich oder Päpstlich ?   ...Eduard Jost.
Constance Sherwood : an autobiography of the sixteenth century.
    ...Georgiana C. L. G. Fullerton.
Darnley ; or, Field of the cloth of gold.   ...George Payne Rainsford
    James.

CATHOLICISM — *Continued.*

Father Clement.   ...Grace Kennedy.
First Fleet family.   ...Louis Becke and Walter Jeffery.
For the Master's sake: a story of the days of Queen Mary.   ...Emily Sarah Holt.
Foster sisters.   ...Lucy Ellen Guernsey.
Gloria.   ...Benito Peréz Galdós.
Hildebrand and the excommunicated emperor.   ...Joseph Sortain.
House of Yorke.   ...Mary A. Tincker.
Isoult Barry of Wynscote: a tale of Tudor times.   ...Emily Sarah Holt.
Jesuit of to-day.   ...Orange McNeill.
John Inglesant.   ...Joseph Holt Ingraham.
Jovinian; or, Early days of Papal Rome.   ...William Henry Giles Kingston.
Kenilworth.   ...Sir Walter Scott.
Königin Bertha.   ...Joseph E. C. Bischoff.
Lady and the priest.   ...Katherine C. Maberley.
Leon Roch.   ...Benito Peréz Galdós.
Miner's daughter: a Catholic tale.   ...Cecilia M. Caddell.
Mixed marriage.   ...Amabel Kerr.
Monastery.   ...Sir Walter Scott.
Money god; or, Empire and the papacy.   ...Abel Quinton.
Onoqua.   ...Frances C. Sparhawk.
Priest and the Huguenot; or, Persecution in the age of Louis XV.   ...Laurence Louis Félix Bungener.   English translation.
Protestant.   ...Anna Eliza Bray.
Remember the Alamo.   ...Amelia Edith Barr.
Robin Tremayne of Bidmin: a tale of the Marian persecution.   ...Emily Sarah Holt.
Romance of a French parsonage.   ...M. B. Edwards.
Rome.   ...Émile Zola.
Rome.   ...Émile Zola.   English translation.
Russian priest.   ...I. N. Potapenko.   English translation.
Three hundred years ago.   ...William Henry Giles Kingston.
Tor Hill.   ...Horace Smith.   (Pseud., Paul Chatfield.)
Trois sermons sous Louis XV.   ...Laurence Louis Félix Bungener.
Truce of God: a tale of the eleventh century.   ...George H. Miles.
Twice crowned: a story of the days of Queen Mary.   ...Harriet B. MacKeever.
Uline's escape; or, Hid with the nuns.   ...Mrs. Alexander S. Orr.
See also **PROTESTANTISM** and **REFORMATION**.

CAVALIERS.

Amyas Egerton, cavalier.   ...Maurice H. Hervey.
Brambletye House.   ...Horace Smith.   (Pseud., Paul Chatfield.)

CAVALIERS — *Continued.*
Cavaliers.   ...S. R. Keightley.
Cavaliers and Roundheads.   ...John George Edgar.
Cavaliers of England.   ...Henry W. Herbert.   (Pseud., Frank
    Forester.)
Century too soon : a story of Bacon's Rebellion.   ...John R. Musick.
Colonial cavalier ; or, Southern life before the revolution.   ...Maud
    Wilder Goodwin.
Last of the cavaliers.   ...Rose Piddington.
Memoirs of a cavalier.   ...Daniel Defoe.
Scottish cavalier.   ...James Grant.
Spanish cavalier.   ...Charlotte Tucker.   (Pseud., A. L. O. E.)
Virginia cavalier.   ...Molly Elliot Seawell.

CAXTON, WILLIAM. *Lived 1422–1491.*
Earl printer : times of Caxton.   ...C. M. M.

CENTRAL AMERICA.   See **AMERICA, Central.**

CHARLEMAGNE. *Reigned 800–814.*
Les chevaliers du Cygne ; ou, La cour de Charlemagne.   ...Sté-
    phanie Félicité, Comtesse de Genlis.
Four sons of Aymon.   ...Anon.
Knights of the swan ; or, Court of Charlemagne.   ...Stéphanie Fé-
    licite, Comtesse de Genlis.   English translation.
Legends of Charlemagne.   ...Thomas Bulfinch.
Legends of Charlemagne.   ...Edward Everett Hale.
Magic runes.   ...Emma Leslie.
Passe Rose.   ...Arthur Hardy.
Pépin et Charlemagne.   ...Alexandre Dumas.
Pepin and Charlemagne.   ...Alexandre Dumas.   English translation.
Romance of history.   ...Leitch Ritchie.
Romances relating to Charlemagne.   ...George Ellis.
Story of Roland.   ...James Baldwin.

CHARLES I. OF ENGLAND. *Reigned 1625–1649.*
Alice Bridge of Norwich : a tale of the time of Charles the First.
    ...Andrew Reed.
Alice Leighton ; or, Good name is rather to be chosen than riches.
    ...Ann J. Cupples (Mrs. George Cupples).
Alice Lisle : a tale of Puritan times.   ...Richard King.
Andrew Marvel and his friends.   ...Marie S. Hall.
Arrah Neil.   ...George Payne Rainsford James.
Benjamin Holbeck : a story of the civil war.   ...Margaret A. Paull.
Boy cavaliers.   ...Henry C. Adams.
Brambletye House ; or, Cavaliers and Roundheads.   ...Horace Smith.
    (Pseud., Paul Chatfield.)

4

CHARLES I. OF ENGLAND — *Continued.*

Buccaneer.   ...Anna Maria Hall.

Carewes : a tale of the civil wars.   ...Mary Gillies.

Castle Cornet : or, Island's troubles in the troublous times.   ...Louisa Hawtrey.   `

Cavalier.   ...George Payne Rainsford James.

Cavaliers.   ...S. R. Keightley.

Cavaliers and Roundheads ; or, Stories of the great civil war.   ...John George Edgar.

Cavaliers of England : times of the great revolution of 1642 and 1688.   ...Henry W. Herbert.   (Pseud., Frank Forester.)

Charmouth Grange : a tale of the seventeenth century.   ...John P. Groves.

Children of the New Forest.   ...Frederick Marryat.

Civil war in Hampshire : a story of Basing House.   ...George N. Godwin.

Clare of Claresmede.   ...Charles Gibbon.

Le Comte de Strafford.   ...François T. M. B. d'Arnaud.

Cost what it may : a story of cavaliers and roundheads.   ...Emma E. Hornibrook.

Courtenay of Walreddon : a romance of the West.   ...Anna Eliza Bray.

Cromwell ; or, Protector's oath.   ...J. Frederick Smith.

Donning Castle : a royalist story.   ...George H. Colomb.

Dorothy's dilemma.   ...Caroline Austin.

Draytons and the Davenants : a story of the civil wars.   ...Elizabeth Charles.

Fairleigh Hall : a tale of Oxfordshire during the great rebellion. ...Augustine D. Crake.

Fair maid of Taunton.   ...Elizabeth M. Alford.

For king and Kent : a true story of the great rebellion.   ...George H. Colomb.

Friends though divided : a tale of the civil war.   ...George Alfred Henty.

Gabriel : a tale of Wichnor Wood.   ...Mary Howitt.

Mandeville : a tale of the seventeenth century in England.   ...William Godwin.

Harry Ogilvie ; or, Black dragoons.   ...James Grant.

Hayslope Grange : a tale of the civil war.   ...Emma Leslie.

Heir of Sherborne ; or, Attainder.   ...Anon.

Henry Masterton ; or, Adventures of a young cavalier.   ...George Payne Rainsford James.

Her Majesty the Queen.   ...John Esten Cooke.

Hide and seek : a story of the New Forest in 1647.   ...Mrs. Frank Cooper.

CHARLES I. OF ENGLAND — *Continued.*

Holmby House: a tale of old Northamptonshire. ...George John Whyte Melville.

In Colston's days: a story of old Bristol. ...Emma Marshall.

Isabel Saint Clair: a romance of the seventh century. ...Julia Addison.

John Inglesant. ...Joseph Henry Shorthouse.

Lady Betty's governess; or, Corbet chronicles. ...Lucy Ellen Guernsey.

Lady Shakerlye: the record of the life of a good and noble woman. ...Anon.

Lady Willoughby; or, Passages from the diary of a wife and mother in the seventeenth century. ...Hannah M. Rathbone.

Langley Grange: a romance of the time of Charles the First. ...Thomas T. Harman.

Leaguer of Lathom: a tale of the civil war in Lancashire. ...William Harrison Ainsworth.

Lettice: a tale of the siege of Chester. ...Pauline Biddulph.

Love the leveller: a tale of the great rebellion. ...Angus Comyn.

Markhams of Ollerton: a tale of the civil war. ...Elizabeth Glaister.

Maudeville: a tale of the seventeenth century in England. ...William Godwin.

Memoirs of a cavalier. ...Daniel Defoe.

Minister Lovell: a story of the days of Laud. ...Emily Sarah Holt.

Nonpareil House: or, Fortunes of Julian Mountjoy. ...Henry Curling.

Old Bristol: a story of Puritan times. ...Anon.

Old Noll; or, Days of the Ironsides. ...Frederick William Robinson.

Oliver Cromwell; or, England's great protector. ...Henry W. Herbert. (Pseud., Frank Forester.)

Ovington Grange: a tale of the South Downs. ...William Harrison Ainsworth.

Prince and the pedler; or, Siege of Bristol. ...Ellen Pickering.

Regicides: a tale of early colonial times. ...Frederick Hull Cogswell.

Reginald Hastings: a tale of the troubles of 164–. ...Eliot B. G. Warburton.

Royalists and Roundheads. ...Elizabeth M. Stewart.

Rosamond Fane; or, Prisoners of Saint James. ...Mary and Catherine Lee.

Scarlet and buff: a tale of Winchester, during the great rebellion. ...J. E. Corbière.

Settlers at home. ...Harriet Martineau.

Siege of Lichfield. ...William Gresley.

Sir Henry Appleton; or, Essex during the great rebellion. ...William E. Heygate.

CHARLES I. OF ENGLAND — *Continued.*

Spanish match ; or, Charles Stuart in Madrid.   ...William Harrison Ainsworth.

Splendid spur.   ...Arthur T. Q. Couch.   (Pseud., Q.)

Stanch for the king; or, Chamber of Koning Hall : a story of the civil wars.   ...Arthur Brown.

Stanfield Hall.   ...J. Frederick Smith.

Strafford : a romance.   ...Henry B. Baker.

Tales of the civil wars.   ...Henry C. Adams.

Three judges : a story of the men who beheaded their king.   ...Israel Putnam Warren.

Two knights of Delany Castle.   ...Mary M. Sherwood.

Under Salisbury spire, in the days of George Herbert.   ...Emma Marshall.

Under the storm ; or, Steadfast's charge.   ...Charlotte Mary Yonge.

Warleigh ; or, Fatal oak : a legend of Devon.   ...Anna Eliza Bray.

Washingtons : a tale of a country parish in the seventeenth century.   ...John N. Simpkinson.

When Charles the First was king.   ...J. S. Fletcher.

White gauntlet.   ...Mayne Reid.

Whitehall ; or, Days and times of Oliver Cromwell.   ...Jane Robinson.

With the king at Oxford : a tale of the great rebellion.   ...John Alfred Church.

Wizard of Windshaw : a tale of the seventeenth century.   ...Anon.

Young Oxford maid in the days of the king and parliament.   ...Henrietta Keddie.   (Pseud., Sarah Tytler.)

See also **ENGLISH HISTORY**.

CHARLES II. OF ENGLAND.   *Reigned 1660–1685.*

Agnes Beaumont : a true story of the year 1670.   ...Marian Caldecott.

Andrew Golding : a tale of the great plague.   ...Anne E. Keeling.

Anne of Argyle.   ...G. Eyre Todd.

Aphra Behn.   ...Clara M. Mundt.   (Pseud., Louise Mühlbach.)

Aphra Behn.   ...Clara M. Mundt.   (Pseud., Louise Mühlbach.)   English translation.

At the sign of the Blue Boar : a story of the reign of Charles II.   ...Emma Leslie.

Beyond the seas.   ...Oswald Crawfurd.

Boscobel ; or, Loyal oak : a tale of the year 1651.   ...William Harrison Ainsworth.

Caleb Field : a tale of the Puritans.   ...Margaret O. W. Oliphant.

Captain Jacques : a romance of the time of the plague.   ...Edward Fitzgibbon.   (Pseud., Somerville Gibney.)

Captain of the guard.   ...James Grant.

CHARLES II. OF ENGLAND — *Continued.*

Pattie Durant: a tale of 1662. ...Ellen Clacy. (Pseud., Cycla.)

Peter, the apprentice : a tale of the restoration. ...Emma Leslie.

Peveril of the peak. ...Sir Walter Scott.

Le prophète irlandois. ...Charles de Saint-Evremond.

Puritan's grave. ...William P. Scargill.

Robber. ...George Payne Rainsford James.

Das Roggenhaus Komplott. ...Georg Hiltl.

Rupert Aubrey of Aubrey Chase : an historical tale of 1681. ...Thomas J. Potter.

Russell : a tale of the reign of Charles II. ...George Payne Rainsford James.

Rye House plot. ...George W. M. Reynolds.

Saint Dunstan's clock : a story of 1666. ...E. Ward.

Saint Valentine's day. ...Elbridge S. Brooks. (In his Storied holidays.)

Shepherd of Grove Hall : a tale of 1662. ...Anon.

Silas Verney : being the story of his adventures in the days of King Charles the Second. ...Edgar Pickering.

Sir Ralph Esher; or, Memoirs of a gentleman in the court of Charles the Second. ...Leigh Hunt.

Stanfield Hall. ...J. Frederick Smith.

Talbot Harland. ...William Harrison Ainsworth.

Through unknown ways; or, Journal-books of Dorothea Trundel. ...Lucy Ellen Guernsey.

Traitor or patriot ? a tale of the Rye House plot. ...Mary C. Roswell.

True hero; or, Story of William Penn. ...William Henry Giles Kingston.

Truth ; or, Persis Clareton. ...Charles B. Tayler.

Two swords : being a story of old Bristol. ...Emma Marshall.

Wearyholm; or, Seedtime and harvest : a tale of the restoration of Charles II. ...Emily Sarah Holt.

Whitefriars; or, Times and days of Charles the Second. ...Jane Robinson.

Winchester Meads in the time of Thomas Ken. ...Emma Marshall.

Woodstock ; or, Cavalier : a tale of the year 1651. ...Sir Walter Scott.

See also **ENGLISH HISTORY, Seventeenth Century.**

CHARLES (THE BOLD) OF BURGUNDY. *Reigned 1467–1477.*

Anne of Geierstein. ...Sir Walter Scott.

Charles le Téméraire. ...Alexandre Dumas.

Charles the Bold. ...Alexandre Dumas. English translation.

Cloister and the hearth. ...Charles Reade.

Marie de Bourgogne. ...Mme. de Saint-Venant.

CHARLES (THE BOLD) OF BURGUNDY — *Continued.*
Mary of Burgundy; or, Revolt of Ghent. ...George Payne Rainsford James.
Quentin Durward. ...Sir Walter Scott.

CHARLES V. OF FRANCE. *Reigned 1364–1380.*
Bertrand Du Guesclin. ...Céline Fallet.
Charles-le-Mauvais; ou, La cour de Navarre. ...Élisabeth Guénard.
Chronique de Bertrand Du Guesclin. ...Cuvelier.
Du Guesclin et Jeanne d'Arc; ou, La France aux XIVe et XVe siècles. ...Léopold Favre.
Le jeune Du Guesclin. ...René de Mont-Louis.
White company. ...Arthur Conan Doyle.

CHARLES VI. OF FRANCE. *Reigned 1390–1422.*
Agincourt. ...George Payne Rainsford James.
De Foix: a romance of Béarn of the fourteenth century. ...Anna Eliza Bray.
Henry of Monmouth; or, Field of Agincourt. ...Major Michel.
Isabel de Bavière. ...Alexandre Dumas.
Isabel of Bavaria; or, Chronicles of France for the reign of Charles VI. ...Alexandre Dumas. English translation.
Joan the maid, deliverer of France and England. ...Elizabeth Charles.
King of a day. ...Florence Wilford.
Lily of Paris; or, King's nurse. ...John P. Simpson.
Phélippa: souvenirs du règne de Charles VI. ...C. Guénot.
Provost of Paris: a tale. ...William S. Browning.
Sword of De Bardwell: a tale of Agincourt. ...C. M. Katherine Phipps.

CHARLES VII. OF FRANCE. *Reigned 1422–1461.*
Agnes Sorel. ...George Payne Rainsford James.
Agnès Sorel; ou, La cour de Charles VII. ...Élisabeth Guénard.
Les écorcheurs; ou, L'usurpation et la peste, fragmens historiques. 1418. ...Victor, Vicomte d'Arlincourt.
Le fratricide; ou, Gilles de Bretagne. ...Joseph A. Welsh.
Personal recollections of Joan of Arc. ...Samuel Langhorne Clemens. (Pseud., Mark Twain.)
Le prince de Bretagne. ...François T. M. B. d'Arnaud.
See also **JOAN OF ARC.**

CHARLES VIII. OF FRANCE. *Reigned 1483–1498.*
Histoire de Jean de Paris, roi de France. ...Anon.

CHARLES IX. OF FRANCE. *Reigned 1560–1574.*
Astrologer's daughter. ...Rose E. Temple.
Le capitaine muet. ...Adolphe Racot.
Catherine de' Medici. ...Honoré de Balzac. English translation.

CHARLES IX. OF FRANCE — *Continued.*

CHARLES X. OF FRANCE. *Reigned 1824-1830.*

CHARLES V. OF GERMANY AND I. OF SPAIN. *Reigned 1519–1556.*

Christian prince, Wolfgang, prince of Anhalt.   ...Franz Hoffmann. English translation.

Constancia's household : a story of the Spanish reformation. ...Emma Leslie.

Fürst Wolfgang : historische Erzählung.   ...Franz Hoffmann.

Kaiser Carl·des Fünften erste Jugendliebe.   ...Ludwig A. Arnin.

Karl von Spanien.   ...Ludwig Storch.

Lichtenstein.   ...Wilhelm Hauff.

Maurice, the elector of Saxony.   ...Katharine Colquhoun.

Moritz von Sachsen.   ...F. C. Schlenkert.

Saint Leon : a tale of the sixteenth century.   ...Caleb Williams Godwin.

Die Schlacht bei Drakenburg.   ...Werner Bergman.

Sibylle von Cleve.   ...Julius Bacher.

Tabithe von Geyersberg.   ...Amalie Schoppe.

CHARLES XII. OF SWEDEN. *Reigned 1697–1718.*

Der Eisenkopf : eine historische Erzählung.   ...Franz Hoffmann.

Fältskärns berättelser.   ...Zacharias Topelius.   (6 volumes.)

Der Freibeuter.   ...Ludwig Storch.

Iron Head.   ...Franz Hoffmann.   English translation.

Last of the free-booters.   ...P. Sparre.

Stiff-necked king.   ...Franz Hoffmann.   English translation.

Surgeon's stories.   ...Zacharias Topelius.   English translation.   (6 volumes.)

Ur Karl XII's ungdom.   ...Johan Börjesson.

CHARLESTON, SOUTH CAROLINA.

Earth trembled.   ...Edward Payson Roe.

In old Saint Stephen's.   ...Jeanie Drake.

Mount Benedict.   ...Peter McCorry.

Partisan : a romance of the revolution.   ...William Gilmore Simms.

Scout.   ...William Gilmore Simms.

CHARTISM.

Alton Locke.   ...Charles Kingsley.

Convict.   ...George Payne Rainsford James.

Gaythorne Hall.   ...John M. Fothergill.

Love and a quiet life.   ...Walter Raymond.

Sybil.   ...Benjamin Disraeli.

CHICAGO, ILLINOIS.

Barriers burned away.   ...Edward Payson Roe.

Cliff-dwellers.   ...Henry B. Fuller.

Foiled by a lawyer.   ...Anon.

CHICAGO, ILLINOIS — Continued.
George's mother.   ...Stephen Crane.
Hardscrabble ; or, Fall of Chicago.   ...John Richardson.
Lucky number : a book of stories of the Chicago slums.   ...I. K. Friedman.
Wau-nan-gee ; or, Massacre at Chicago : a romance of the American revolution.   ...John Richardson.
With the procession.   ...Henry B. Fuller.

CHILDREN'S CRUSADES.   See **CRUSADES, Children's.**

CHILD-LIFE.
Adventures of Huckleberry Finn.   ...Samuel Langhorne Clemens. (Pseud., Mark Twain.)
Bimbi.   ...Louise De la Ramé.   (Pseud., Ouida.)
Bushy: a romance founded on fact.   ...Cynthia M. Westover (Mrs. I. Alden).
Captain January.   ...Laura E. Richards.
Child life in New England.   ...Sarah L. Hall.
Child world.   ...Mary Abigail Dodge.   (Pseud., Gail Hamilton.)
Child's classics of prose.   ...Mary R. Fitch Pierce, Compiler.
Clovernook children.   ...Alice Cary.
Five little Peppers and how they grew.   ...Margaret Sidney.
Giovanni and the other children who have made stories.   ...Frances Hodgson Burnett.
Helen's babies.   ...John Habberton.
Italian child life.   ...Marietta Ambrosi.
Jolly good times ; or, Child life on a farm.   ...Mary P. Smith. (Pseud., P. Thorne.)
Jo's boys.   ...Louisa May Alcott.
Läsning för Barn.   ...Zacharias Topelius.
Little girl of long ago.   ...Eliza Orne White.
Little Lord Fauntleroy.   ...Frances Hodgson Burnett.
Little men.   ...Louisa May Alcott.
Little Saint Elizabeth.   ...Frances Hodgson Burnett.
Little women.   ...Louisa May Alcott.
Long ago : a year of child life.   ...Louise T. Cragin.
Loyal little red-coat.   ...Ruth Ogden.
Marm Liza.   ...Kate Douglas Wiggin (Mrs. George Riggs).
Mary of Lorraine : an historical romance.   ...James Grant.
No heroes.   ...Blanche Willis Howard.
Old curiosity shop.   ...Charles Dickens.
Old-fashioned girl.   ...Louisa May Alcott.
One I knew best of all.   ...Frances Hodgson Burnett.
One of the McIntyres.   ...Amelia Holbrook.
Reading for children.   ...Zacharias Topelius.   English translation.

CHILD-LIFE — *Continued.*
Robin's recruit. ...A. G. Plympton.
Rocky Fork. ...Mary Hartwell Catherwood.
Sara Crewe. ...Frances Hodgson Burnett.
Sentimental Tommy: the story of his boyhood. ...James Matthew
  Barrie.
Story of a baby. ...Ethel S. Turner.
Story of a child. ...Margaret Deland.
Story of Patsy. ...Kate Douglas Wiggin (Mrs. George Riggs).
Tell me a story. ...Mary L. Molesworth. (Pseud., Ennis Graham.)
Three Greek children. ...Alfred John Church.
Timothy's quest. ...Kate Douglas Wiggin (Mrs. George Riggs).
Uncle Bob's baby. ...Wilbur Fisk Brown.
When Molly was six. ...Eliza Orne White.
Wreck of the golden fleece : the story of a North sea fisher-boy.
  ...Robert Leighton.
Young castaways. ...Florence C. Dixie.

CHINA.
War-tiger: a tale of the conquest of China. ...Anon.

CHOPIN, FREDERIC. *Lived 1810–1849.*
Lucrezia Floriani Lavinia. ...Amantine L. A. D. Dudevant.
  (George Sand, Pseud.) (Prince Carol personates Chopin.)

CHRIST, LIVES OF.
Ahasvérus. ...Edgar Quinet.
Am Kreuz : ein Passionsroman aus Oberammergau. ...Wilhelmine
  von Hillern.
Antipas, son of Chuza, and others whom Jesus loved. ...Louisa
  Seymour Houghton.
Asa of Bethlehem and his household. ...Mary E. Jennings.
Barabbas. ...Marie Corelli.
Ben Hur : a tale of the Christ. ...Lew Wallace.
Doom of the holy city. ...Lydia Hoyt Farmer.
Emmanuel : the story of the Messiah. ...William Forbes Cooley.
Golden age. ...Kenneth Grahame.
Julian. ...William Ware.
On the cross. ...Wilhelmine von Hillern. English translation.
Philochristus. ...Edwin A. Abbott.
Prince of the House of David; or, Three years in the Holy City :
  Scenes in the life of Jesus. ...Joseph Holt Ingraham.
Quiet King. ...Caroline Atwater Mason.
Trooper Peter Halket of Mashonaland. ...Olive Schreiner. (Pseud.,
  Ralph Iron.)
Story of the other wise man. ...Henry Van Dyke.

CHRIST, LIVES OF — *Continued.*
Salathiel the immortal.   ...George Croly.
See also **BIBLE, Special History and Characters**, and **CHURCH HISTORY, First Century.**

CHRISTIAN SCIENCE AND FAITH HEALING.
Faith doctor.   ...Edward Eggleston.
Lourdes.   ...Émile Zola.
Lourdes.   ...Émile Zola.   English translation.
Siegfried the Mystic.   ...Ida Worden Wheeler.
God's light as it came to me.   ...Anon.

CHRISTMAS.
At last ! Christmas in the West Indies.   ...Charles Kingsley.
Bachelor's Christmas, and other stories.   ...Robert Grant.
Birds' Christmas carol.   ...Kate Douglas Wiggin (Mrs. George Riggs).
Breaking up of the ice : a Christmas tale.   ...Anon.
Christmas at Narragansett.   ...Edward Everett Hale.   ·
Christmas at sea.   ...E. Shippen.
Christmas at Surf-Point.   ...Willis B. Allen.
Christmas at the Pole; or, God everywhere : a tale for the young.   ...Laurence Louis Félix Bungener.   English translation.
Christmas books.   ...Charles Dickens.
Christmas cake in four quarters.   ...Anon.
Christmas carol.   ...Charles Dickens.
Christmas evergreens.   ...Willis B. Allen.
Christmas eve and Christmas day.   ...Edward Everett Hale.
Christmas fantasy.   ...Thomas Bailey Aldrich.   (In his Two bites of a cherry.)
Christmas in 1574.   ...Elbridge S. Brooks.   (In his Storied holidays.)
Christmas stories.   ...Charles Dickens.
Christmas stories.   ...Benjamin Leopold Farjeon.
Christmas story of a little church.   ...Grace King.
Christmas week at Bigler's mill.   ...Dora E. W. Spratt.
Claudia and Pudens; or, Early Christmas in Gloucester.   ...Samuel Lysons.
Colonel's Christmas dinner.   ...Charles King.
Cricket on the hearth.   ...Charles Dickens.
Elsie : a Christmas story.   ...Alexander L. Kjelland.   English translation.
First Christmas of New England.   ...Harriet Beecher Stowe.
Good old days ; or, Christmas under Queen Elizabeth. ...Esmé Stuart.
Good spirit : a story for the Christmas fireside.   ...G. Abbott.
Last of the fairies : a Christmas tale.   ...George Payne Rainsford James.

CHRISTMAS — *Continued.*
Mr. Bixby's Christmas visitor.   ...Charles S. Gage.
Our Christmas in a palace.   ...Edward Everett Hale.
Parson's miracle, and My grandmother's grandmother's Christmas candle.   ...Hezekiah Butterworth.
Polly : a Christmas recollection.   ...Thomas Nelson Page.
Red and white.   ...Edward Everett Hale.
Salamander.   ...E. O. Smith.
Santa Claus on a lark.   ...Washington Gladden.
Solomon Crow's Christmas pockets, and other tales.   ...Ruth McEnery Stuart.
Solomon Isaacs : a Christmas story.   ...Benjamin Leopold Farjeon.
Three Christmas days.   ...Caroline E. Davis.
2894 ; or, Fossil man.   ...Walter Browne.
Young patroon.   ...P. H. Myers.

CHRYSOSTOM.   See **SAINT CHRYSOSTOM.**

CHURCH HISTORY.   *A Selection.*
First Century.
Agathocles.   ...Caroline Pichler.
Arius the Libyan.   ...Nathan C. Kouns.
Burning of Rome.   ...Alfred John Church.
Daybreak in Britain.   ...Charlotte Tucker. (Pseud., A. L. O. E.)
Dorcas, the daughter of Faustina.   ...Nathan C. Kouns.
Early Christianity.   ...Charlotte Tucker. (Pseud., A. L. O. E.)
Early dawn; or, Sketches of Christian life in England in the olden times.   ...Elizabeth Charles.
Edol the Druid.   ...William Henry Giles Kingston.
Gaudentius.   ...Gerald S. Davies.
Glaucia.   ...Emma Leslie.
Helena's household : a tale of Rome in first century.   ...James De Mille.
Lapsed but not lost.   ...Elizabeth Charles.
Light from the catacombs.   ...E. L. M.
Narcissus.   ...William B. Carpenter.
Neither Rome nor Judæa.   ...E. Hoven.
Onesimus : memoirs of a disciple of Saint Paul.   ...Edwin A. Abbott.
Pomponia.   ...Mrs. J. B. Peploe (Webb).
Prince of the House of David; or, Three years in the Holy City : Scenes in the life of Jesus.
Quiet King.   ...Caroline Atwater Mason.
Quo Vadis.   ...Henryk Sienkiewicz.
Quo Vadis : a narrative of the time of Nero.   ...Henryk Sienkiewicz. English translation.
Saint Paul in Greece.   ...Gerald S. Davies.

CHURCH HISTORY — Continued.

Salathiel, the immortal.   ...George Croly.
Story of the other wise man.   ...Henry Van Dyke.
To the lions.   ...Alfred John Church.
Triumphs of the cross.   ...John Mason Neale.   (Pseud., Aurelius Gratianus.)
Valeria.   ...Ballydear and Bowden.
Valerius : a Roman story.   ...John Gibson Lockhart.
Vestal.   ...Mme. de La Grange.
Victory of the vanquished.   ...Elizabeth Charles.
Work while ye have light.   ...Lyeff Tolstoi.   English translation.

Third Century.

Aurelian.   ...William Ware.
Callista : a sketch of the third century.   ...John Henry Newman.
Child martyr and early Christians at Rome.   ...Anon.
Theban legion : a story of the times of Diocletian.   ...William M. Blackburn.

Fourth Century.

Alypius of Tagaste : a tale of the early church.   ...Mrs. J. B. Peploe (Webb).
Conquering and to conquer : story of Rome in the days of Saint Jerome.   ...Elizabeth Charles.
Erling; or, Days of Saint Olaf.   ...F. Scarlett Potter.
Gathering clouds.   ...Frederick William Farrar.
Out of the mouth of the lions.   ...Anon.
Quadratus : a tale of the world in the Church.   ...Emma Leslie.

Fifth Century.

Alypius of Tagaste : a tale of the early Church.   ...Mrs. J. B. Peploe (Webb).
Conquering and to conquer.   ...Elizabeth Charles.
Fabiola; or, Church of the catacombs.   ...Nicholas Patrick Stephen Wiseman.
Hypatia.   ...Charles Kingsley.
Maid and Cleon.   ...Elizabeth Charles.
Pearl of Antioch : a picture of the East at the end of the fourth century.   ...Marc A. Bayle.
Quadratus : a tale of the world in the church.   ...Emma Leslie.

Tenth Century.

Adventures of Olaf Tryggveson, king of Norway.   ...Pamelia M. Reed (Mrs. Joseph J. Reed).
Erling; or, Days of Saint Olaf.   ...F. Scarlett Potter.
Hakon Jarl.   ...Oehlenschlager.
Heroes of the North; or, Stories from Norwegian Chronicle.   ...F. Scarlett Potter.

CHURCH HISTORY — *Continued.*

Eleventh Century.

Ivo and Vereno.' ...Anon.

Fifteenth Century.

Before the dawn : a story of Paris and the Jacquerie. ...George Dulac.

Conrad. ...Emma Leslie.

Coulying Castle ; or, Knight of the olden days. ...Agnes Giberne.

Dearer than life : a tale of the times of Wycliffe. ...Emma Leslie.

For or against. ...Frances M. Wilbraham.

Geoffrey the Lollard. ...Mrs. D. C. Knevels. (Pseud., Frances Eastwood.)

Gilbert Wright, the gospeller. ...F. S. Merryweather.

Gladys of Herleck. ...Anon.

Hubert Ellerdale. ...W. O. Rhind.

In Wicliff's days. ...G. Stebbing.

Jack of the mill. ...William Howitt.

John de Wycliffe, the first reformer. ...Emily Sarah Holt.

Knight of Dilham. ...Arthur Brown.

Lollard. ...Minnie K. Davis.

Lollard priest. ...Henry C. Adams.

Lollards. ...Thomas Gaspey.

Margery's son : a fifteenth century tale of the court of Scotland. ...Emily Sarah Holt.

Mistress Margery : a tale of the Lollards. ...Emily Sarah Holt.

Richard Hunne : a story of old London. ...George E. Sargent.

White rose of Langley : a story of the court of England in the olden time. ...Emily Sarah Holt.

Wycliffites ; or, England in the fifteenth century. ...Margaret Mackay.

See also **WYCLIFFE, JOHN.**

Sixteenth and Seventeenth Centuries.

Bolsover Castle : a tale from Protestant history of the sixteenth century. ...Anon.

Chief's daughter. ...Anon.

Dark year of Dundee. ...Deborah Alcock.

Darnley ; or, Field of the cloth of gold. ...George Payne Rainsford James.

For the Master's sake : a story of the days of Queen Mary. ...Emily Sarah Holt.

Isoult Barry of Wynscote : a tale of Tudor times. ...Emily Sarah Holt.

Kenilworth. ...Sir Walter Scott.

Lettice Eden : a tale of the last days of Henry the Eighth. ..Emily Sarah Holt.

CHURCH HISTORY — *Continued.*

Magdalen Hepburn.   ...Margaret O. W. Oliphant.

No cross, no crown.   ...Caroline C. Davis.

Robin Tremayne of Bodmin : a tale of the Marian persecution. ...Emily Sarah Holt.

Steadfast.   ...Rose Terry Cooke.

Three hundred years ago.   ...William Henry Giles Kingston.

Twice crowned : a story of the days of Queen Mary.   ...Harriet B. MacKeever.

Uline's escape; or, Hid with the nuns.   ...Mrs. Alexander S. Orr.

Westminster Abbey.   ...Emma Robinson.

Within sea walls ; or, How the Dutch kept the faith.   ...E. H. Walshe and George E. Sargent.

See also **REFORMATION.**

Eighteenth Century.

Ashcliffe Hall.   ...Emily Sarah Holt.

Nineteenth Century.

Tithe-proctor.   ...William Carleton.

See also **PERSECUTIONS.**

CIVIL WAR. *A Selection.*

England. 1625.

Alice Bridge of Norwich : a tale of the time of Charles the First. ...Andrew Reed.

Alice Leighton : or, Good name is rather to be chosen than riches. ...Ann J. Cupples (Mrs. George Cupples).

Alice Lisle : a tale of Puritan times.   ...Richard King.

Andrew Marvel and his friends : a story of the siege of Hull. ...Marie S. Hall.

Armourer's daughter ; or, Border rivals.   ...Anon.

Arrah Neil.   ...George Payne Rainsford James.

Buccaneer.   ...Anna Maria Hall.

Castle Cornet ; or, Island's troubles in troublous times.   ...Louisa Hawtrey.

Children of the New Forest.   ...Frederick Marryat.

Cromwell ; or, Protector's oath.   ...J. Frederick Smith.

Draytons and Davenports : a story of the civil wars.   ...Elizabeth Charles.

Harry Ogilvie ; or, Black dragoons.   ...James Grant.

Henry Masterton ; or, Adventures of a young cavalier.   ...George Payne Rainsford James.

Her Majesty the Queen.   ...John Esten Cooke.

Hide and seek : a story of the New Forest in 1647.   ...Mrs. Frank Cooper.

CIVIL WAR — *Continued.*

Holmby House; a tale of old Northamptonshire. ...George John Whyte Melville.

John Inglesant. ...Joseph Henry Shorthouse.

John Milton and his times. ...Max Ring. English translation.

John Milton und seine Zeit. ...Max Ring.

Lady Betty's governess; or, Corbet chronicles. ...Lucy Ellen Guernsey.

Lady Shakerlye : the record of the life of a good and noble woman. ...Anon.

Lady Willoughby ; or, Passages from the diary of a wife and mother in the seventeenth century. ...Hannah Manning Rathbone.

Leaguer of Lathom : a tale of the civil war in Lancashire. ...William Harrison Ainsworth.

Lettice : a tale of the siege of Chester. ...Pauline Biddulph.

Life of Colonel Jack. ...Daniel Defoe.

Love the leveller : a tale of the great rebellion. ...Angus Comyn.

Markhams of Ollerton : a tale of the civil war. ...Elizabeth Glaister.

Maudeville : a tale of the seventeenth century in England. ...William Godwin.

Memoirs of a cavalier. ...Daniel Defoe.

Memoirs of troublous times. ...Emma Marshall.

Old Noll ; or, Days of the Ironsides. ...Frederick William Robinson.

Oliver Cromwell ; or, England's great protector. ...Henry W. Herbert. (Pseud., Frank Forester.)

On both sides of the sea: a story of the commonwealth and the restoration. ...Elizabeth Charles.

Ovingdean Grange : a tale of the South Downs. ...William Harrison Ainsworth.

Reginald Hastings : a tale of the troubles of 1649. ...Eliot B. G. Warburton.

Rosamond Fane ; or, Prisoners of Saint James. ...Mary and Catherine Lee.

Saint George and Saint Michael. ...George Macdonald.

Scholar and the trooper. ...William E. Heygate.

Settlers at home. ...Harriet Martineau.

Siege of Colchester. ...George F. Townshend.

Siege of Lichfield. ...William Gresley.

Sir Henry Appleton ; or, Essex during the great rebellion. ...William E. Heygate.

Stanfield Hall. ...J. Frederick Smith.

Two knights of Delany Castle. ...Mary M. Sherwood.

Washingtons : a tale of a country parish in the seventeenth century. ...John N. Simpkinson.

White gauntlet. ...Mayne Reid.

5

CIVIL WAR — *Continued.*

Whitehall : or, Days and times of Oliver Cromwell.   ...Jane Robinson.

Young Castellan : a tale of the English civil war.   ...George Manville Fenn.

- See also **CHARLES I. OF ENGLAND.**

United States. 1861.

Aboard the Atlantic.   ...Henry Frith.

American mail bag ; or, Tales of the war.   ...Anon.

Among the camps.   ...Thomas Nelson Page.

Among the cotton thieves.   ...Edward Bacon.

Among the guerillas.   ...James Roberts Gilmore.   (Pseud., Edmund Kirke.)

Angel of the battlefield.   ... Wesley Bradshaw.

" As we went marching on."   ...George W. Hosmer.

At anchor : a story of our civil war.   ...Anon.

Baby Rue : her friends and her enemies.   ...Charlotte M. Clark.   (Pseud., Charles M. Clay.)

Bertha the beauty : a story of the Southern revolution.   ...Sarah J. C. Whittlesey.

Between the lines.   ...Charles King.

Black angel : a tale of the American civil war.   ... William Stephens Hayward.

Bloody chasm.   ...John William De Forest.

Bloody junto ; or, Escape of John Wilkes Booth.   ...R. H. Crozier.

Border war : a tale of disunion.   ...John B. Jones.

Brave old salt ; or, Life on the quarter deck.   ... William Taylor Adams.   (Pseud., Oliver Optic.)

Bristling with thorns : a story of war and reconstruction.   ...O. T. Beard.

Brother soldiers.   ...Mary S. Robinson.

Brownings.   ...Jane G. Fuller.

Cameron Hall : a story of the civil war.   ...Mary A. Cruse.

Chattanooga.   ...John Jolliffe.

Colonel Dunwoddie, millionaire.   ...William Mumford Baker.   (Pseud., George H. Harrington.)

Colonel's daughter ; or, Winning his spurs.   ...Charles King.

Confederate flag on the ocean : a tale of the cruises of the Sumter and Alabama.   ...William H. Peck.

Confederate spy.   ...R. H. Crozier.

Conspirators.   ...William H. Peck.

Cotton stealing.   ...Anon.

Daisy Swain, the flower of the Shenandoah.   ...Anon.

Days of Shoddy : a novel of the great rebellion in 1861.   ...Henry Morford.

CIVIL WAR — *Continued.*

Dora Darling ; the daughter of the regiment. ...Jane Goodwin Austin.

'89 ; or, Grand master's secret. ...Edgar Henry.

Elopement : a tale of the Confederate States of America. ...L. Fairfax.

Etowah : a romance of the Confederacy. ...Francis Fontaine.

Fairfax : a chronicle of the valley of Shenandoah. ...John Esten Cooke.

Fiery cross : a tale of the great American war. ...William Stephens Hayward.

Fighting Joe ; or, Fortunes of a staff officer. ...William Taylor Adams. (Pseud., Oliver Optic.)

Fighting Quakers : a true story of our war for our union. ...Augustine J. H. Duganne.

Fitz-Hugh Saint Clair, the South Carolina rebel boy. ...Sallie F. Chapin.

Fort Lafayette ; or, Love and secession. ...Benjamin Wood.

Forward with the flag. ...Mary S. Robinson.

Gilead guards : a story of war times in a New England town. ...Mrs. O. W. Scott.

Gray and the blue. ...Edward E. Roe.

Great battle year. ...Mary S. Robinson.

Gun-bearer. ...Edward A. Robinson.

Hammer and rapier. ...John Esten Cooke.

Heroine of the Confederacy. ...Florence J. O'Connor.

Hilt to hilt ; or, Days and nights in the Shenandoah in the autumn of 1864. ...John Esten Cooke.

His sombre rivals. ...Edward Payson Roe.

Home scenes during the Rebellion. ...Margaret Roberts.

Hospital sketches, and camp and fireside stories. ...Louisa May Alcott.

Inside : a chronicle of secession. ...William Mumford Baker.

In war time. ...S. Weir Mitchell.

In war times at La Rose Blanche. ...M. E. M. Davis.

John Charáxes : a tale of the civil war in America. ...George Ticknor Curtis.

Katy of Catochin. ...George A. Townsend.

Kernwood ; or, After many days. ...Virginia L. French.

Little bugler. ...George M. Royce.

Little regiment, and other episodes of the American civil war. ...Stephen Crane.

Love and rebellion. ...Martha C. Keller.

Lucia Dare. ...Sarah A. Dorsey. (Pseud., Filia.)

McDonalds ; or, Ashes of Southern homes : a tale of Sherman's march. ...William H. Peck.

CIVIL WAR — *Continued.*

Macpherson, the great Confederate philosopher. ...Alfred C. Hills.

Marcy, the blockade-runner. ...Charles A. Fosdick. (Pseud., Harry Castlemon.)

Millicent Halford: a tale of the dark days of Kentucky in 1861. ...Martha Remick.

" Miss Lou." ...Edward Payson Roe.

Miss Ravenel's conversion from secession to loyalty. ...John William De Forest.

Modern Hagar. ...Charlotte M. Clark. (Pseud., Charles M. Clay.)

Mohun; or, Last days of Lee and his paladins. ...John Esten Cooke.

On the blockade. ...William Taylor Adams. (Pseud., Oliver Optic.)

On the plantation: a story of a Georgia boy's adventures during the war. ...Joel Chandler Harris.

Original belle. ...Edward Payson Roe.

Other fools and their doings; or, Life among the freedmen. ...Harriet N. K. Goff.

Out of the foam. ...John Esten Cooke.

Patriotism at home; or, Young invincibles. ...I. H. Anderson.

Poor white; or, Rebel conscript. ...Emily C. Pearson.

Prince Paul: the freedman soldier. ...Emily C. Pearson.

Princess of Peele. ...William Westall.

Rattlesnake. ...Edward Z. C. Judson. (Pseud., Ned Buntline.)

Rebel fiend; or, Scout of secession. ...W. D. Reynolds.

Red acorn. ...John McElroy.

Red badge of courage. ...Stephen Crane.

Refugee; or, Union boys of '61. ...Paul Pritchard.

Remy Saint Remy; or, Boy in blue. ...Mrs. C. H. Gildersleeve.

Reunited: a story of the civil war. ...Alfred R. Calhoun.

Rival volunteers; or, Black plume rifles. ...Mary A. Howe.

Robert Warren; or, Texan refugee. ...Anon.

Rodney, the partisan. ...Charles A. Fosdick. (Pseud., Harry Castlemon.)

Roebuck. ...C. W. Russell.

Roland Blake. ...S. Weir Mitchell.

Romance of the republic. ...Lydia Maria Child.

Rose Mather: a tale of the war. ...Mary J. Holmes.

Running the blockade. ...William H. Thomes.

Sailor boy. ...William Taylor Adams. (Pseud., Oliver Optic)

Sailor boys of '61. ...James R. Soley.

Sanctuary: a story of the civil war. ...George W. Nichols.

Shelby's expedition to Mexico. ...John N. Edwards.

Shoulder straps: a novel of New York and the army. ...Henry Morford.

CIVIL WAR — *Continued.*

Siege of Spolete : a camp tale of Arlington Heights. ...Michael J. A. McCaffery.

Soldier boy. ...William Taylor Adams. (Pseud., Oliver Optic.)

Die Spionin. ...Adolph Schermer.

Stand by the Union. ...William Taylor Adams. (Pseud., Oliver Optic.)

Star of the South. ...W. S. Hayward.

Stars and bars ; or, Reign of terror in Missouri. ...I. Kelso.

Stories of the civil war. ...Albert F. Blaisdell, ed.

Surry of Eagle's nest ; or, Memoirs of a staff officer serving in Virginia. ...John Esten Cooke.

Taken by the enemy. ...William Taylor Adams. (Pseud., Oliver Optic.)

Tale of New Orleans life and of the present war. ...Clarimonde.

Thinking bayonet. ...James K. Hosmer.

Tiger-lilies. ...Sidney Lanier.

Throckmorton. ...Molly E. Seawell.

Tobias Wilson : a tale of the great rebellion. ...Jeremiah Clemens.

Tom Clifton ; or, Western boys in Grant's and Sherman's army. ...W. L. Goss.

Tried and true ; or, Love and loyalty. ...Bella Z. Spencer.

True to his colors. ...Charles A. Fosdick. (Pseud., Harry Castlemon.)

Two college friends. ...Frederick W. Loring.

Two little confederates. ...Thomas Nelson Page.

Uncle Daniel's story of Tom Anderson and twenty great battles. ...Anon.

Union : a story of the great rebellion. ...John R. Musick.

Unofficial patriot. ...Helen H. Gardener.

Valerie Aylmer. ...Frances C. Fisher.

Victor. ...Ellery Sinclair.

Virginia Graham, the spy of the grand army. ...Justin Jones.

Waiting for the verdict. ...Rebecca Harding Davis.

War-time wooing. ...Charles King.

Wearing of the gray. .. John Esten Cooke.

Weiss und Schwarz. ...Friedrich W. Arming.

When the war broke out. ...Edward A. Rand.

Wild work. ...Mary Edwards Bryan.

Winning his way. ...Charles Carleton Coffin.

With gauge and swallow. ...Albion Winegar Tourgee.

Within the enemy's lines. ...William Taylor Adams. (Pseud., Oliver Optic.)

With Lee in Virginia. ...George Alfred Henty.

Without blemish : to-day's problem. ...J. H. Walworth.

CIVIL WAR — *Continued.*
Women ; or, Chronicles of the late war. ...Mary T. Magill.
Work of the two great captains. ...Mary S. Robinson.
Yankee middy ; or, Adventures of a naval officer. ...William Taylor Adams. (Pseud., Oliver Optic.)
Year of wreck. ...George C. Benham.
Young lieutenant. ...William Taylor Adams. (Pseud., Oliver Optic.)

CLOVIS I. OF FRANCE. *Reigned 489-511.*
Ierne of Armorica. ...J. C. Bateman.

COLERIDGE, SAMUEL TAYLOR. *Lived 1772-1834.*
Days of Lamb and Coleridge. ...Alice E. Lord.

COLLEGE AND UNIVERSITY LIFE. *A Selection.*
Adventures of Mr. Verdant Green. ...Edward Bradley.
Babe. B. A. ...Edward F. Benson.
Cambridge freshman. ...Martin Legrand.
Chapters of a college romance. ...Isaac Butts.
Charlie Lucken at school and college. ...Henry C. Adams.
College days at Oxford. ...Henry C. Adams.
College days ; or, Harry's career at Yale. ...John S. Wood.
College girls. ...Abbe C. Goodloe.
College life. ...Joseph T. J. Hewlett.
Dashwood Priory. ...Emily Juliana May.
Donald Marcy. ...Elizabeth Stuart Phelps Ward (Mrs. Herbert D. Ward).
Fair Harvard. ...William T. Washburn.
Forbes of Harvard : a story of college life in the early fifties. ...Elbert Hubbard.
Hammersmith : his Harvard days. ...Mark Sibley Severance.
Harvard stories : sketches of the undergraduate. ...Waldron K. Post.
Julian Home : a tale of college life. ...Frederick William Farrar.
Junior Dean. ...Alan Saint Aubin.
Lost manuscript. ...Gustav Freytag. English translation.
New Academe. ...Edward Hartington.
Princetonian : a story of undergraduate life at the College of New Jersey. ...James Barnes.
Reginald Dalton : a story of English university life. ...John Gibson Lockhart.
School-boy honour : a tale of Talminster College. ...Henry C. Adams.
Student life at Harvard. ...George H. Tripp.
Tales of college life. ...Edward Bradley.
Thing that hath been. ...A. H. Gilkes.

COLLEGE AND UNIVERSITY LIFE — *Continued.*
Two college boys. ...Edward A. Rand.
Two college friends. ...Frederick W. Loring.
Two college girls. ...Helen D. Brown.
Die verlorene Handschrift. ...Gustav Freytag.
Wooden spoon. ...Theron Brown.
With the best intentions : a tale of undergraduate life at Cambridge (Eng). ...John Bickerdyke.
Yale yarns. ...John Seymour Wood.

COLORADO.
At the end of the rainbow. ...Julia A. Sabin.
Doctor Grattan. ...William A. Hammond.
In the heart of the Rockies : a story of adventure in Colorado. ...George Alfred Henty.
Lal. ...William A. Hammond.
Naulahka. ...Rudyard Kipling.
Nelly's silver mine. ...Helen Hunt Jackson.
Peak and prairie. ...Anna Fuller.
Silver caves : a mining story. ...Ernest Ingersoll.
Story of a cañon. ...Beveridge Hill.
Zeph. ...Helen Hunt Jackson.

COLUMBUS, CHRISTOPHER. *Lived 1436-1506.*
Christophe Colomb. ...A. Dousseau.
Columbia : a story of the discovery of America. ...John R. Musick.
Columbus and Beatriz. ...Constance Goddard DuBois.
Cristoforo Colombo. ...Ludwig August Frankl.
Diccon the Bold. ...John Russell Coryell.
Diego Pinzon and the fearful voyage he took into the unknown ocean, A. D. 1492. ...John Russell Coryell.
Die heroischen Epopeën, Gustav Wasa und Columbus. ...Franz Michael Franzen.
Ismael ben Kaissar. ...J. F. Denis.
Legend of Christopher Columbus. ...Joanna Baillie.
Mercedes of Castile ; or, Voyage to Cathay. ...James Fenimore Cooper.
El nuevo mundo, descubierto por Cristoval Colon. ...L. F. de Vega-Carpio.
Out of the sunset sea. ...Albion Winegar Tourgee.
Story of Columbus. ...Elizabeth Eggleston Seeley.
Tales from American history, containing the principal facts in the life of Christopher Columbus. ...Eliza Robbins.
With the admiral of the ocean sea. ...C. P. MacKie.

COMMONWEALTH OF ENGLAND. *1649–1660.*

Andrew Marvel and his friends : a story of the siege of Hull. ...Maria S. Hall.

Armourer's daughter ; or, Border rivals. ...Anon.

Boscobel ; or, Royal oak : a tale of the year 1651. ...William Harrison Ainsworth.

Brambletye House ; or, Cavaliers and roundheads. ...Horace Smith (Paul Chatfield).

Chevalier's daughter. ...Lucy Ellen Guernsey.

Commonwealth of Oceana. ...James Harrington.

Constance Aylmer : a story of the 17th century. ...Helen F. Parker.

Cromwell ; or, Protector's oath. ...J. Frederick Smith.

Days of Laud and of the Commonwealth. ...Mrs. Newton Crosland. (Pseud., Camilla Toulmin.)

Draytons and Davenants : a story of the civil wars. ...Elizabeth Charles.

Hanley Castle : an episode of the civil wars and the battle of Worcester. ...William S. Symonds.

Henry Masterton. ...George Payne Rainsford James.

In bonds but fetterless. ...Richard Cuninghame.

Ironsides. ...Anon.

John Milton and his times. ...Max Ring. English translation.

John Milton und seine Zeit. ...Max Ring.

King's namesake. ...Catherine M. Phillimore.

King's ransom. ...Anon.

Last of the fairies : a Christmas tale. ...George Payne Rainsford James.

Life of Colonel Jack. ...Daniel Defoe.

Maiden and married life of Mary Powell, afterwards Mistress Milton. ...Anne Manning.

Marmaduke Wyvil ; or, Maid's revenge. ...Henry W. Herbert. (Pseud., Frank Forester.)

Memoirs of a cavalier. ...Daniel Defoe.

Memoirs of troublous times. ...Emma Marshall.

Old Noll ; or, Days of the Ironsides. ...Frederick William Robinson.

On both sides of the sea : a story of the commonwealth and the restoration. ...Elizabeth Charles.

Ovingdean Grange : a tale of the South Downs. ...William Harrison Ainsworth.

Pigeon pie : a tale of roundhead times. ...Charlotte Mary Yonge.

Reginald Hastings : a tale of the troubles of 164-. ...Eliot B. G. Warburton.

Saint George and Saint Michael. ...George Macdonald.

Saxby : a tale of the commonwealth time. ...Emma Leslie.

Scapegrace Dick. ...Frances M. Peard.

Whitehall ; or, Days of Oliver Cromwell. ...Jane Robinson.

COMMONWEALTH OF ENGLAND — *Continued.*
Woodstock; or, Cavalier : a tale of the year 1651. ...Sir Walter Scott.
See also **CROMWELL, OLIVER**, and **MILTON, JOHN**.

CONGRESS OF VIENNA, Nineteenth Century.
Auf dem Wiener Kongress. ...Julius Bacher.
Der Congress zu Wien. ...Eduard Breier.
Die Geheimnisse des Waldschlosses. ...Reinhard Edmund Hahn.
See also **VIENNA**.

CONNECTICUT.
Famous victory. ...Anon.
Kinley Hollow. ...Gideon H. Hollister.
Orange Grove : a tale of the Connecticut. ...Anon.
Summer in Oldport Harbor. ...W. H. Metcalf.

CONQUEST OF GRANADA.
Alhambra. ...Washington Irving.
Conquest of Granada. ...Washington Irving. [wer-Lytton.
Leila ; or, Siege of Granada. ...Edward George Earle Lytton Bul-
Tales from Spanish history. ...Elizabeth J. Brabazon.

CONQUEST OF MOGUL.
Akbar. ...A. S. Van Limburg-Bronner.

CONSPIRACIES AND PLOTS.
Adventures of Rob Roy. ...James Grant.
Beatrice Tyldesley. ...William Harrison Ainsworth. (Lancashire,
1694.)
Black dwarf. ...Sir Walter Scott. (Jacobite, 1708.)
Le Chevalier d'Harmental. ...Alexandre Dumas.
Cinq-Mars; ou, Conjuration sous Louis XIII. ...Alfred Victor,
Comte de Vigny.
Cinq-Mars : a conspiracy under Louis XIII. ...Alfred Victor de
Vigny. English translation.
Coombe Abbey. ...Selina Bunsbury. (Gunpowder, 1605.)
La conspiration de Walstein. ...Jean François Sarrasin.
Conspirators. ...William H. Peck.
Conspirators; or, Chevalier d'Harmental. ...Alexandre Dumas.
English translation.
Father Darcy. ...Anne Marsh.
For liberty's sake : the story of Robert Ferguson. ...John B. Marsh.
(Ryehouse)
Gowrie ; or, King's plot. ...George Payne Rainsford James.
Guy Fawkes. ...William Harrison Ainsworth.
John Deane of Nottingham. ...William Henry Giles Kingston.
Katy of Catoctin. ...George Alfred Townsend. (Lincoln's assas-
sination.)
King's highway. ...George Payne Rainsford James. (Jacobite.)

CONSPIRACIES AND PLOTS — *Continued.*
Peveril of the Peak.   ...Sir Walter Scott.
Redgauntlet.   ...Sir Walter Scott.
Das Roggenhaus Komplott.   ...George Hiltl.   (Ryehouse, 1685.)
Royal police.   ...John H. Robinson.   (Boston, 1715.)
Ryehouse plot.   ...G. W. Reynolds.
Traitor or patriot ?   ...M. C. Rowsell.   (Ryehouse, 1683.)
Yemasee : a romance of California.   ...William Gilmore Simms.
   (Indian, 1715.)

CONSTANTINE (THE GREAT). *Lived 272-337.*
Cellene.   ...Sneyd.
Constantine.   ...Edmund Spencer.
Evanus : a tale of Constantine the Great.   ...Augustine D. Crake.

CONSTANTINOPLE, TURKEY.
Blue and green.   ...Sir Henry Pottinger.
Count Robert of Paris.   ...Sir Walter Scott.
Dog of Constantinople.   ...Izora C. Chandler.
Prince of India ; or, Why Constantinople fell.   ...Lew Wallace.
Story of the times of Scanderbeg.   ...Anon.
Theodora Phranza ; or, Fall of Constantinople.   ...John Mason Neale.
   (Pseud., Aurelius Gratianus.)

CORDAY, CHARLOTTE. *Lived 1768-1795.*
L'ami du peuple.   ...J. M. Gassier Saint Amand.
Charlotte Corday.   ...Henri Alphonse Esquiros.
Charlotte Corday: an historical novel.   ...Henri Alphonse Esquiros.
   English translation.
Charlotte Corday.   ...Carl W. T. Frenzel.
Charlotte Corday.   ...Rose E. Temple.
Dream-Charlotte.   ...Matilda Barbara Betham-Edwards.
Ingénue.   ...Alexandre Dumas.   [lation.
Ingenue ; or, Death of Marat.   ...Alexandre Dumas.   English trans-
Maid of Normandy.   ...Edmund J. Eyre.

CORNWALL, ENGLAND.   [lantyne.
Deep down : a tale of the Cornish mine.   ...Robert Michael Bal-
Delectable duchy: stories, studies, and sketches.   ...Arthur T. Q.
   Couch.   (Pseud., Q.)
Menhardoc.   ...George Manville Fenn.
Pendower : a story of Cornwall in the time of Henry the Eighth.
   ...Marianne Filluel.
Singer of the sea.   ...Amelia Edith Barr.
True Cornish maid.   ...G. Norway.
Vicar's people.   ...George Manville Fenn.
Watchers on the Longships.   ...James F. Cobb.
Wreckers and Methodists.   ...H. D. Lowry.

COVENANTERS.

Andrew Gillon : a tale of the Scottish covenanters. ...John Tod.
  · (Pseud., John Strathesk.)
Beacon light : a tale of the covenanters. ...Ellen E. Guthrie.
Brownie of Bodsbeck. ...James Hogg.
Fiery cross. ...B. Hutton.
Journal of Lady Beatrix Graham. ...Jane M. F. Smith.
Helen of the glen. ...Robert Pollok.
Hunted and harried : a tale of the Scottish covenanters. ...Robert
  Michael Ballantyne.
In the smoke of war. ...Walter Raymond.
Janet and her father. ...Mary E. Bamford.
Legend of Montrose. ...Sir Walter Scott.
Men of the Moss-Hags. ...Samuel Rutherford Crockett.
Old Mortality. ...Sir Walter Scott.
Peden, the prophet : a tale of the covenanters. ...Andrew M. Brown.
Persecuted family : a tale of the covenanters. ...Robert Pollok
Philip Colville : a covenanter's story. ...Grace Kennedy.
Scottish cavalier. ...James Grant.
Tales of the covenanters. ...Ellen E. Guthrie.
Tales of the covenanters. ...J. W. Wilson.
Tales of the wars of Montrose. ...James Hogg.

CRIMEAN WAR. *1853–1856.*

Beatrice and Benedick : a romance of the Crimea. ...Hawley Smart.
Children of the great king. ...Matilda Horsburgh.
Fall of Sebastopol ; or, Jack Archer in the Crimea. ...George
  Alfred Henty.
Frederick Gordon ; or, Storming of the Redan. ...Anon.
Henri de la Tour ; or, Comrades in arms. ...J. Frederick Smith.
Interpreter : a tale of the Crimean war. ...George John Whyte
  Melville.
Jack Archer : a tale of the Crimea. ...George Alfred Henty.
Lady Wedderburn's wish : a tale of the Crimean war. ...James Grant.
Laura Everingham ; or, Highlanders of Glen Ora. ...James Grant.
Lord Hermitage. ...James Grant.
Mabel Ashton : a tale of the Crimean war. ...Elizabeth Allnatt.
Martyrs to circumstance. ...Maria Thérèse Longworth Yelverton.
One of the six hundred. ...James Grant.
Ravenshoe. ...Henry Kingsley.
Sebastopol. ...Lyeff Nikolaievich Tolstoi. English translation from
  the Russian.
Under the Red Dragon. ...James Grant.
Vera ; or, Russian princess and English earl. ...Charlotte L. H.
  Dempster.
Young Brown ; or, Law of inheritance. ...Grenville Murray.

CRIMINOLOGY. *A Selection.*

Adventures of Sherlock Holmes.   ...Arthur Conan Doyle.
Another crime.   ...Julian Hawthorne.
Artist in crime.   ...Rodrigues Ottolengui.
Book of strange sins.   ...Coulson Kernahan.
Chronicles of Martin Hewitt.   ...Anon.
Conflict of evidence.   ...Rodrigues Ottolengui.
La corde au cou.   ...Émile Gaboriau.
Le crime d'Orcival.   ...Émile Gaboriau.
Crime and punishment.   ...Fedor M. Dostoevsky.   English translation.
Le crime du vieux Blas.   ...C. Mendès.
Crime of Sylvestre Bonnard.   ...Anatole France.
Crime of the century.   ...Rodrigues Ottolengui.
Crime of the 'Liza Jane.   ...Fergus W. Hume.
Criminal.   ...Johann C. Friedrich von Schiller.   English translation.
La dégringolade.   ...Émile Gaboriau.
Le dossier No. 113.   ...Émile Gaboriau.
For fifteen years.   ...Louis Ulbach.   English translation.
Good people of Pawlocz.   ...Coloman Mikszath.
Lady of quality.   ...Frances Hodgson Burnett.
Le marteau d'acier.   ...Louis Ulbach.
Memoirs of Jane Cameron.   ...Frederick William Robinson.
Memoirs of Sherlock Holmes.   ...Arthur Conan Doyle.
M. Lecoq.   ...Émile Gaboriau.
Modern wizard.   ...Rodrigues Ottolengui.
Moondyne.   ...John Boyle O'Reilly.
Mystery of Orcival.   ...Émile Gaboriau.   English translation.
Mystery of Paul Chadwick.   ...J. W. Postgate.
Quality of crime.   ...William Dean Howells.
Quality of mercy.   ...William Dean Howells.
Quinze ans de bagne.   ...Louis Ulbach.
Scarlet letter.   ...Nathaniel Hawthorne.
Section 558.   ...Julian Hawthorne.
Shadow of a crime.   ...Thomas Henry Hall Caine.
Silent witness.   ...Mrs. J. H. Walworth.
Sketches of life and character taken from the police court, Bow Street.   ...Edwin Hodder.   (Pseud., Old Merry.)
Somebody's Ned.   ...Mrs. A. M. Freeman.
Steel hammer.   ...Louis Ulbach.   English translation.
Story of John Coles.   ...M. E. Kenyon.
Study in scarlet.   ...Arthur Conan Doyle.
Veiled hand : a novel of the sixties, seventies, and eighties.   ...Frederick Wicks.

CRIMINOLOGY — *Continued.*

What will he do with it? ...Edward George Earle Lytton Bulwer-Lytton.

Who poisoned Hetty Duncan? ...J. E. Muddock.

See also **PRISONS AND PRISONERS, DETECTIVE STORIES**, and **SIN**.

CROMWELL, OLIVER. *Lived 1599–1658.*

Andrew Marvel and his friends: a story of the siege of Hull. ...Maria S. Hall.

Brambletye House; or, Cavaliers and roundheads. ...Horace Smith. (Pseud., Paul Chatfield.)

Cavaliers. ...S. R. Keightley.

Cavaliers and roundheads. ...John George Edgar.

Chances of war. ...A. Whitelock.

Confederate chieftains. ...Mary A. Sadlier (Mrs. J. Sadlier).

Cromwell; or, Protector's oath. ...J. Frederick Smith.

Ethne. ...Mrs. E. M. Field.

Father John; or, Cromwell in Ireland. ...S. E. A.

Henry Masterton; or, Adventures of a young cavalier. ...George Payne Rainsford James.

John Milton and his times. ..Max Ring. English translation.

John Milton und seine Zeit. ...Max Ring.

Last of the Corbes. ...John Wright.

Life of Colonel Jack. ...Daniel Defoe.

Little wizard. ...Stanley J. Weyman.

Memoirs of troublous times. ...Emma Marshall.

Mistress Spitfire. ...J. S. Fletcher.

Nellie Netterville. ...Cecilia M. Caddell.

Old Noll; or, Days of the Ironsides. ...Frederick William Robinson.

Oliver Cromwell; or, England's great protector. ...Henry W. Herbert. (Pseud., Frank Forester.)

On both sides of the sea; a story of the commonwealth and the restoration. ...Elizabeth Charles.

Silk of the kine. ...L. McManus.

Whitehall; or, Days of Oliver Cromwell. ...Jane Robinson.

Woodstock; or, Cavalier: a tale of the year 1651. ...Sir Walter Scott.

See also **COMMONWEALTH OF ENGLAND** and **ENGLISH HISTORY**.

CRUSADES. *General and Special.*

La bannière bleue, aventures d'un musulman, d'un chrétien et d'un Païen à l'époque des croisades et de la conquête mongole. ...Léon Cahun.

Betrothed. ...Sir Walter Scott. (Third crusade.)

Blue banner. ...Léon Cahun. English translation

CRUSADES — *Continued.*

Boy crusaders. ...John George Edgar. (Seventh crusade.)
Brothers in arms : a story of the crusades. ...F. Bayford Harrison.
Cheshire pilgrims. ...Frances M. Wilbraham.
Count Robert of Paris. ...Sir Walter Scott.
Daventrys. ...Julia Pardoe.
Elfrica. ...Charlotte G. Boger.
Die Fahrten Thiodolfs des Isländers. ...Friedrich H. C. La Motte
    Fouqué.
Fighting the Saracens; or, Boy knight. ...George Alfred Henty.
Florine, princess of Burgundy. ...William B. MacCabe. (First
    crusade.)
Foure prentises of London. ...Thomas Heywood.
Heroines of the crusades. ...Charles A. Bloss.
I Lombardi alla prima crociata. ...Tommaso Grossi.
In the brave days of old: the story of the crusades. ...Henry Frith.
Ivanhoe. ...Sir Walter Scott. (Third crusade.)
Lady Sybil's choice : a tale of the crusades. ...Emily Sarah Holt.
    (Third crusade.)
Longbeard, Lord of London. ...Charles Mackay.
Mathilde, ou mémoires tirés de l'histoire des croisades. ...Sophie
    Risteau Cottin.
Matilda, Princess of England. ...Sophia Risteau Cottin. English
    translation.
Maud and Miriam; or, Fair crusader. ...Harriet B. MacKeever.
    (Third crusade.)
Orphans of Alsace. ...Anon.
Parcival. ...Albert E. Brachvogel.
Philip Augustus ; or, Brothers in arms. ...George Payne Rainsford
    James. (Third crusade.)
Prince and the page : the last crusade. ...Charlotte Mary Yonge.
    (Eighth crusade.)
Ransom : a tale of the thirteenth century. ...Laura Jewry.
Raymond de Saint Gilles at the crusades. ...Henriette de Witt.
    (First crusade.)
Richard Cœur de Lion. ...James White.
Richard the Lion-hearted. ...Robert Tomes.
Saracen. ...Sophie Risteau Cottin.
Saxon's daughter : a tale of the crusades. ...Major Michel.
Shadow of the ragged stone. ...C. F. Grindrod.
Sir Walter's ward. ...William Everard.
Stories of the crusades. ...William E. Dutton.
Stories of the crusades. ...John Mason Neale. (Pseud., Aurelius
    Gratianus.)
Sword and scimetar. ...Alfred Trumble.

CRUSADES — *Continued.*

Tales of the early ages. ...Horace Smith. (Pseud., Paul Chatfield.)
Talisman. ...Sir Walter Scott. (Third crusade.)
Under Bayard's banner. ...Henry Frith.
Westminster cloisters : the story of a life's ambition. ...M. Bidder.
Winning his spurs. ...Elijah Kellogg.
See also RICHARD I. OF ENGLAND.

Children's Crusade, Thirteenth Century.

Crusaders and captives. ...George E. Merrill.
With script and staff. ...Elia W. Peattie.
Young prophetess : a tale of the children's crusade. ...R. Leighton
Gerhart.

CUBA.

Caoba, the guerilla chief : a real romance of the Cuban rebellion.
...P. H. Emerson.
Conspiracy : a Cuban romance. ...Adam Badeau.
Cruise of the Midge. ...Michael Scott.
Daughter of Cuba. ...Helen M. Bowen.
Free flag of Cuba. ...H. M. Hardimann.
Enriqueta Faber. ...Andrés Clemente Vazquez.
Humbled pride : a story of the Mexican war. ...John R. Musick.
Juanita : a romance of real life in Cuba fifty years ago. ...Margaret
Mann.
El Mulato Sab. ...Gertrudes Gomez de Avellaneda.
Otilia : episodio de la guerra de Cuba. ...V. Aguilar.
Rover's secret. ...William J. C. Lancaster. (Pseud., Harry Col-
lingwood.)
Vasconselos : a romance of the new world. ...William Gilmore
Simms.
With Maceo in Cuba : adventures of a Minnesota boy. ...Franc R.
E. Woodward.

DELAWARE.

Conspirators. ...Eliza Ann Dupuy.
Koningsmarke ; or, Old times in the new world. ...James Paulding.
(Pseud., Lancelot Langstaff, Esq.)
Miriam's heritage. ...A. Calder.

DENMARK.

Description, History, Manners, and Customs.

Erik menveds Barndom. ...Bernhard Severin Ingemann.
Eventyr og Fortoellinger. ...Bernhard Severin Ingemann.
King Eric V. ...Bernhard Severin Ingemann. English translation.
Kong Erik. ...Bernhard Severin Ingemann.
Lovells : a story of the Danish war. ...Mrs. J. B. Peploe (Webb),

DENMARK — *Continued.*
Prinds Otto af Danmark.   ...Bernhard Severin Ingemann.
Valdemar Seier.   ...Bernhard Severin Ingemann.
DE SOTO, FERDINANDO.   *Lived 1496–1542.*
Vasconselos: a romance of the new world.   ...William Gilmore
Simms.
DETECTIVE STORIES.   *A Selection.*
Adventures of Sherlock Holmes.   ...Arthur Conan Doyle.
American penman.   ...Julian Hawthorne.
Another crime.   ...Julian Hawthorne.
Baffling quest.   ...Richard Dowling.
Behind closed doors.   ...Anna Katharine Green.
Captain Shannon.   ...Coulson Kernahan.
Chronicles of Martin Hewitt.   ...Arthur Morrison.
Clues from a detective's camera.   ...Headon Hill.
Clues from the note-book of Zambra, the detective.   ...Headon Hill.
Conflict of evidence.   ...Roderigues Ottolengui.
La corde au cou.   ...Émile Gaboriau.
Le crime d'Orcival.   ...Émile Gaboriau.
Cynthia Wakeman's money.   ...Anna Katharine Green.
Dangerous ground.   ...Lawrence L. Lynch, Pseud.
La dégringolade.   ...Émile Gaboriau.
Detective triumphs.   ...J. E. Muddock.
Le dossier No. 113.   ...Émile Gaboriau.
Dugdale millions.   ...William Hudson.   (Pseud., Barclay North.)
Excellent knave.   ...J. Fitzgerald Molloy.
Experience of Loveday Brooke, lady detective.   ...C. L. Pirkis.
For fifteen years.   ...Louis Ulbach.   English translation.
From clue to capture.   ...J. E. Muddock.
From information received.   ...J. E. Muddock.
Great bank robber.   ...Julian Hawthorne.
Great Porter Square.   ...Benjamin Leopold Farjeon.
In the grip of the law.   ...J. E. Muddock.
Leavenworth case: a lawyer's story.   ...Anna Katharine Green.
Link by link.   ...J. E. Muddock.
Long arm.   ...Mary Wilkins and J. E. Chamberlain.
M. Lecoq.   ...Émile Gaboriau.
Man from Manchester.   ...J. E. Muddock.
Man hunter.   ...J. E. Muddock.
Marked "Personal."   ...Anna Katharine Green.
Le marteau d'acier.   ...Louis Ulbach.
Memoirs of Sherlock Holmes.   ...Arthur Conan Doyle.
Mill mystery.   ...Anna Katharine Green.
Modern wizard.   ...Rodrigues Ottolengui.

DETECTIVE STORIES — *Continued.*
Mystery of Orcival.   ...Émile Gaboriau.   English translation.
Mystery of Paul Chadwick.   ...J. W. Postgate.
Mystery of the Patrician Club.   ...Albert D. Vandam.
Personal adventures of a detective.   ...Lieutenant Carmichael.
Quinze ans de bagne.   ...Louis Ulbach.
Reminiscences of a chief inspector.   ...Chief Inspector Littlefield.
Section 558 ; or, Fatal letter.   ...Julian Hawthorne.
7-12 : a detective story.   ...Anna Katharine Green.
Shadowed by three.   ...Lawrence L. Lynch, Pseud.
Silent witness.   ...Mrs. J. H. Walworth.
Steel hammer.   ...Louis Ulbach.   English translation.
Strange disappearance.   ...Anna Katharine Green.
Study in scarlet.   ...Arthur Conan Doyle.
Suspicions aroused.   ...J. E. Muddock.
Sword of Damocles : a story of New York life.   ...Anna Katharine
    Green.
Tracked to doom.   ...J. E. Muddock.
Trial of Gideon.   ...Julian Hawthorne.
Wedderburn's will.   ...T. Cobb.
Who poisoned Hetty Duncan ?   ...J. E. Muddock.
X. Y. Z. : a detective story.   ...Anna Katharine Green.
See also **CRIMINOLOGY**.

DIETRICH OF BERN.
Didrik af Barns Saga.   ...Carl C. Rafn.
Dietrichs Bratfahrt von Albrecht von Kemenat.   ...F. H. von Hagen.
Dietrichs erste Ausfahrt.   Herausg. von D. Franz Stark.
Dietrich und seine Gesellen aus der Heidelberger Handschrift.
    ...Anon.
Dietrich von Bero.   ...Adolf Wechsler.
Saga Dietrichs Konungs af Bern.   ...C. R. Unger.

DOGMAS.
Autobiography of Mark Rutherford.   ...William Hale White.
    (Pseud., Mark Rutherford.)
Bertha ; or, Pope and the emperor.   ...William B. MacCabe.
Bez Dogmatic.   ...Henryk Sienkiewicz.
Born player.   ...Mary West.
Der christlichen, königlichen Fürsten Herculiscus und Herculadisla
    Wundergeschichte.   ...Andreas E. Buchholtz.
Damnation of Theron Ware.   ...Harold Frederick.
Dark year of Dundee.   ...Deborah Alcock.
Dedora Heywood.   ...Gertrude Smith.
Heresy of Mehetabel Clark.   ...Annie T. Slosson.
Hildebrand and the excommunicated emperor.   ...Joseph Sortain.

DOGMAS — *Continued.*

King Henry and his times. ...Clara M. Mundt. (Pseud., Louise Mühlbach.) English translation.

König Heinrich VIII. und sein Hof. ...Clara M. Mundt. (Pseud., Louise Mühlbach.)

Legend of Thomas Didymus. ...James Freeman Clarke.

Magdalen Hepburn. ...Margaret O. W. Oliphant.

Mariam's schooling. ...William Hale White. (Pseud., Mark Rutherford.)

Mark Rutherford, dissenting minister. ...William Hale White. (Pseud., Mark Rutherford.)

Mark Rutherford's deliverance. ...William Hale White. (Pseud., Mark Rutherford.)

Micah Clarke and his statement as made to his three grandchildren. ...Arthur Conan Doyle.

Minister's wooing. ...Harriet Beecher Stowe.

Modern instance. ...William Dean Howells.

New England conscience. ...Bella C. Greene.

New minister. ...Kenneth Paul.

No cross, no crown. ...Caroline E. Davis.

Old Town folks. ...Harriet Beecher Stowe.

Orthodox. ...Dorothea Gerard.

Out of step. ...Maria Louise Pool.

Parson Jones. ...Florence Marryat Lean (Mrs. Francis Lean).

Revolution in Tanner's Lane. ...William Hale White. (Pseud., Mark Rutherford.)

Soul of the bishop. ...Henrietta E. V. Stannard. (Pseud., John Strange Winter.)

Stories of a sanctified town. ...Lucy S. Furman.

Truce of God: a tale of the eleventh century. ...George H. Miles.

Two Salomes. ...Maria Louise Pool.

Without a dogma. ...Henryk Sienkiewicz. English translation.

Wolverton. ...D. A. Reynolds.

Won and not one. ...Emily Lucas Blackall.

DRUIDS.

Daughter of the Druids. ...A. K. Hopkins.

Edol, the Druid. · ...William Henry Giles Kingston.

DUDLEY, ROBERT.

See **LEICESTER, ROBERT DUDLEY, EARL OF.**

DÜRER, ALBERT. *Lived 1471–1528.*

Artist's married life. ...Leopold Schefer. English translation.

Dürer in Venedig. ...Adolf Stern.

Franz Sternbald's Wanderungen. ...J. Ludwig Tieck.

DÜRER, ALBERT — *Continued.*
Die Künstlerehe. ...Leopold Schefer.
Norica. ...August Hagen.
See also **ART.**

DUTCH CHARACTER. *A Selection.*
Baroness : a Dutch story. ...Frances M. Peard.
Black tulip. ...Alexandre Dumas. English translatiòn.
Bow of orange ribbon : a romance of New York. ...Amelia Edith
Barr.
By England's aid ; or, Freeing the Netherlands. ...George Alfred
Henty.
By pike and dyke : a tale of the rise of the Dutch republic.
...George Alfred Henty.
Fisherman's daughter. ...Hendrik Conscience. English translation.
Galama ; or, Beggars. ...John B. de Liefde.
Gideon Florenoz. ...Anna L. G. Bosboom.
Hans Brinker ; or, Silver skates : a story of life in Holland. ...Mary
Mapes Dodge.
Hotspur : a tale of the old Dutch manor. ...Mansfield Tracy
Walworth.
Leycester in Nederland. ...Anna L. G. Bosboom.
Liberators of Holland. ...Elizabeth Charles. (In her Martyrs of
Spain.)
Old maid's love : a Dutch tale told in English. ...J. van der
Poorsen-Schwartz. (Pseud., Maarten Maartens.)
Question of taste. ...J. van der Poorsen-Schwartz. (Pseud.,
Maarten Maartens.)
Siska van Roosemael de ware geschiedenis van eene juffer die
nog leeft. ...Hendrik Conscience.
Summer in a Dutch country house. ...Margaret H. Traherne.
La tulipe noire. ...Alexandre Dumas.
De vrouwen van het Leycestersche tijdoak. ...Anna L. G. Bos-
boom.
Within sea walls ; or, How the Dutch kept the faith. ...Elizabeth
H. Walshe and George Sargent.

EDWARD THE BLACK PRINCE. Titles are with **EDWARD III.
OF ENGLAND.**

EDWARD I. OF ENGLAND. *Reigned 1272-1307.*
Coberly Hall : a Gloucestershire tale. ...Robert Hughes.
Edwin the boy outlaw : or, Dawn of freedom in England. ...J.
Frederick Hodgetts.
Famous chronicle of Edward I. ...George Peele.
Lily of Saint Paul's. ...William D. Watson.

EDWARD I. OF ENGLAND — *Continued.*
Lord of Dynevor.   ...Evelyn Everett Green.
Merchant and the friar.   ...Francis Palgrave.
Our little lady.   ...Emily Sarah Holt.
Prentice Hugh.   ...Frances M. Peard.
Rothelan : a romance of the English histories.   ...John Galt.
Stories from old English poetry.   ...Abby S. Richardson.
Vendgiag; or, Blessed One : a tale of the thirteenth century.
   ...Anon.

EDWARD II. OF ENGLAND. *Reigned 1307–1327.*
Anecdotes de la cour et du règne d'Édouard II.   ...Claudine A. G.
   Tencin.
Berkley Castle.   ...G. F. Berkley.
Days of Bruce.   ...Grace Aguilar.
In all time of our tribulation : the story of the Piers Gavestone.
   ...Emily Sarah Holt.
Margam Abbey : an historical romance.   ...Anon.

EDWARD III. OF ENGLAND. *Reigned 1327–1377.* (Includes also
   his son Edward the Black Prince.)
Border lances : a romance of the northern marches in the reign of
   Edward Third.   ...E. L. S.
Cressy and Poictiers ; or, Story of the Black Prince's page.   ...John
   George Edgar.
Edward the Black Prince ; or, Feudal days.   ...Pierce Egan, Jr.
Garter : a romance of English history.   ...H. Neale.
In the days of chivalry : a tale of the times of the Black Prince.
   ...Evelyn Everett Green.
King Edward III.   ...John Bancroft.
Lances of Lynwood.   ...Charlotte Mary Yonge.
Saint George for England.   ...George Alfred Henty.
Story of Edward the Black Prince.   ...Meredith Jones.
Vision and the creed of Piers Ploughman.   ...William Langland.
Well in the desert : an old legend of the house of Arundel.
   ...Emily Sarah Holt.
White company.   ...Arthur Conan Doyle.

EDWARD IV. OF ENGLAND. *Reigned 1461–1483.*
Evenings at Haddon Hall.   ...Anon.
Goldsmith's ward : a tale of London city in the fifteenth century.
   ...Mrs. R. H. Reade.
Jane Shore : or, Goldsmith's wife.   ...Mary Bennett.
Last of the barons.   ...Edward George Earle Lytton Bulwer-
   Lytton.
See also **WAR OF THE ROSES.**

EDWARD VI. OF ENGLAND. *Reigned 1547–1553.*

All for the best; or, Bernard Gilpin's motto. ...Emily Sarah Holt.
Cecily : a tale of the English reformation. ...Emma Leslie.
Colloquies of Edward Osborne. ...Anne Manning.
Constable of the tower. ...William Harrison Ainsworth.
Diary and hours of the Lady Adalie. ...Charlotte Pepys.
Fall of Somerset. ...William Harrison Ainsworth.
Heir of Treherne : a tale of the Devonshire rebellion. ...Augustine D. Crake.
Jack and the tanner of Wymondham : a tale of the time of Edward the Sixth. ...Anne Manning.
Lady Jane Grey. ...Thomas Miller.
Lady Jane Grey und ihre Zeit. ...Louise M. Grafin von Robiano.
Mistress Hazelwode. ...Frederick H. More.

EGYPT.

Description, History, Manners, and Customs.

Eine Aegyptische Königstochter. ...Georg Moritz Ebers.
Alexandrians. ...Anon.
Beyond recall. ...Adeline Sergeant.
Bride of the Nile. ...Georg Moritz Ebers. English translation.
Chapter of adventures ; or, Through the bombardment of Alex·andria. ...George Alfred Henty.
Cleopatra. ...Georg Moritz Ebers. English translation.
Daughter of an Egyptian king. ...Georg Moritz Ebers. English translation.
Egyptian tales. ...W. M. Flinders Petrie.
Ephraim and Helah : a story of the Exodus. ...Edwin Hodder.
Epicurean. ...Thomas Moore.
Hassan; or, Child of the Pyramids. ...Charles A. Murray.
Homo sum. ...Georg Moritz Ebers.
Homo sum. ...Georg Moritz Ebers. English translation.
Hypatia. ...Charles Kingsley.
Der Kaiser. ...Georg Moritz Ebers.
King's treasure house : a romance of ancient Egypt. ...Wilhelm Walloth. English translation.
Kismet. ...Julia C. Fletcher.
Kleopatra : Historischer Roman. ...Georg Moritz Ebers.
Maid and the Cleon. ...Elizabeth Charles.
Mr. Potter of Texas. ...Archibald Clavering Gunter.
Die Nilbraut. ...Georg Moritz Ebers.
Out of Egypt : stories from the threshold of the East. ...Percy Hemingway.
Patriarchal times. ...C. M. O'Keefe.
Pearl of Antioch : a picture of the East at the end of the fourth century. ...Marc A. Bayle.

EGYPT — Continued.
Per aspera.   ...Georg Moritz Ebers.
Per aspera: a thorny path.   ...Georg Moritz Ebers.   English translation.
Pillar of fire; or, Israel in bondage.   ...Joseph Holt Ingraham.
Queen Moo and the Egyptian sphinx.   ...C. L. Augustus Plongeon.
Rameses.   ...Edward Upham.
Rescued from Egypt.   ...Charlotte Tucker.   (Pseud., A. L. O. E.)
Das Schatzhaus des Königs: eine Erzählung aus dem Alten Aegypten.   ...Wilhelm Walloth.
Der Scheik von Alessandria und seine Sklaven.   ...Wilhelm Hauff.
Die Schwestern.   ...Georg Moritz Ebers.
Serapis.   ...Georg Moritz Ebers.
Serapis.   ...Georg Moritz Ebers.   English translation.
Sheik of Alexandria and his slaves.   ...Wilhelm Hauff.   English translation.
Sisters.   ...Georg Moritz Ebers.   English translation.
Strange journey; or, Pictures from Egypt and the Soudan.   ...Christina Georgiana Rossetti.
Uarda: Roman aus dem alten Aegypten.   ...Georg Moritz Ebers.
Uarda: a romance of ancient Egypt.   ...Georg Moritz Ebers.   English translation.
Zorah.   ...Elizabeth Balch.   (Pseud., D. T. S.)

ELIZABETH OF ENGLAND. *Reigned 1558-1603.*
By England's aid; or, Freeing of the Netherlands.   ...George Alfred Henty.
By little and little: a tale of the Spanish armada.   ...Emma Leslie.
Clare Avery: a story of the Spanish armada.   ...Emily Sarah Holt.
Constance Sherwood: autobiography of the sixteenth century.   ...Georgiana C. L. G. Fullerton.
Eventide light; or, Passages in the life of Dame Margaret Hoby, only child and sole heir of Sir Arthur Dakyns.   ...Emma Marshall.
For queen and king: or, Loyal 'prentice: a story of old London.   ...Henry Frith.
Fritz of Fritz-Ford.   ...Anna Eliza Bray.
Gloucester cathedral; or, Last days of the Tudors.   ...J. R. Clarke.
Good old days.   ...Esmé Stuart.
Hildebrand; or, Days of Queen Elizabeth.   ...Anon.
Huguenot family in an English village.   ...Henrietta Keddie. (Pseud., Sarah Tytler.)
Joyce Morrell's harvest.   ...Emily Sarah Holt.
Kenilworth.   ...Sir Walter Scott.

ELIZABETH OF ENGLAND — *Continued.*

Last earl of Desmond: a historical romance. ...Charles B. Gibson.

Lord Mayor: a tale of London in 1584. ...Emily Sarah Holt.

Lost evidence. ...Hannah D. Burdon.

Loyal hearts. ...Evelyn Everett Green.

May Lane: a story of the sixteenth century. ...Anon.

Michaelmas in 1574. ...Eldridge S. Brooks. (In his Storied holidays.)

Nut-brown maids; or, First hosier and his hosen. ...Henrietta Keddie. (Pseud., Sarah Tytler.)

Pirates' fort: a tale of the sixteenth century. ...L. MacNally.

Ralf Redman's atonement: a tale of the black death of 1598. ...Herbert V. Mills.

Recess; or, Tale of other times. ...Sophia Lee.

Royal merchant; or, Events in the days of Sir Thomas Gresham. ...William Henry Giles Kingston.

Sir Lundar: a story of the days of the great Queen Bess. ...Talbot B. Reed.

Spae-wife; or, Queen's secret: a tale of the days of Elizabeth. ...John Boyce.

Stories of the wars. ...J. Tillotson.

Story of Arthur Penreath, sometime gentleman to Sir Walter Raleigh. ...Verney L. Cameron.

Story of John Marbeck. ...Emma Marshall.

Sword and pen: reign of Queen Elizabeth. ...William H. D. Adams.

Sydney the knight. ...Ella T. Disosway.

Tablette booke of Lady Mary Keyes. ...Anon.

Three hundred years ago; or, Martyrs of Brentwood. ...William Henry Giles Kingston.

Through storm and stress. ...J. S. Fletcher.

Twin heroes: the separatists of the time of Queen Elizabeth. ...F. A. Reed.

Under Drake's flag: a tale of the Spanish main. ...George Alfred Henty.

Virgin queen; or, Romance of royalty. ...J. Frederick Smith.

Westward Ho! or, Voyages and adventures of Sir Amyas Leigh. ...Charles Kingsley.

Wild times: a tale of the days of Queen Elizabeth. ...Cecilia M. Caddell.

See also **ENGLISH HISTORY**.

ENGLAND. *A Selection.*

Description, Manners, and Customs.

Abigel Rowe. ...Lewis S. Wingfield.

Aims and obstacles. ...George Payne Rainsford James.

ENGLAND — *Continued.*

Alton Locke. ...Charles Kingsley.

Anne Hathaway ; or, Shakespeare in Love. ...Edward Severn.

As in a looking-glass. ...F. C. Phillips.

Behind the veil : a tale of the days of William the Conqueror. ...Emily Sarah Holt.

Black arrow ; a tale of the two roses. ...Robert Louis Stevenson.

Bledisloe; or, Aunt Pen's American nieces. ...Ada M. Trotter. (Pseud., Lawrence Severn.)

Brave Dame Mary; or, Siege of Corfe Castle. ...Louisa Hawtrey.

Camp on the Severn. ...Augustine D. Crake.

Cathedral courtship. ...Kate Douglas Wiggin (Mrs. George Riggs).

Cavaliers. ...S. R. Keightley.

Children of the New Forest. ...Frederick Marryat.

Clarissa Harlowe. ...Samuel Richardson.

Coningsby. ...Benjamin Disraeli.

Constable of the Tower. ...William Harrison Ainsworth.

Cornet of horse : a tale of Marlborough's wars. ...George Alfred Henty.

Craddock Nowell : tale of the New Forest. ...Richard Doddridge Blackmore.

David's loom : a story of Rochdale life in the nineteenth century. ...John T. Clegg.

Daybreak in Britain. ...Charlotte Tucker. (Pseud., A. L. O. E.)

Delectable duchy : stories, studies, and sketches. ...Arthur T. Q. Couch. (Pseud., Q.)

Demos : a story of English socialism. ...George R. Gissing.

Dodo. ...E. F. Benson.

Dove in the eagle's nest. ...Charlotte Mary Yonge.

Early dawn ; or, Sketches of Christian life in England in the olden time. ...Elizabeth Charles.

Edol the Druid. ...William Henry Giles Kingston.

Eldrick the Saxon. ...A. S. Bride.

Endymion. ...Benjamin Disraeli.

English squire. ...Christabel R. Coleridge.

Evelyn Manwaring : a tale of Hampton Court. ...Greville J. Chester.

First chronicle of Aescendun. ...Augustine D. Crake.

Fitch of bacon. ...William Harrison Ainsworth.

Forest of Arden. ...William Gresley.

Foure prentises of London. ...Thomas Heywood.

Goldsmith's ward : a tale of London city in the fifteenth century. ...Mrs. R. H. Reade.

Harold, the boy earl. ...J. Frederick Hodgetts.

Held fast for England : a tale of the siege of Gibraltar. ...George Alfred Henty.

ENGLAND — *Continued.*

Henrietta Temple.   ...Benjamin Disraeli.

Hetty Hyde's lovers.   ...James Grant.

Hide and seek : a story of the New Forest in 1647.   ...E. E. Cooper (Mrs. Frank Cooper).

His bad angel.   ...Richard Harding Davis.

Historical tales ; or, Romance of reality.   ...Charles Morris.

History of David Grieve.   ...Mary Augusta Ward (Mrs. Humphry Ward).

History of Tom Jones, a foundling.   ...Henry Fielding.

Imogen : a story of the mission of Augustine.   ...Emily Sarah Holt.

Ivanhoe.   ...Sir Walter Scott.

Life, adventures, and piracies of the famous Captain Singleton.   ...Daniel Defoe.

Little wizard.   ...Stanley J. Weyman.

Lothair.   ...Benjamin Disraeli.

Marcella.   ...Mary Augusta Ward (Mrs. Humphry Ward).

Miss Angel.   ...Anna Isabella Thackeray (Mrs. Richmond Ritchie).

Mustard leaves.   ...E. Balch.

My novel.   ...Edward George Earle Lytton Bulwer-Lytton.

North and South.   ...Elizabeth C. Gaskell.

Pastor's fireside.   ...Jane Porter.

Pictures of cottage life in the West of England. ...Margaret E. Poole.

Le prophète irlandois.   ...Charles de Saint-Evremond.

Punch's letters to his son.   ...Douglas W. Jerrold.

Rival apprentices.   ...Anon.

Ronald Morton ; or, Fireships.   ...William Henry Giles Kingston.

Rubicon.   ...E. F. Benson.

Saint Cedd's cross.   ...Edward Lewes Cutts.

Sea-kings of England.   ...Edwin Atherstone.

Second chronicle of Aescendun.   ...Augustine D. Crake.

Secret passion.   ...Robert Folk Williams.

Shakespeare and his friends.   ...Robert Folk Williams.

Shakespeare : the poet, the lover, the actor, and the man.   ...Henry Curling.

Shepherd of Grove Hall.   ...Anon.

Stories of England and foreign life.   ...William and Mary Howitt.

Stories of old England.   ...George E. Sargent.

Sybil; or, Two nations.   ...Benjamin Disraeli.

Tales and legends of national origin or widely current in England from the earliest times.   ...William C. Hazlett.

Tales of kings and queens of England.   ...Stephen Percy.

Tales of old English life.   ...William F. Collier.

Tales of the Saxons.   ...Emily Taylor.

True Cornish maid.   ...G. Norway.

ENGLAND — *Continued.*

Twice crowned . a story of the days of Queen Mary.   ...Harriet B. MacKeever.

Venetia.   ...Benjamin Disraeli.

Wager of battle.   ...William Henry Herbert.

Westminster Abbey.   ...Emma Robinson.

Westward Ho ! or, Voyages and adventures of Sir Amyas Leigh. ...Charles Kingsley.

When London was burned.   ...J. Finnemore.

White gauntlet.   ...Mayne Reid.

Woodlanders.   ...Thomas Hardy.

Wreckers and Methodists.   ...H. D. Lowry.

Wulf the Saxon.   ...Ralph Peacock.

See also **ENGLISH HISTORY.**

ENGLISH HISTORY. *A Selection.*

Sixth Century.

Rivals.   ...Gerald Griffin (in his Tales of the Munster festivals, third series).

Ninth Century.

Athelwold.   ...Amélie Rives Chanler.

Chronicle of Ethelfled.   ...Anne Manning.

Danes in England.   ...Alfred H. Engelbach.

Dragon and the raven ; or, Days of King Alfred.   ...George Alfred Henty.

Glastonbury.   ...A. Payne.

Sea-kings of England.   ...Edwin Atherstone.

Eleventh Century.

Andreds-weald ; or, House of Michelham : a tale of the Norman Conquest.   ...Augustine D. Crake.

Behind the veil : a tale of the days of William the Conqueror. ...Emily Sarah Holt.

Bishop's daughter.   ...Anon.

Camp of refuge.   ...Charles MacFarlane.

Diane.   ...Katharine S. Macquoid.

Dutch in the Medway.   ...Charles MacFarlane.

Harold, the boy earl.   ...J. Frederick Hodgetts.

Harold, the last of the Saxon kings.   ...Edward George Earle Lytton Bulwer-Lytton.

Hereward the Wake.   ...Charles Kingsley.

Legend of Reading Abbey.   ...Charles Knight.

Rufus ; or, Red king.   ...James Grant.

Siege of Norwich Castle.   ...M. M. Blake.

Stanfield Hall.   ...J. Frederick Smith.

ENGLISH HISTORY — *Continued.*

Westminster cloisters: the story of a life's ambition.   ...M. Bidder.
See also **CRUSADES** and **HOOD, ROBIN.**

Thirteenth Century.

Cheshire pilgrims.   ...Frances M. Wilbraham.
Coberly Hall: a Gloucestershire tale.   ...Robert Hughes.
De Montfort; or, Old English nobleman.   ...Anon.
Earl Hubert's daughter; or, Polishing of the pearl.   ..Emily Sarah Holt.
Edwin the outlaw ; or, Dawn of freedom in England.   ...J. Frederick Hodgetts.
Famous chronicle of Edward I.   ...George Peele.
Gaston de Blondeville; or, Court of Henry III.   ...Anna Radcliffe.
How I won my spurs.   ...John George Edgar.
In all time of our tribulation : the story of Piers Gavestone.   ...Emily Sarah Holt.
Lily of Saint Paul's.   ...William D. Watson.
Lord of Dynevor.   ...Evelyn Everett Green.
Merchant and the friar.   ...Francis Palgrave.
Osbert of Aldgate and the troubadour · a tale of the Goldsmiths' Company.   ...Elizabeth M. Stewart.
Our little lady.   ...Emily Sarah Holt.
Prentice Hugh.   ...Frances M. Peard.
Rothelan : a romance of the English histories.   ...John Galt.
Royston Gower.   ...Thomas Miller.
Runnymede and Lincoln fair : a story of the great charter.   ...John George Edgar.
Siege of Kenilworth.   ...L. S. Stanhope.
Sir Michael Scott.   ...Allan Cunningham.
Stephen Langton ; or, Days of King John.   ...Martin F. Tupper.
Stories from old English poetry.   ...Abby S. Richardson.
Stories of the city of London.   ...Mrs. Newton Crosland.   (Pseud., Camilla Toulmin.)
Troublesome raigne and lamentable death of Edward II. King of England.   ...Christopher Marlowe.
Vendgiad ; or, Blessed one : a tale of thirteenth century.   ..Anon.

Fourteenth Century.

Abbess of Shaftesbury ; or, Days of John of Gaunt.   ...Anon.
Anecdotes de la cour et du règne d'Edouard II.   ..Claudine A G. Tencin.
Berkeley Castle.   ...G. F Berkeley.
Bondman : a tale of the times of Wat Tyler   ..Anon.
Border lances : a romance of the Northern marches in the reign of Edward the Third.   ..E. L S.

ENGLISH HISTORY — *Continued.*

Boy foresters. ...Anne Bowman.

Castle Dangerous. ...Sir Walter Scott.

Chronicles of Yate Court : a narrative of 1399. ...Charles Charlton.

Claribel, the sea maid : a tale of the Fishmongers' Company. ...Elizabeth M. Stewart.

Cressy and Poictiers ; or, Story of the Black Prince's page. ...John George Edgar.

Days of Bruce. ...Grace Aguilar.

Dick Delver : a story of the peasants' revolt of the fourteenth century. ...Henrietta E. Burch.

Edward the Black Prince ; or, Feudal days. ...Pierce Egan, Jr.

Eleanor and I : a tale of the days of King Richard II. ...Mary E. Bamford.

Forest days : a romance of old days. ...George Payne Rainsford James.

Garter : a romance of English history. ...H. Neele.

Geoffrey the Lollard. ...Mrs. D. C. Knevels. (Pseud., Frances Eastwood.)

Hubert Ellis : a story of Richard the Second's days. ...F. Davenant.

Idol of the clownes ; or, Insurrection of Wat the Tyler in the 4th yeare of Richard II. ...John Cleaveland.

In convent walls : the story of the Despensers. ...Emily Sarah Holt.

In the days of chivalry : a tale of the times of the Black Prince· ...Evelyn Everett Green.

John of Gaunt. ...James White.

John Standish ; or, Harrowing of London. ...Edward Gilliat.

King Edward III. ...John Bancroft.

Lances of Lynwood. ...Charlotte Mary Yonge.

Last wolf : a tale of England in the fourteenth century. ...Anne Mercier.

Maid Marian. ...Thomas L. Peacock.

Maid Marian and Robin Hood. ...J. E. Muddock.

Margan Abbey : a historical romance. ...Anon.

Mediation of Ralph Hardelot. ...W. Minto.

Merry England; or, Nobles and serfs. ...William Harrison Ainsworth.

Mistress Margery : a tale of the Lollards.· ...Emily Sarah Holt.

Otterbourne. ...Edward Duras.

Queen Phillippa and the hurrer's daughter : a tale of the Haberdashers' Company. ...Elizabeth M. Stewart.

Robin Hood. ...Pierce Egan, Jr.

Saint George for England : a tale of Cressy and Poictiers. ...George Alfred Henty.

Story of Edward the Black Prince. ...Meredith Jones.

ENGLISH HISTORY — *Continued.*

Stories from English history during the middle ages.   ...Maria Hack

True stories of the times of Richard II.   ...Henry P. Dunster.

Vision and creed of Piers Ploughman.   ...William Langland.

Vox clamantis.   ...John Gower.

Wardship of Steepcoombe.   ...Charlotte Mary Yonge.

Wat Tyler.   ...Pierce Egan, Jr.

Well in the desert : an old legend of the house of Arundel.   ...Emily Sarah Holt.

White company.   ...Arthur Conan Doyle.

Whittington and the Knight Sans-Terre.   ...Elizabeth M. Stewart.

See also **WYCLIFFE, JOHN**.

Fifteenth Century.

Agincourt.   ...George Payne Rainsford James.

Armourer's daughtei ; or, Border rivals.   ...Anon.

At ye Grene Griffin; or, Mrs. Treadwell's cook.   ...Emily Sarah Holt.

Aylmere ; or, Bondman of Kent.   ...Robert T. Conrad.

Berkeley Castle.   ...G. F. Berkeley.

Bosworth-field.   ...John Beaumont.

Bosworth-field; or, Fate of the Plantagenet.   ...Paul Leicester.

Caged lion.   ...Charlotte Mary Yonge.

Captain of the Wight : a romance of Carisbrooke Castle in 1488.   ...Frank Cowper.

Castle Harcourt.   ...L. F. Winter.

Chantry priest of Barnet: a tale of the two Roses.   ...Alfred John Church.

Chronicle history of Perkin Warbeck.   ...John Ford.

Civile wars between the two houses of Lancaster and Yorke.   ...Samuel Daniel.

Coulyng Castle ; or, A knight of the olden days.   ...Agnes Giberne.

Earl Printer: times of Caxton.   ...C. M. M.

Fortress : an historical tale of the fifteenth century.   ...Anon.

Fortunes of Perkin Warbeck.   ...Mary Wollstonecraft Shelley.

Goldsmith's ward : a tale of London city in the fifteenth century.   ...Mrs. R. H. Reade.

Henry of Monmouth.   ...Major Michel.

Henry Seventh.   ...Mary Wollstonecraft Shelley.

Historical tales of the Lancastrian times.   ...Henry P: Dunster.

In the wars of the Roses.   ...Evelyn Everett Green.

Jack of the mill.   ...William Howitt.

Jane Shore; or, Goldsmith's wife.   ...Margaret Bennett.

Joan the maid deliverer of France and England.   ...Elizabeth Charles.

ENGLISH HISTORY — *Continued.*

Judged by appearances : a tale of the civil wars. ...Eleanor Lloyd.

Last of the Barons. ...Edward George Earle Lytton Bulwer-Lytton.

Last of the Plantagenets. ...William Hazeltine.

Lollard priest. ...Henry C. Adams.

Maid of Warsaw. ...Ernst Jones.

Malvern Chase : an episode of the wars of the Roses and the battle of Tewkesbury. ...W. S. Symonds.

Midsummer eve, A. D. 1503. ...Elbridge S. Brooks. (In his Storied holidays.)

Noble purpose nobly won. ...Anne Manning.

Old English baron : a Gothic story. ...Clare Reeve.

Owen Tudor. ...Jane Robinson.

Queen's badge. ...Frances M. Wilbraham.

Red and white : a tale of the war of the Roses. ...Emily Sarah Holt.

Richard de Lacy : a tale of the later Lollards. ...C. Edmund Maurice.

Richard of York ; or, White rose of England. ...Anon.

Richard Plantagenet. ...J. Frederick Hodgetts.

Rival roses : a romance of English history. ...Eliza S. Francis.

Rocking stone : chronicle of the times of the wars of the Roses. ...William Henry Giles Kingston.

Siballa the sorceress; or, Flower girl of London : a tale of the days of Richard III. ...W. H. Peck.

Stormy life. ...Georgiana C. L. G. Fullerton.

Stanley : a tale of the fifteenth century. ...Anon.

Sword of De Bardwell. ...Lucy Ellen Guernsey.

Tangled webb : a tale of the fifteenth century. ...Emily Sarah Holt.

Two penniless princesses. ...Charlotte Mary Yonge.

Walter Leukam. ...Elizabeth M. Stewart.

War of the Roses. ...John George Edgar.

Ward of the crown. ...H. D. Wolfensberger (formerly Burdon).

White rose of Langley : a tale of the court of England. ...Emily Sarah Holt.

Whittington and the knight Sans-terre. ...Elizabeth M. Stewart.

Woodman. ...George Payne Rainsford James.

York and Lancaster rose. ...Annie Keary.

See also **WYCLIFFE** and the **LOLLARDS.**

Sixteenth Century.

Abbot. ...Sir Walter Scott.

Agnes Martin ; or, Fall of Cardinal Wolsey. ...Anon.

Alice Sherwin : a tale of the days of Sir Thomas More. ...C. J. M.

All for the best ; or, Bernard Gilpin's motto. ...Emily Sarah Holt.

Anne Boleyn. ...Mrs. K. Thompson.

Anne Boleyn; or, Suppression of the religious houses. ...Anon.

ENGLISH HISTORY — *Continued.*

Anne Hathaway; or, Shakespeare in love. ...Edward Severn.

Bolsover Castle : a tale from Protestant history of the sixteenth century. ...Anon.

Bride of Bucklersbury : a tale of the Grocers' Company. ...Elizabeth M. Stewart.

By England's aid; or, Freeing the Netherlands. ...George Alfred Henty.

By little and little : a tale of the Spanish main. ...Emma Leslie.

Captain Cobbler ; or, Lincolnshire rebellion. ...Thomas Cooper.

Cardinal Pole ; or, Days of Philip and Mary. ...William Harrison Ainsworth.

Cecily : a tale of the English reformation. ...Emma Leslie.

Chronicles of Camber Castle : a tale of the reformation. ...Anon.

Citation and examination of William Shakespeare. ...Walter Savage Landor.

Clare Avery : a story of the Spanish armada. ...Emily Sarah Holt.

Cloister and the hearth. ...Charles Reade.

Colloquies of Edward Osborne. ...Anne Manning.

Constable of the tower. ...William Harrison Ainsworth.

Constance Sherwood : autobiography of the sixteenth century. ...Georgiana C. L. G. Fullerton.

Crichton. ...William Harrison Ainsworth.

Darnley ; or, Field of the cloth of gold. ...George Payne Rainsford James.

Diary and hours of the Lady Adalie. ...Charlotte Pepys.

Drake and the Dons. ...Richard Lovett.

England's daybreak : narratives of the reformation. ...Edward Bickersteth.

Eventide light ; or, Passages in the life of Dame Margaret Hoby, only child and sole heiress · of Sir Arthur Dakyns. ...Emma Marshall.

Fall of Somerset. ...William Harrison Ainsworth.

Fitz of Fitz-Ford. ...Anna Eliza Bray.

For God and gold. ...Julian Corbett.

Forest of Arden. ...William Gresley.

Forest youth ; or, Shakespere as he lived. ...Henry Curling.

For the Master's sake : a story of the days of Queen Mary. ...Emily Sarah Holt.

Freston Tower : a tale of the times of Wolsey. ...Richard Cobbold.

Geraldine Maynard ; or, Abduction : a tale of the days of Shakespere. ...Henry Curling.

Good old days ; or, Christmas under Queen Elizabeth. ...Esmé Stuart.

Heir of Treherne : a tale of the Devonshire rebellion. ...Augustine D. Crake.

7

ENGLISH HISTORY — *Continued.*

Richard Hunne : a story of old London,   ...George E. Sargent.

Robin Tremayne of Bodmin : a tale of the Marian persecution. ...Emily Sarah Holt.

Royal merchant ; or, Events in the days of Sir Thomas Gresham, during the reigns of Queens Mary and Elizabeth.   ...William Henry Giles Kingston.

Sir Ludar : a story of the days of the great Queen Bess.   ...Talbot B. Reed.

Secret passion.   ...Robert Folk Williams.

Shakespeare and his friends ; or, Golden age of Merry England. ...Robert Folke Williams.

Shaksperc : the poet the lover, the actor, and the man.   ...Henry Curling.

Sir Ludar.   ...Talbot B. Reed.

Spae-wife ; or, Queen's secret : a tale of the days of Elizabeth. ...John Boyce.

Star of the court ; or, Maid of honor and the Queen of England, Anne Boleyn.   ...Selina Bunbury.

Stolen mask ; or, Mysterious cash-box.   ...Wilkie Collins.

Stories of the wars.   ...J. Tillotson.

Story of John Heywood.   ...Charles Bruce.

Story of John Marbeck.   ...Emma Marshall.

Tablette booke of Lady Mary Keyes.   ...Anon.

Three hundred years ago ; or, Mysteries of Brentwood.   ...William Henry Giles Kingston.

Tor hill.   ...Horace Smith.   (Pseud., Paul Chatfield.)

Tower hill.   ...William Harrison Ainsworth.

Tower of London.   ...William Harrison Ainsworth.

Tudors and Stuarts.   ...Anon.

Twice crowned : a story of the days of Queen Mary.   ...Harriet B. MacKeever.

Twin heroes : the Separatists of the time of Queen Elizabeth. ...F. A. Reed.

Uline's escape ; or, Hid with the nuns.   ...Mrs. Alexander S. Orr

Victor : a tale of the great persecution.   ...George G. Perry.

Virgin queen ; or, Romance of royalty.   ...J. Frederick Smith.

Westminster Abbey.   ...Emma Robinson.

Westward Ho ! or, Voyages and adventures of Sir Amyas Leigh. ...Charles Kingsley.

Wetherden Hall : an historical story of the days of Queen Mary. ...Arthur Brown.

Wild times.   ...Cecilia M. Caddell.   •

William Shakespeare.   ...Heinrich J. Koenig.   English translation.

William Shakespeare : culturgeschichtlich-biographischer Roman. ...Heribert Rau.

ENGLISH HISTORY — *Continued.*
Williams Dichter und Trachten.   ...Heinrich J. Koenig.
Windsor Castle.   ...William Harrison Ainsworth.
Youth of Shakespeare.   ...Robert Folk Williams.
See also **ELIZABETH OF ENGLAND; HENRY VIII. OF ENG-
LAND; MARY QUEEN OF SCOTS,** and **SPANISH HISTORY.**

Seventeenth Century.

Afloat and ashore with Sir Walter Raleigh.   ...Janet Hardy.
Agnes Beaumont : a true story of the year 1670.   ...Marian Caldecott.
Aimée : a tale of the days of James the Second.   ...Agnes Giberne.
Alice Bridge of Norwich : a tale of the time of Charles the First.
    ...Andrew Reed.
Alice Leighton ; or, Good name is rather to be chosen than riches.
    ...Ann J. Cupples (Mrs. G. Cupples).
Andrew Golding : a tale of the great plague.   ...Annie E. Keeling.
Andrew Marvel and his friends : a story of the siege of Hull.
    ...Marie S. Hall.
Aphra Behn.   ...Clara M. Mundt.   (Pseud., Louise Mühlbach.)
Aphra Behn.   ...Clara M. Mundt.   (Pseud., Louise Mühlbach.)
    English translation.
Arabella Stuart.   ...George Payne Rainsford James.
Armourer's daughter ; or, Border rivals.   ...Anon.
Arthur Arundel.   ...Horace Smith.   (Pseud., Paul Chatfield.)
At the sign of the Blue Boar : a story of the reign of Charles II.
    ...Emma Leslie.
Baldearg O'Donnell.   ...A. S. G. Canning.
Barony.   ...Anna Maria Porter.
Beatrice Tyldesley.   ...William Harrison Ainsworth.
Beyond the seas : being the surprising adventures and ingenious
    opinions of Ralph, Lord St. Keyne.   ...Oswald Crawfurd.
Boscobel ; or, Loyal oak : a tale of the year 1651.   ...William Har-
    rison Ainsworth.
Boscobel : a narrative of the adventures of Charles II. after the
    battle of Worcester.   ...George W. Dodd.
Brambletye House ; or, Cavaliers and roundheads.   ...Horace Smith.
    (Pseud., Paul Chatfield.)
Brave men of Eyam ; or, Tale of the great plague.   ...Edward N.
    Hoare.
Buccaneer.   ...Anna Maria Hall.
Caged lion.   ...Charlotte Mary Yonge.
Caleb Field : a tale of the Puritans.   ...Margaret O. W. Oliphant.
Campion Court : the days of the ejectment.   ...Emma J. Worboise.
Captain Jacques : a romance of the time of the plague.   ...Edward
    Fitzgibbon.   (Pseud., Somerville Gibney.)

ENGLISH HISTORY — *Continued.*

Captain of the guard.   ...James Grant.

Castle Cornet; or, Island's troubles in the troublous times.   ...Louisa Hawtrey.

Cavaliers of England; or, Times of the revolutions of 1642 and 1648. ...Henry W. Herbert. (Pseud., Frank Forester.)

Carved cartoon.   ...Austin Clare.

Cherry and violet: a tale of the great plague. ...Anne Manning.

Chevalier's daughter: being one of the Stanton chronicles.   ...Lucy Ellen Guernsey.

Children of the New Forest.   ...Frederick Marryat.

Christmas, 1611.   ...Elbridge S. Brooks. (In his Storied holidays.)

Claude Duval: a romance of the days of Charles II.   ...Henry D. Miles.

Commonwealth of Oceana.   ...James Harrington.

Constance Aylmer: a tale of the seventeenth century.   ...Helen F. Parker.

Coombe Abbey.   ...Selina Bunsbury.

Cost what it may: a story of cavaliers and roundheads.   ...Emma E. Hornibrook.

Court at Tunbridge in 1664.   ...Catherine G. F. Gore.

Courtnay of Wabreddon: a romance of the West.   ...Anna Eliza Bray.

Courtier of the days of Charles II.   ...Catherine G. F. Gore.

Courtier of the days of Charles II.   ...Mrs. Gordon.

Courtship of Morrice Buckler.   ...Alfred Edward Woodley Mason.

Cromwell; or, Protector's oath.   ...J. Frederick Smith.

Dame Rebecca Berry; or, Court scenes in the reign of Charles the Second.   ...Elizabeth I. Spencer.

Danvers papers.   ...Charlotte Mary Yonge.

Darien; or, Merchant Prince.   ...Eliot B. G. Warburton.

Daughter of Tyrconnell.   ...Mary A. Sadlier (Mrs. James Sadlier).

Days of Laud and the commonwealth.   ...Mrs. Newton Crosland. (Pseud., Camilla Toulmin.)

Deborah's diary.   ...Anne Manning.

Diary of Mary Powell.   ...Anne Manning.

Dog-fiend; or, Snarley-yow.   ...Frederick Marryat.

Dorothy Arden.   ...J. M. Callwell.

Draytons and Davenants: a tale of the civil wars.   ...Elizabeth Charles.

Drifted and shifted: a domestic chronicle of the seventeenth century.   ...Miss McLaren.

La duchesse de Châtillon.   ...François F. M. B. d'Arnaud.

Duke of Monmouth.   ...Gerald Griffin.

Dutch in the Medway.   ...Charles Macfarlane.

ENGLISH HISTORY — *Continued.*

Edgar Nelthrope ; or, Fair maids of Taunton.   ...Andrew Reed, Jr.

Fate : a tale of stirring times.   ...George Payne Rainsford James.

Father Darcy : an historical romance.   ...Anne Marsh.

For liberty's sake : the story of Robert Ferguson.   ...John B. Marsh.

For queen and king ; or, Royal 'prentice.   ...Henry Frith.

Fortunes of Nigel.   ...Sir Walter Scott.

Guy Fawkes ; or, Gunpowder · treason.   ...William Harrison Ainsworth.

Harry Ogilvie ; or, Black dragoons.   ...James Grant.

Henry Masterton ; or, Adventures of a young cavalier.   ...George Payne Rainsford James.

Her majesty the queen.   ...John Esten Cooke.

Hero in the strife.   ...Louisa C. Silke.

Hide and seek : a story of the New Forest in 1647.   ...Mrs. Frank Cooper.

History of the great plague in London in 1665.   ...Daniel Defoe.

Holmby House : a tale of old Northamptonshire.   ...George John Whyte Melville.

Ida Vane : a tale of the restoration.   ...Andrew Reed, Jr.

In the East country with Sir Thomas Brown.   ...Emma Marshall.

In the golden days.   ...Ada E. Bayly.   (Pseud., Edna Lyall.)

In the service of Rachel, Lady Russell.   ...Emma Marshall.

Ironsides : a tale of the English commonwealth.   ...Anon.

It might have been ; or, Story of the gunpowder plot.   ...Emily Sarah Holt.

James the Second ; or, Revolution of 1688.   ...William Harrison Ainsworth.

Janet and her father.   ...Mary E. Bamford.

Jannett Cragg, the Quakeress.   ...Maria Wright.

John Deane of Nottingham.   ...William Harrison Ainsworth.

John Inglesant.   ...Joseph Henry Shorthouse.

John Milton and his times : historical novel.   ...Max Ring.   English translation.

John Milton und seine Zeit.   ...Max Ring.

King's highway.   ...George Payne Rainsford James.

King's namesake : a tale of Carisbrooke Castle.   ...Catherine M. Phillimore.

Lady Betty.   ...Christabel R. Coleridge.

Lady Betty's governess ; or, Corbet chronicles.   ...Lucy Ellen Guernsey.

Lady Clancarty ; or, Wedded and wooed : a tale of the assassination plot of 1696.   ...Thomas Taylor.

Lady Shakerlye : the record of the life of a good and noble woman.   ...Anon.

ENGLISH HISTORY — *Continued.*

Lady Willoughby; or, Passages from the diary of a wife and mother in the seventeenth century.   ...Hannah M. Rathbone.

Langham rebels.   ...Lucy Ellen Guernsey.

Last of the cavaliers.   ...Rose Piddington.

Last of the fairies : a Christmas tale.   ...George Payne Rainsford James.

Leaguer of Lathom : a tale of the civil war in Lancashire.   ...William Harrison Ainsworth.

Life of Colonel Jack.   ...Daniel Defoe.

Lord Montagu's page : an historical romance of the seventeenth century.   ...George Payne Rainsford James.

Lorna Doone : a romance of Exmoor.   ...Richard Doddridge Blackmore.

Maid of honor.   ...Catherine G. F. Gore.   (In her Romances of real life.)

Man's foes.   ...E. H. Strain.

Markhams of Ollerton : a tale of the civil war.   ...Elizabeth Glaister.

Mary Bunyan, the dreamer's blind daughter.   ...Sallie Rochester Ford.

Mary Hollis : a romance of the days of Charles II. and William, Prince of Orange.   ...Hendrik J. Schimmel.   English translation.

Masque of Ludlow.   ...Anne Manning.

Maudeville : a tale of the seventeenth century in England.   ...William Godwin.

Memoirs of a cavalier.   ...Daniel Defoe.

Memoirs of a lady-in-waiting.   ...J. D. Fenton.

Memoirs of troublous times.   ...Emma Marshall.

Merry monarch ; or, England under Charles II.   ..William H. D. Adams.

Micah Clarke, his statement as made to his three grandchildren.   ...Arthur Conan Doyle.

Miser's secret; or, Days of James I.   ...Anon.

Mistress Spitfire.   ...J. S. Fletcher.

Mistress Dorothy Marvin.   ...J. C. Snaith.

Oak staircase ; or, Stories of Lord and Lady Desmond.   ...Mary and Catherine Lee.

Old Noll; or, Days of the Ironsides.   ...Frederick William Robinson.

Old Saint Paul's : a tale of the plague and the fire.   ...William Harrison Ainsworth.

Oliver Cromwell; or, England's great protector.   ...Henry W. Herbert.   (Pseud., Frank Forester.)

ENGLISH HISTORY — *Continued.*

Oliver Wyndham; or, Tale of the great plague. ...Mrs. J. B. Peploe (Webb).

On both sides of the sea : a story of the commonwealth and the restoration. ...Elizabeth Charles.

Orange and green. ...George Alfred Henty.

Outlaw. ...Anna Maria Hall.

Ovingdean grange : a tale of the South Downs. ...William Harrison Ainsworth.

Owain Goch : a tale of the revolution. ...Thomas Roscoe.

Pattie Durant : a tale of 1662. ...Ellen Clacy. (Pseud., Cycla.)

Peter the apprentice : a tale of restoration. ...Emma Leslie.

Peveril of the peak. ...Sir Walter Scott.

Prince and the pedler; or, Siege of Bristol. ...Ellen Pickering.

Le prophète irlandois. ...Charles de Saint-Évremond.

Puritan's grave. ...William P. Scargill.

Reginald Hastings : a tale of the troubles of 164–. ...Eliot B. G. Warburton.

Robber. ...George Payne Rainsford James.

Roger Willoughby; or, Times of Benbow. ...William Henry Giles Kingston.

Das Roggenhaus-Komplott. ●...Georg Hiltl.

Rosamond Fane; or, Prisoners of Saint James. ...Mary and Catharine Lee.

Rosemary : a tale of the fire of London. ...Georgiana C. L. G. Fullerton.

Rupert Aubrey of Aubrey Chase : an historical tale of 1681. ...Thomas John Potter.

Russell : a tale of the reign of Charles II. ...George Payne Rainsford James.

Rye House plot. ...George W. M. Reynolds.

Saint Clair of the Isles. ...E. Helme.

Saint Dunstan's clock : a story of 1666. ...E. Ward.

Saint George and Saint Michael. ...George Macdonald.

Saint Valentine's day, 1664. ...Elbridge S. Brooks. (In his Storied holidays.)

Secret foe. ...Ellen Pickering.

Settlers at home. ...Harriet Martineau.

Shepherd of Grove Hall : a tale of 1662. ...Anon.

Siege of Colchester. ...George F. Townsend.

Siege of Lichfield. ...William Gresley.

Silas Verney : being the story of his adventures in the days of King Charles the Second. ...Edgar Pickering.

Sir Henry Appleton; or, Essex during the great rebellion. ...William E. Heygate.

ENGLISH HISTORY — *Continued.*

Sir Ralph Esher; or, Memoirs of a gentleman of the court of Charles the Second.   ...Leigh Hunt.

Sir Ralph Willoughby.   ...Sir Samuel Egerton Brydges.

Spanish match; or, Charles Stuart at Madrid.   ...William Harrison Ainsworth.

Stanfield Hall.   ...J. Frederick Smith.

Star-chamber.   ...William Harrison Ainsworth.

Stronges of Netherstronge: a tale of Sedgemoor.   ...Edith J. May.

Talbot Harland.   ...William Harrison Ainsworth.

Through unknown ways; or, Journal-books of Dorothea Trundel.   ...Lucy Ellen Guernsey.

Traitor or patriot? a tale of the Rye House plot.   ...Mary C. Rowsell.

True-born gentleman.   ...Daniel Defoe.

True hero; or, Story of William Penn.   ...William Henry Giles Kingston.

Truth; or, Persis Clareton.   ...Charles B. Tayler.

Two knights of Delany Castle.   ...Mary M. Sherwood.

Two penniless princesses.   ...Charlotte Mary Yonge.

Two swords: being a story of old Bristol.   ...Emma Marshall.

Walter Colyton: a tale of 1688.   ...Horace Smith.   (Pseud., Paul Chatfield.)

Washingtons: a tale of a country parish in the seventeenth century.   ...John N. Simpkinson.

Wearyholme; or, Seed time and harvest: a tale of the restoration of Charles II.   ...Emily Sarah Holt.

White gauntlet.   ...Mayne Reid.

Whitefriars; or, Times and days of Charles II.   ...Jane Robinson.

Whitehall; or, Days and times of Oliver Cromwell.   ...Jane Robinson.

William, Prince of Orange; or, King and the hostage.   ...Titus M. Merriman.

Willitoft; or, Days of James I.: a tale.   ...Anon.

Winchester Meads in the times of Thomas Ken.   ...Emma Marshall.

Winifred; or, English maiden in the seventeenth century.   ...Lucy Ellen Guernsey.

Woodstock; or, Cavalier: a tale of the year 1651.   ...Sir Walter Scott.

See also **CHARLES I. OF ENGLAND, COMMONWEALTH OF ENGLAND**, and **JAMES II. OF ENGLAND**.

Eighteenth Century.

Adventures of Roderick Random.   ...George Tobias Smollett.

Against the stream: the story of the heroic age in England.   ...Elizabeth Charles.

Ashcliffe Hall: a tale of the last century.   ...Emily Sarah Holt.

ENGLISH HISTORY — *Continued.*

Life, adventures, and piracies of the famous Captain Singleton.
...Daniel Defoe.

Lord Harry Bellaîre : a tale of the last century.   ...Anne Manning.

Lord Mayor of London.   ...William Harrison Ainsworth.

Lucy Arden ; or, Hollywood Hall.   ...James Grant.

Maid of honor ; or, Court of George I.   ...Robert Folk Williams.

Maiden's lodge : a tale of the times of Queen Anne.   ...Emily Sarah
Holt.

Miser's daughter.   ...William Harrison Ainsworth.

Miss Angel.   ...Anna Isabella Thackeray (Mrs. Richmond Ritchie).

New voyage around the world.   ...Daniel Defoe.

Old Chelsea Bun-house.   ...Anne Manning.

Old manor house.   ...Charlotte Smith.

Oliver Ellis ; or, Fusiliers.   ...James Grant.

Pastor's fireside.   ...Jane Porter.

Peg Woffington.   ...Charles Reade.

Phantom regiment ; or, Stories of "Ours."   ...James Grant.

Preston fight ; or, Insurrection of 1715.   ...William Harrison Ainsworth.

Queen's jewel.   ...M. P. Blyth.

Ralf Skirlaugh, the Lancashire squire.   ...Edward Peacock.

Rival apprentices.   ...Anon.

Romance of war ; or, Highlanders in Spain.   ...James Grant.

Ronald Morton ; or, Fire ships : a story of the last naval war.
...William Henry Giles Kingston.

Rye House plot.   ...George M. Reynolds.

Saint James's ; or, Court of Queen Anne.   ...William Harrison
Ainsworth.

Saville House : an historical romance of the time of George the
First.   ...Mary M. Hardy.   (Pseud., Addeston Hill.)

Shelley : Biographische Novelle.   ...Wilhelm Ritter von Hamm.

South-sea bubble : a tale of the year 1720.   ...William Harrison
Ainsworth.

Spiritual Quixote ; or, Summer Ramble of Mr. Geoffry Wildgoose.
...Richard Graves.

Strawberry Hill.   ...Robert Folk Williams.

Surgeon's daughter.   ...Sir Walter Scott.

Tale of two cities.   ...Charles Dickens.

Tapestried chamber.   ...Sir Walter Scott.

Traitor or patriot ? a tale of the Rye House plot.   ...Mary C. Rowsell.

Treasure trove.   ...Samuel Lover.

Tremaine ; or, Man of refinement.   ...Robert P. Ward.

Vicar of Wakefield.   ...Oliver Goldsmith.

Virginians : a tale of the last century.   ...William Makepeace
Thackeray.

ENGLISH HISTORY — *Continued.*

Waverley. ...Sir Walter Scott.

When George III. was king. ...Anon.

Wilhelm Hogarth : Historische Roman. ...Albert E. Brachvogel.

William Hogarth : a romance. ...Albert E. Brachvogel. English translation.

World went very well then. ...Walter Besant.

Young buglers : a tale of the Peninsular War. ...George Alfred Henty.

Youth and manhood of Cyril Thornton. ...Thomas Hamilton.

See also **REBELLIONS. Old and Young Pretender's.**

Nineteenth Century.

Abigel Rowe : a chronicle of the regency. ...Lewis S. Wingfield.

Aims and obstacles. ...George Payne Rainsford James.

Alton Locke. ...Charles Kingsley.

Convict. ...George Payne Rainsford James.

Diana's crescent. ...Anne Manning.

Four Georges. ...William Makepeace Thackeray.

Frederick Gordon ; or, Storming of the Sedan. ...Anon.

Gaythorne Hall. ...John M. Fothergill.

Henri de la Tour ; or, Comrades in arms. ...J. Frederick Smith.

Interpreter : a tale of the Crimean War. ...George John Whyte Melville.

Lady Wedderburn's wish : a tale of the Crimean war. ...James Grant.

Laura Everingham ; or, Highlanders of Glen Ora. ...James Grant.

Lord Hermitage. ...James Grant.

Mainstone's housekeeper. ...Eliza Meteyard. (Pseud., Silver pen.)

Manchester strike. ...Harriet Martineau.

Martyrs to circumstances. ...Maria Thérèsa Longworth Yelverton.

Mary Barton : a tale of Manchester life. ...Elizabeth C. Gaskell.

North and South. ...Elizabeth C. Gaskell.

One of the " Six hundred." ...James Grant.

Phantom regiment ; or, Stories of " Ours." ...James Grant.

Pride and prejudice. ...Jane Austen.

Ravenshoe. ...Henry Kingsley.

Romance of war ; or, Highlanders in Spain. ...James Grant.

Shadow of the sword. ...Robert W. Buchanan.

Subaltern. ...George Robert Gleig.

Sybil ; or, Two nations. ...Benjamin Disraeli.

Under the Red Dragon. ...James Grant.

Vera ; or, Russian Princess and English Earl. ...Charlotte L. H. Dempster.

ENGLISH HISTORY — *Continued.*

Weird of the Wentworths : a tale of George IV.'s time.... Johannes Scotus. Pseud.

Wenderholm : a story of Lancashire and Yorkshire.   ...Philip Gilbert Hamerton.

Whitehall ; or, Days of George IV.   ...William Maginn.

Young buglers : a tale of the Peninsular war.   ...George Alfred Henty.

ERIC V. OF DENMARK. *Reigned 1259–1286.*

King Eric V.   ...Bernhard Severin Ingemann.   English translation.

Kong Erik.   ...Bernhard Severin Ingemann.

EUGENE DE BEAUHARNAIS, SON OF JOSEPHINE OF FRANCE. *Lived 1789–1824.*

Courtship by command.   ...M. M. Blake.

Prinz Eugen.   ...Albert Pulitzer.

Romance of Prince Eugene.   ...Albert Pulitzer.   English translation.

EUGENE, PRINCE OF SAVOY. *Lived 1663–1736.*

Prince Eugene and his times.   ...Clara M. Mundt. (Pseud., Louise Mühlbach.)   English translation.

Prinz Eugen und seine Zeit.   ...Clara M. Mundt. (Pseud., Louise Mühlbach.)

Prinz Eugen unter Kaiser Leopold.   ...Johann Georg Ludwig Hesekiel.

Prinz Eugen von Saxoyen.   ...Carl T. Griesinger.

Prinz Eugenius von Savoyen.   ...Emil von Boxberger.

EUROPE, HISTORY OF MEDIÆVAL AND MODERN. *A Selection.*

Eighth Century.

Ingraban.   ...Gustav Freytag.

Ingraban.   ...Gustav Freytag.   English translation.

Ninth Century.

Albrecht.   ...Arlo Bates.

Athelwold.   ...Amélie Rives Chanler.

Chronicle of Ethelfled.   Anne Manning.

Danes in England.   ...Alfred H. Engelbach.

Glastonbury.   ...A. Payne.

Legends of Charlemagne.   ...Thomas Bulfinch.

Passe Rose.   ...Arthur Sherburne Hardy.

Romances relating to Charlemagne.   ...George Ellis.

Sea kings of England.   ...Edwin Atherstone.

EUROPE, HISTORY OF MEDIÆVAL AND MODERN — *Continued.*

Tenth Century.

Ekkehard. ...Joseph Victor von Scheffel.
Erling; or, Days of Saint Olaf. ...F. Scarlett Potter.
Heroes of the North; or, Stories from Norwegian chronicle. ...F. Scarlett Potter.
Ogine: Chronique du Xᵉ siècle. ...Armand Garreau.
Popular tales from the Norse. ...George Webbe Dasent.
Vikings of the Baltic. ...George Webbe Dasent.

Eleventh Century.

Bertha: a historical romance of the time of Henry IV. of Germany. ...Joseph E. C. Bischoff. English translation.
Bertha; or, Pope and the emperor. ...William B. MacCabe.
Boy crusaders. ...John George Edgar.
Count Robert of Paris. ...Sir Walter Scott.
Dutch in the Medway. ...Charles MacFarlane.
Die Fahrten Thiodolfs des Isländers. ...Friedrich H. C. La Motte Fouqué.
Fighting the Saracens; or, Boy knight. ...George Alfred Henty.
Florine, princess of Burgundy. ...William B. MacCabe.
Foure prentises of London, with the conquest of Jerusalem. ...Thomas Heywood.
Harold, the last of the Saxon kings. ...Edward George Earle Lytton Bulwer-Lytton.
Hereward, the Saxon. ...Charles Knight.
Hereward, the Wake. ...Charles Kingsley.
Heroines of the crusades. ...C. A. Bloss.
Hildebrand and the excommunicated emperor. ...Joseph Sortain.
I Lombardi alla prima crociata. ...Tommaso Grossi.
In the brave days of old: the story of the crusades. ...Henry Frith.
Ivo and Vereno. ...Anon.
Kaiser Friedrich genannt Barbarossa. ...Carl J. Simrock.
Königin Bertha. Historischer Roman. ...Joseph E. C. Bischoff.
Lady and the priest. ...Katherine C. Maberley.
Legend of Reading Abbey. ...Charles Knight.
Mathilde; ou, Mémoires tirés de l'histoire des croisades. ...Sophie Risteau Cottin.
Parcival: Roman. ...Albert E. Brachvogel.
Prince and the page, the last crusade. ...Charlotte Mary Yonge.
Raymond de Saint Gilles at the crusade. 1095-1099. ...Henriette de Witt.
Rufus; or, Red king. ...James Grant.

EUROPE, HISTORY OF MEDIÆVAL AND MODERN — *Continued.*
Saracen ; or, Matilda and Malek Adkel : a crusade romance.
...Sophie Risteau Cottin.   ...English translation.
Sir Walter's ward : a tale of the crusades.   ...William Everard.
Stories of the crusades.   ...William E. Dutton.
Sword and scimetar : the romance of the crusades.   ...Alfred
    Trumble.
Tales of the early ages.   ...Horace Smith. (Pseud., Paul Chatfield.,
Truce of God : a tale of the eleventh century.   ...George H. Miles.
Wulf, the Saxon.   ...Ralph Peacock.

Twelfth Century.

L'Abbaye de Typhaines.   ...Joseph Arthur, Comte de Gobineau.
Arthur of Brittany.   ...Peter Leicester.
Barbarossa : historischer Roman.   ...Joseph E. C. Bischoff.
Betrothed.   ...Sir Walter Scott.
Blondel : a historical fancy.   ...George E. Rice.
Boy's adventure in the Baron's war.   ...John George Edgar.
Fair Rosamond : Days of King Henry II.   ...Thomas Miller.
Friedrichs des ersten letzte Lebenstage.   ...Julius Bacher.
Heinrich der Löwe.   ...Carl L. Haeberlin.
Heinrich der Löwe.   ...Carl C. F. Riedmann.
Heinrich der sechste teutscher Kaiser.   ...Carl F. A. Buchner.
Heinrich von England und sein Söhne.   ...Franziska C. J. F. Tarnow.
Ivanhoe.   ...Sir Walter Scott.
John of England.   ...Henry Curling.
John o' London.   ...Edward Fitzgibbon. (Pseud., Somerville Gibney.)
King and the troubadour.   ...Julia Corner.
King John and the brigand's daughter.   ...Anon.
Lady Sybil's choice : a tale of the crusades.   ...Emily Sarah Holt.
Leper-house of Janval.   ...Catherine C. F. Gore.
Maude and Miriam.   ...Harriet B. MacKeever.
Merry adventures of Robin Hood.   ...Howard Pyle.
Richard Cœur de Lion.   ...James White.
Royston Gower.   ...Thomas Miller.
Runnymede and Lincoln fair.   ...John George Edgar.
Sir Guy de Lusignan.   ...E. Cornelia Knight.
Talisman.   ...Sir Walter Scott.
Typhaines : a tale of the twelfth century.   ...Joseph Arthur, Comte
    de Gobineau. English translation.
See also **JOHN OF ENGLAND** and **HOOD, ROBIN**.

Thirteenth Century.

Castle of Ehrenstein : its lords, spiritual and temporal ; its inhabi-
    tants, earthly and unearthly. ...George Payne Rainsford James.

EUROPE, HISTORY OF MEDIÆVAL AND MODERN — *Continued.*

Cheshire pilgrims. ...Frances M. Wilbraham.

Coberly Hall: a Gloucestershire tale.

Crusaders and captives: a tale of the children's crusade. ...George E. Merrill.

Days of Bruce. .. Grace Aguilar.

Earl Hubert's daughter; or, Polishing of the pearl. ...Emily Sarah Holt.

Gaston de Blondeville; or, Court of Henry III. ...Anne Radcliffe.

Heinrich von Ofterdingen. . .Friedrich L. von Hardenberg.

Henry of Ofterdingen. ...Friedrich L. von Hardenberg. English translation.

How I won my spurs. ...John George Edgar.

Knights of the lion: a romance of the thirteenth century. ...Anon.

Our little lady. ...Emily Sarah Holt.

Philip Augustus; or, Brothers in arms. ...George Payne Rainsford James.

Prentice Hugh. ...Frances M. Peard.

Rothelan: a romance of the English histories. ...John Galt.

Royston Gower. ...Thomas Miller.

Rudolf von Habsburg. ...Friedrich C. Schlenkert.

Runnymede and Lincoln fair. ...John George Edgar.

Scottish chiefs. ...Jane Porter.

Siege of Kenilworth. ...L. S. Stanhope.

Sir Michael Scott. ...Allan Cunningham.

Stephen Langton; or, Days of King John. ...Martin Farquhar Tupper.

Stories of the city of London. ...Mrs. Newton Crosland. (Pseud., Camilla Toulmin.)

Vendgiad; or, Blessed one: a tale of the 13th century. ...Anon.

With script and staff: a tale of the children's crusade. ...Elia W. Peattie.

Wolfram von Eschenbach. ...Ludwig Lang.

Fourteenth Century.

Abbess of Shaftesbury; or, Days of John of Gaunt. ...Anon.

Le bâtard de Mauléon. ...Alexandre Dumas.

Berkeley castle. ...G. F. Berkeley.

Boy foresters. ...Anne Bowman.

Castle dangerous. ...Sir Walter Scott.

Chronicles of Yate Court: a narrative of 1399. Charles Charlton.

Cressy and Poictiers; or, Story of the Black Prince's page. ...John George Edgar.

Der falsche Wöldemar. ...Georg Wilhelm Heinrich Haering.

EUROPE, HISTORY OF MEDIÆVAL AND MODERN — *Continued.*

Forest days: a romance of old days.   ...George Payne Rainsford
    James.
Foster brothers.   ...Thornton Hunt.
Geoffrey the Lollard.   ...Mrs. D. C. Knevels.   (Pseud., Frances
    Eastwood.)
Half brothers; or, Head and the hand.   ...Alexandre Dumas.
    English translation.
Hubert Ellis: a story of Richard II.'s days.   ...F. Davenant.
John of Gaunt.   ...James White.
Der Klosterjäger.   ...Ludwig A. Ganghofer.
Knight of Mauléon.   ...Alexandre Dumas.   English translation
    (same as Half brothers).
Lances of Lynwood.   ...Charlotte Mary Yonge.
Maid Marian.   ...Thomas L. Peacock.
Maid Marian and Robin Hood.   ...J. E. Muddock.
Marco Visconti.   ...Tommaso Grossi.
Margam Abbey: an historical romance.   ...Anon.
Merry England; or, Nobles and serfs.   ...William Harrison
    Ainsworth.
Mistress Margery: a tale of the Lollards.   ...Emily Sarah Holt.
Otterbourne.   ...Edward Duras.
Provost of Paris: a tale.   ...William S. Browning.
Rienzi, the last of the Roman Tribunes.   ...Edward George Earle
    Lytton Bulwer-Lytton.
Robin Hood.   ...Pierce Egan, jr.
Sword of De Bardwell: a tale of Agincourt.   ...C. M. Katherine
    Phipps.
Tower of the hawk: some passages in the history of the House of
    Hapsburg.   ...Jane L. Willyams.
Valperga.   ...Mary Wollstonecraft Shelley.
Well in the desert: an old legend of the house of Arundel.   ...Emily
    Sarah Holt.
White company.   ...Arthur Conan Doyle.

Fifteenth Century.

Die Abenteuer der Herzoga Christoph von Bayern genannt der
    Kampfer.   ...Franz Trautmann.
Agnes of Sorrento.   ...Harriet Beecher Stowe.
Agincourt.   ...George Payne Rainsford James.
L'Amazone française Jeanne d'Arc.   ...Marie Thérèse Peroux
    d'Albany.
Anne of Geierstein.   ...Sir Walter Scott.
Before the dawn: a story of Paris and the Jacquerie.   ...George
    Dulac.

EUROPE, HISTORY OF MEDIÆVAL AND MODERN — *Continued.*

Briefe eines Frauenzimmers aus dem XV. Jahrh. nach alten Urschriften. ...Paul von Stetten.

Das Buch von den Wienern. ...Michael Beheim.

Caged lion. ...Charlotte Mary Yonge.

Captain of the guard. ...James Grant.

Cardinal Ximenes. ...Jean Bertheroy. English translation.

Conrad. ...Emma Leslie.

Coulyng Castle; or, Knight of the olden days. ...Agnes Giberne.

Days of yore. ...Henrietta Keddie. (Pseud., Sarah Tytler.)

Dearer than life : a tale of the times of Wycliffe. ...Emma Leslie.

Dolores. ...Gertrúdes Gomez de Avellaneda.

Dove in the eagle's nest. ...Charlotte Mary Yonge.

Dürer in Venedig. ...Adolf Stern.

Earl printer : times of Caxton. ...C. M. M.

Ernest and Albert ; or, Story of the stolen princes. ...Anon.

Ettore Fieramosca ; or, Challenge of Barletta. ...Massimo Taparelli, Marchese d' Azeglio. English translation.

Ettore Fieramosca : ossia, La disfida di Barletta. ...Massimo Taparelli, Marchese d' Azeglio.

Fair maid of Perth. ...Sir Walter Scott.

For, or against. ...Frances M. Wilbraham.

Fortunes of Perkin Warbeck. ...Mary Wollstonecraft Shelley.

Franz Sternbald's Wanderungen. ...Johan Ludwig Tieck.

Geoffrey, the Lollard. ...Mrs. D. C. Knevels. (Pseud., Frances Eastwood.)

Die Gred. Roman aus dem alten Nüremberg. ...Georg Moritz Ebers.

Gutenberg. ...Emily C. Pearson. (Pseud., Ervie.)

House of Fiesole. ...Catherine Shaw.

Hubert Ellerdale. ...W. O. Rhind.

Hermann von Unna. ...Christiane Benedicte Eugenie Naubert.

Henry VII. ...Mary Wollstonecraft Shelley.

Historical tales of the Lancastrian times. ...H. P. Dunster.

In Wycliffe's days. ...G. Stebbing.

Jack of the mill. ...William Howitt.

Jeanne d'Arc ; ou, La France au XVᵉ siècle. ...Jules Foussette.

Jew. ...Carl Spindler. English translation.

Joan the maid, deliverer of France and England. ...Elizabeth Charles.

Johannes Gutenberg. ...Adolf Stern.

John de Wycliffe, the first reformer. ...Emily Sarah Holt.

Der Jude. ...Carl Spindler.

Knight of Dilham : a story of the Lollards. ...Arthur Brown.

Die Künstlerehe. ...Leopold Schefer.

Kunz von Kauffungen. ...Ludwig Storch.

EUROPE, HISTORY OF MEDIÆVAL AND MODERN — *Continued.*
White rose of Langley: a story of the court of England in olden times. ...Emily Sarah Holt.
Wycliffites; or, England in the fifteenth century. ...Margaret Mackay.
Ximénès. ...Jean Bertheroy.
Yellow frigate. ...James Grant.
See also **ANNE BOLEYN.**

Sixteenth Century.

L'Abbesse de Montmartre. ...Henri Augu.
Abbot. ...Sir Walter Scott.
Agnes de Mansfeldt. ...Thomas C. Grattan.
Alice Sherwin : a tale of the days of Sir Thomas More. ...C. J. M.
Anne Hathaway; or, Shakespeare in love. ...E Severn.
Ascanio. ...Alexandre Dumas.
Ascanio. ...Alexandre Dumas. English translation.
Aus dem sechzehnten Jahrhundert. ...Wilhelm Jensen.
Barbara Ittenhausen. ...Emma Wuttze-Biller.
Baron of Hertz. ...Albert de Labadye. English translation.
Le Baron de Hertz. ...Albert de Labadye.
Die Belagerung von Gotha. ...Auguste W. Lorenz.
Die Belagerung von Leipzig in den Jahren 1546-7. ...F. Lohmann.
Die Belagerung Magdeburg. ...Friedrich L. Schmidt.
Bolsover castle : a tale from the Protestant history of the sixteenth century. ...Anon.
Bothwell; or, Days of Mary Queen of Scots. ...James Grant.
Braes of Yarrow. ...H. Gibbon.
Brigand ; or, Corse de Leon. ...George Payne Rainsford James.
Das Buch von Doktor Luther. ...Hermann O. Nietschmann.
Burgomaster's wife. ...Georg Moritz Ebers. English translation.
By England's aid; or, Freeing the Netherlands. ...George Alfred Henty.
Cæsar Borgia. ...E. Robinson.
Cardinal Pole ; or, Days of Philip and Mary. ...William Harrison Ainsworth.
Chaplet of pearls; or, White and black Ribaumont. ...Charlotte Mary Yonge.
Christlich oder Päpstlich ? ...Eduard Jost.
Christian or Romanist ? ...Eduard Jost. English translation.
Chronicles of the Schönberg-Cotta family. ...Elizabeth Charles.
Clare Avery: a story of the Spanish armada. ...Emily Sarah Holt.
Cloister and the hearth. ...Charles Reade.
Clytia: a romance of the sixteenth century. ...Adolf Hausrath. English translation.

EUROPE, HISTORY OF MEDIÆVAL AND MODERN — *Continued.*

Colloquies of Edward Osborne.   ...Anne Manning.

Constable of the tower.   ...William Harrison Ainsworth.

Constancia's household : a story of the Spanish reformation. ...Emma Leslie.

Cooper of Nuremberg.   ...Ernest Theodor Wilhelm Hoffmann. English translation.

Count Arensberg.   ...Joseph Sortain.

Count Ulrich von Lindburg : a tale of the reformation.   ...William Henry Giles Kingston.

Crichton.   ...William Harrison Ainsworth.

Dark year of Dundee.   ...Deborah Alcock.

Darnley ; or, Field of the cloth of gold.   ...George Payne Rainsford James.

David Rizzio.   ...William H. Ireland.

Doctor Luther's book.   ...Hermann O. Nietschmann.   English translation.

Drake and the Dons.   ...R. Lovett.

Duke Christian of Luneburgh.   ...Jane Porter.

Duke Christopher : a story of the reformation.   ...Frances H. Christopher.

Der Fall von Kounstanz.   ...Otto Mueller.

Fall of Somerset.   ...William Harrison Ainsworth.

Fatal revenge ; or, Family of Montorio.   ...Charles Robert Maturin. (Pseud., Dennis Jasper Murphy.)

Fate of Castle Löwengard : a story of the days of Luther.   ...Esmé Stuart.

Florence betrayed ; or, Last days of the republic.   ...Massimo Taparelli, Marchese d'Azeglio.   English translation.

Florian Geyer.   ...Wilhelm R. Heller.

Forest of Arden.   ...William Gresley.

Forester's daughter : a tale of the reformation.   ...Anon.

For the master's sake : a story of the days of Queen Mary.   ...Emily Sarah Holt.

Die Frau Bürgemeisterin.   ...Georg Moritz Ebers.

From dark to dawn in Italy.   ...Anon.

Der Fürstentag.   ...Ludwig Bechstein.

Gabriele die Waise auf Rosenburg.   ...Johann B. Fastenau.

Der Geächtete.   ...Gustav. A. von Heeringen.

Gentleman of France.   ...Stanley J. Weyman.

Georg Podiebrad, König von Bohmen und Markgraf zur Lausitz. ...Christian R. Koestlin.

Good old days ; or, Christmas under Queen Elizabeth.   ...Esmé Stuart.

Good old times : a tale of Auvergne.   ...Anne Manning.

EUROPE, HISTORY OF MEDIÆVAL AND MODERN — *Continued.*

Hans Sachs. ...Adolf Friedrich Furschau.

Der Heiland von der Rhön. ...Paul Lippert.

Histoire vom Ritter Götz von Berlichingen mit der eisernen Hand. ...Heinrich Doering.

Household of Sir Thomas More. ...Anne Manning.

House of the wolf: a romance. ...Stanley J. Weyman.

Im blauen Hecht. ...Georg Moritz Ebers.

Im Zwing und Bann. ...Wilhelm Jensen.

In the Blue Pike. ...Georg Moritz Ebers. English translation.

In the olden time. ...Margaret Roberts.

Isoult Barry of Wynscote : a tale of Tudor times. ...Emily Sarah Holt.

Jack and the tanner of Wymondham : a tale of the time of Edward the Sixth. ...Anne Manning.

Jane Seton. ...James Grant.

Joachim Slüter. Julius von Wickede.

Joyce Morrell's harvest. ...Emily Sarah Holt.

Kaiser Karl des Fünften erste Jugendliebe. ...Ludwig A. Arnim.

Katharine von Schwarzburg. ...Carl G. Berneck.

Kenilworth. ...Sir Walter Scott.

Klytia. Adolf Hausrath.

Knight of Saint John. ...Anna M. Porter.

King Henry and his court. ...Clara M. Mundt. (Pseud., Louise Mühlbach.) English translation.

König Heinrich VIII. und sein Hof. ...Clara M. Mundt. (Pseud., Louise Mühlbach.)

Die Kronenwächter. Ludwig A. von Arnim.

Lady Jane Grey. ...Thomas Miller.

Lady Jane Grey und ihre Zeit. ...Louise M. Gräfin von Robiano.

Lady Rosamond's book. ...Lucy Ellen Guernsey.

Lancashire witches : a romance of Pendle Forest. ...William Harrison Ainsworth.

Lettice Eden : a tale of the last days of Henry the Eighth. ...Emily Sarah Holt.

Die letzten Humanisten. ...Adolf Stern.

Lichtenstein. ...William Hauff.

Lichtenstein ; or, Swabian league. ...William Hauff. English translation.

Die Lichtensteiner. ...Carl F. van der Velde.

Lichtensteins : a tale of the thirty years' war. ...Carl F. van der Velde. English translation.

Lincolnshire tragedy. ...Anne Manning.

Lukas Cranach: historischer Roman. ...Hermann Klencke.

Luther and the Cardinal. ...Hermann O. Nietschmann. English translation.

EUROPE, HISTORY OF MEDIÆVAL AND MODERN — *Continued.*

Luther in Rom. Roman. ...Levin Schuecking.

Luther in Rome; or, Corradina, the last of the Hohenstaufen ...Levin Schuecking. English translation.

Luther und die Seinen. ...Franz Lubozatsky.

Luther und seine Zeit. ...Theodor Koenig.

Luther's Brautfahrt. ...Joseph E. C. Bischoff.

Magdalen Hepburn. ...Margaret O. W. Oliphant.

Margaret Roper. ...Agnes M. Stewart.

Margarethe: a tale of the sixteenth century. ...Emma Leslie.

Marie Stuart. ...George W. M. Reynolds.

Martin Luther: historischer Roman. ...Heinrich Eisenlohr.

Martin Luther und Graf Erbach. Hermann O. Nietschmann.

Martyrs of Spain and liberators of Holland. ...Elizabeth Charles.

Mary of Lorraine. ...James Grant.

Maurice, the elector of Saxony. ...Kathárine Colquhoun.

Michael Kohlhaas. ...Heinrich von Kleist.

Mistress Haselwode. ...Frederick II. More.

Monastery. ...Sir Walter Scott.

Monk and knight. ...Frank W. Gunsaulus.

Montchensey: the days of Shakespeare. ...Nathan Drake.

Moritz von Sachsen. ...Friedrich C. Schlenkert.

Mountain patriots: a tale of the reformation in Savoy. ...Mrs. Alexander S. Orr.

Necromancer. ...George W. M. Reynolds.

Niccolò de Lapi; ovvero, I Palleschi e i Piagnoni. ...Massimo Taparelli, Marchese d'Azeglio.

No cross no crown. ...Caroline E. Davis.

Passages in the life of the fair gospeller Mistress Anne Askew. ...Anne Manning.

Pendower. ...M. Filleul.

Philippine Welser, oder vor dreihundert Jahren. ...Adelbert Baudissin.

Professor of Alchemy. ...Percy Ross.

Der Prophet. ...Theodor Muegge.

Protestant: a tale of the reign of Queen Mary. ...Anna Eliza Bray.

Recess; or, Tale of other times. ...Sophia Lee.

Richard Hunne: a story of old London. ...George E. Sargent.

Robin Tremayne of Bodwin: a tale of the Marian persecution. ...Emily Sarah Holt.

Royal merchant; or, Events in the days of Sir Thomas Gresham. ...William Henry Giles Kingston.

Saint Bartholomew's eve: a tale of the Huguenot wars. ...George Alfred Henty.

Die Schlacht bei Drakenburg. ...Werner Bergman.

EUROPE, HISTORY OF MEDIÆVAL AND MODERN — *Continued.*

Die Schlacht bei Hemmingstedt. ...Amalie Schoppe.

Die Schlacht von Marignano. ...Carl A. F. von Witzleben.

Secret passion. ...Robert Folk Williams.

Shakespeare and his friends. ...Robert Folk Williams.

Shakspere the poet. ...Henry Curling.

Shakespeare's youth. ...Robert Folk Williams.

Sibylle von Cleve. ...Julius Bacher.

Sidonia the sorceress. ...Johann Wilhelm Meinhold.

Le Siège de La Rochelle. ...Stéphanie Félicité, Comtesse de Genlis.

Siege of La Rochelle. ...R. C. Dallas.

Der Sohn des Kaisers. ...Adolf Muetzelburg.

Southenan. ...John Galt.

Spae-wife; or, Queen's secret: a tale of the days of Elizabeth. John Boyce.

Spanish brothers : a tale of the sixteenth century. ...Debnal Alcock.

Stories of the wars. ...J. Tillotson.

Struggle in Ferrara. ...William Gilbert.

Tablette booke of Lady Mary Keyes. ...Anon.

Tabithe von Geyersberg. ...Amalie Schoppe.·

Tales from Alsace. ...Anne Manning.

Thomas Müntzer, Roman. ...Theodor Mundt.

Three hundred years ago ; or, Mysteries of Brentwood. ...William Henry Giles Kingston.

Three Musketeers. ...Alexandre Dumas. English translation.

Tor hill. ...Horace Smith. (Pseud., Paul Chatfield.)

Tower Hill. . ...William Harrison Ainsworth.

Tower of London. ...William Harrison Ainsworth.

Les trois mousquetaires. ...Alexandre Dumas.

Trübe Tage. ...Wilhelm Koch.

True as steel. ...George Walter Thornbury.

Twice crowned : a story of the days of Queen Mary. ...Harriet B. MacKeever.

Twin heroes : the Separatists of the time of Queen Elizabeth. ...F. A. Reed.

Uline's escape ; or, Hid with the nuns. ...Mrs. Alexander S. Orr.

Ulric; or, Voices. ...Theodore S. Fay.

Westminster Abbey. ...Emma Robinson.

Westward Ho! or, Voyages and adventures of Sir Amyas Leigh. ...Charles Kingsley.

Wild times. ...Cecilia M. Caddell.

Die Wiellinger. ...Friedrich Wilhelm Arming.

Windsor Castle. ...William Harrison Ainsworth.

Word, only a word. ...Georg Moritz Ebers. English translation.

EUROPE, HISTORY OF MEDIÆVAL AND MODERN — *Continued.*
Ein Wort. ...Georg Moritz Ebers.
Zord Idő. ...Zsigmund Koerner.

Seventeenth Century.

Der abenteuerliche Simplicissimus. ...Hans J. C. von Grummelshausen.
Agnes Beaumont: a true story of the year 1670. ...Marian Caldecott.
Agnes de Mansfeldt. ...Thomas C. Grattan.
Aimée: a tale of the days of James the Second. ...Agnes Giberne.
Alice Bridge of Norwich: a tale of the time of Charles the First. ...Andrew Reed.
Allwin, ein Roman. ...Friedrich H. C. La Motte Fouqué.
Allzeit getreu. ...H. Brand.
Andrew Golding: a tale of the great plague. ...Annie E. Keeling.
Aphra Behn. ...Clara M. Mundt. (Pseud., Louise Mühlbach.)
Aphra Behn. ...Clara Mundt. (Pseud., Louise Mühlbach.) English translation.
Andrew Marvel and his friends: a story of the siege of Hull. ...Marie S. Hall.
Arabella Stuart. ...George Payne Rainsford James.
Armourer's daughter; or, Border rivals. ...Anon.
Arthur Arundel. ...Horace Smith. (Pseud., Paul Chatfield.)
Arthur Blane; or, Hundred cuirassiers. ...James Grant.
At the sign of the Blue Boar: a story of the reign of Charles II. ...Emma Leslie.
Axel. ...Carl F. van der Velde.
Axel: a tale of the thirty years' war. ...Carl F. van der Velde. English translation.
Baron and squire: a story of the thirty years' war. ...Wilhelm Noeldechen. English translation.
Barony. ...Anna Maria Porter.
Beatrice Tyldesley. ...William Harrison Ainsworth.
Berber; or, Mountaineer of the Atlas. ...William S. Mayo.
Bernhard, Herzog zu Sachsen-Weimar. ...Friedrich G. Schlenkert.
Betrothed. ...Alessandro Manzoni. ...English translation.
Beyond the seas. ...Oswald Crawfurd.
Bílá hora, Aneb: Tři léta z třiceti. ...Ludwig Rellstab.
Blameless knights; or, Lützen and La Vendée. ...Alice H. F. Byng, Vicountess Enfield.
Boscobel; or, Loyal oak: a tale of the year 1651. ...William Harrison Ainsworth.
Boscobel: a narrative of the adventures of Charles II. after the battle of Worcester. ...George W. Dodd.

EUROPE, HISTORY OF MEDIÆVAL AND MODERN — *Continued.*

Brambletye house ; or, Cavaliers and roundheads. ...Horace Smith. (Pseud., Paul Chatfield.)

Brave men of Eyam; or, Tale of the great plague. ...Edward N. Hoare.

Brave resolve. ...John B. de Liefde.

Breughel brothers. ..Alexander Ungern-Sternberg. English translation.

Die Brüder; oder, Das Geheimniss. ...Alexander Ungern-Sternberg.

Buccaneer. ...Anna Maria Hall.

Der Buchführer von Lemgo. ...Johann Georg L. Hesekiel.

Der Bürgemeister von Neisse. ...Georg Hartwig.

Das Bürgerweib von Weimar. ...Julius W. Grosse.

Cajus Rungholt. ...Lucian Buerger.

Caleb Field : a tale of the Puritans. ...Margaret O. W. Oliphant.

Campion Court: the days of the ejectment. ...Emma J. Worboise.

Captain Jacques: a romance of the time of the plague. ...Edward Fitzgibbon. (Pseud., Somerville Gibney.)

Captain of the guard. ...James Grant.

Carved Cartoon. ...Austin Clare.

Castle Cornet ; or, Island's troubles in the troublous times. ...Louisa Hawtrey.

Cavaliers of England; or, Times of the revolutions of 1642 and 1648. ...Henry W. Herbert. (Pseud., Frank Forester.)

Charles II. ; or, Merry Monarch. ...John II. Payne.

Cherry and violet : a tale of the great plague. ...Anne Manning.

Chevalier's daughter : being one of the Stanton Corbet Chronicles. ...Lucy Ellen Guernsey.

Children of the New Forest. ...Frederick Marryat.

Der Christlichen königlichen Fürsten Herculiscus und Herculadisla Wundergeschichte. ...Andreas H. Buchholtz.

Des Christlichen teutschen Grossfürsten Hercules und der böhmischen königlichen Fräulein Valisca Wundergeschichte. ...Antreas II. Buchholtz.

Cinq-Mars ; ou, une conjuration sous Louis XIII. ...Alfred Victor, Comte de Vigny.

Cinq-Mars: a conspiracy under Louis XIII. ...Alfred Victor, Comte de Vigny. English translation.

Claude Duval: a romance of the days of Charles II. ...Henry D. Miles.

La conspiration de Walstein. ...Jean François Sarrasin.

Constance Aylmer : a story of the seventeenth century. ...Helen F. Parker.

Constance Sherwood. ...Georgiana C. L. G. Fullerton.

Coombe Abbey. ...Selina Bunbury.

EUROPE, HISTORY OF MEDIÆVAL AND MODERN — *Continued*.

Court at Tunbridge in 1664.   ...Catherine G. F. Gore.

Courtenay of Walreddon : a romance of the West.   ...Anna Eliza Bray.

Courtier in the days of Charles II.   ...Mrs. Gordon.

Courtier of the days of Charles II.   ...Catherine G. F. Gore.

Cromwell ; or, Protector's oath.   ...J. Frederick Smith.

Dame Rebecca Berry; or, Court scenes in the reign of Charles the Second.   ...Elizabeth I. Spencer.

Danvers papers.   ...Charlotte Mary Yonge.

Darien; or, Merchant prince.   ...Eliot B. G. Warburton.

Deborah's diary.   ...Anne Manning.

Der Deutsche Krieg.   ...Heinrich Laube.

Der Deutsch Michel.   ...Johann Georg L. Hesekiel.

Ein deutsches Schneiderlein.   ...Franz Isador Proschko.

Deutschlands Kassandra.   ...Heribert Rau.

Dorothe.   ...W. Alexsis.

Dorothe.   ...Georg Wilhelm Heinrich Haering.

Draytons and Davenants : a tale of the civil wars.   ...Elizabeth Charles.

Der dreissigjährige Krieg.   ...Clara M. Mundt. (Pseud., Louise Mühlbach.)

Die drei Wünsche.   ...Carl A. F. von Witzleben.

Drifted and sifted : a domestic chronicle of the seventeenth century.   ...Miss McLaren.

La duchesse de châtillon.   ...François T. M. d'Arnaud.

Duellists.   ...Anon.

Duke of Monmouth.   ...Gerald Griffin.

Dutch in the Medway.   ...Charles Macfarlane.

Edgar Nelthorpe ; or, Fair maids of Taunton.   ...Andrew Reed, jr.

Every man his own trumpeter.   ...George Walter Thornbury.

Fair Else, Duke Ulrich, and other stories.   ...Margaret Roberts.

Fältskarns Berättelser.   ..Zacharias Topelius.

Fate : a tale of stirring times.   ...George Payne Rainsford James.

Father Darcy: an historical romance.   ...Anne Marsh.

Der Fels von Erz.   ...Albert E. Brachvogel.

Fides: eine Erzählung aus der Zeit des dreissigjährigen Krieges.   ...F. Jordan.

Der Findling.   ...Carl A. F. von Witzleben.

Fire of London; or, Which is which?   ...Georgiana C. L. G. Fullerton.

For liberty's sake : the story of Robert Ferguson.   ...John B. Marsh.

Fortunes of Nigel.   ...Sir Walter Scott.

Frau Schatz Regime.   ...Johann Georg L. Hesekiel.

Gabriel : a story of the Jews in Prague.   ...Salomon Kohn. English translation.

EUROPE, HISTORY OF MEDIÆVAL AND MODERN — *Continued.*

Journal of Lady Beatrix Graham.   ...Jane M. F. Smith.

Karl II. : oder der Lustige Monarch.   ...Johann R. Lenz.

King's highway.   ...George Payne Rainsford James.

King's namesake : a tale of Carisbrooke Castle.   ...Catherine M. Phillimore.

King's service.   ...Anon.

Klosterheim.   ...Thomas De Quincey.

Das Kräuterweible von Wimpfen.   ...Conrad Fron.

Kruitzner ; or German's tale.   ...Harriet Lee.   (In her Canterbury tales.)

Lady Betty.   ...Christabel R. Coleridge.

Lady Betty's governess ; or, Corbet Chronicles.   ...Lucy Ellen Guernsey.

Lady Shakerlye : the record of the life of a good and noble woman.   ...Anon.

Lady Willoughby ; or, Passages from the diary of a wife and mother in the seventeenth century.   ...Hannah M. Rathbone.

Langham rebels.   ...Lucy Ellen Guernsey.

Last of the cavaliers.   ...Rose Piddington.

Last of the fairies : a Christmas tale.   ...George Payne Rainsford James.

Leaguer of Lathom : a tale of the civil war in Lancashire.   ...William Harrison Ainsworth.

Legend of Montrose.   ...Sir Walter Scott.

Leila ; or, Siege of Granada.   ...Edward George Earle Lytton Bulwer-Lytton.

Lichtenstein.   ...Wilhelm Hauff.

Die Lichtensteiner.   ...Carl F. von der Velde.

Lichtenstein ; or, Swabian league.   ...Wilhelm Hauff.   English translation.

Lichtensteins : a tale of the thirty years' war.   ...Carl F. van der Velde.

Die Lieder des dreissigjährigen Krieges.   ...Eduard Weller.

Life of Colonel Jack.   ...Daniel Defoe.

Lion of the North : a tale of the time of Gustavus Adolphus and wars of religion.   ...George Alfred Henty.

Lord Montagu's page : an historical romance of the seventeenth century.   ...George Payne Rainsford James.

Lorna Doone : a romance of Exmore.   ...Richard Doddridge Blackmore.

Maid of honor.   ...Catherine G. F. Gore.

Maid of Stralsund : a story of the thirty years' war.   ...John B. de Liefde.

Maiden and married life of Mary Powell, afterwards Mistress Milton.   ...Anne Manning.

EUROPE, HISTORY OF MEDIÆVAL AND MODERN — *Continued*.

Der Pfarrer von Andouse.   ...Heinrich Moewes.

Phantom ship.   ...Frederick Marryat.

Philip Rollo; or, Scottish musketeers.   ...James Grant.

Prinz Eugen und seine Zeit.   ...Clara M. Mundt.   (Pseud., Louise Mühlbach.)

Prinz Eugen unter Kaiser Leopold.   ...Johann Georg L. Hasekiel.

Prinz Eugen von Savoyen.   ...Carl T. Griesinger.

Prinz Eugenius von Savoyen.   ...Emil von Boxberger.

Prince Eugene and his times.   ...Clara M. Mundt.   (Pseud., Louise Mühlbach.)   English translation.

Le prophète irlandois.   ...Charles de Saint-Évremond.

Puritan's grave.   ...William P. Scargill.

Der Raub Strassburgs in Jahre 1681.   ...Heribert Rau.

Refugees: a tale of two continents.   ...Arthur Conan Doyle.

Reginald Hastings: a tale of the troubles of 164-.   ...Eliot B. G. Warburton.

Die Retter Niederwesels.   ...Philipp F. W. Oertel.

Der Ring.   ...Carl A. F. von Witzleben.   (In his Ausgewählte Schriften.)

Ritterlicher Sinn.   ...Carl A. F. von Witzleben.

Der Rittmeister von Alt-Rosen.   ...Gustav Freytag.

Robber.   ...George Payne Rainsford James.

Roger Willoughby; or, Times of Benbow.   ...William Henry Giles Kingston.

Das Roggenhaus Komplott.   ...Georg Hiltl.

Rosamond Fane: or, Prisoners of Saint James'.   ...Mary and Catherine Lee.

Die Rose von Heidelberg.   ...Louisa M. Robiano.

Rosemary: a tale of the fire in London.   ...Georgiana C. L. G. Fullerton.

Rupert Aubrey of Aubrey Chase: an historical tale of 1681.   ...Thomas John Potter.

Russell: a tale of the reign of Charles II.   ...George Payne Rainsford James.

Ryehouse plot.   ...George W. M. Reynolds.

Saint Dunstan's clock: a story of 1666.   ...E. Ward.

Saint George and Saint Michael.   ...George Macdonald.

Saint Valentine's day, 1664.   ...Elbridge S. Brooks.   (In his Storied holidays.)

Schoolmaster and his son.   ...Carl H. Caspari.   English translation.

Der Schulmeister und sein Sohn : eine Erzählung aus dem dreissigjähr. Kriege.   ...Carl H. Caspari.

Die Schweden in Prag.   ...Caroline Pichler.

Scottish cavalier.   ...James Grant.

EUROPE, HISTORY OF MEDIÆVAL AND MODERN — *Continued.*

Settlers at home. ...Harriet Martineau.

Shepherd of Grovehall : a tale of 1662. ...Anon.

Siege of Colchester. ...George F. Townsend.

Siege of Lichfield. ...William Gresley.

Silas Verney : being the story of his adventures in the days of King Charles the Second. ...Edgar Pickering.

Sir Henry Appleton ; or, Essex during the-great rebellion. ...William E. Heygate.

Sir Ralph Esher; or, Memoirs of a gentleman of the court of Charles the Second. ...Leigh Hunt.

Sir Ralph Willoughby. ...Sir Samuel Egerton Brydges.

Spanish match ; or, Charles Stuart at Madrid. ...William Harrison Ainsworth.

Spinoza. ...Berthold Auerbach.

Stanfield Hall. ...J. Frederick Smith.

Star-chamber. ...William Harrison Ainsworth.

Stronge of Netherstronge : a tale of Sedgemoor. ...Edith J. May.

Story of an abduction in the seventeenth century. ...Jacob van Lennep.

Talbot Harland. ...William Harrison Ainsworth.

Tales of the covenanters. ...J. M. Wilson.

Tales of the wars of Montrose. ...J. Hogg.

Tempest tossed; the story of Seijungfer. ...Margaret Roberts.

Through unknown ways. ...Lucy Ellen Guernsey.

Times of Gustav Adolf. ...Zacharias Topelius. English translation.

Die Tochter des Piccolomini. ...Georg C. R. Herlosssohn.

Too strange not be true. ...Georgiana C. L. G. Fullerton.

Traitor or patriot ? a tale of the Ryehouse plot. ...Mary C. Rowsell.

Trnová koruna. Historicky olraz z třicetilete války. ...Beněs Třebizský. [ton.

True hero ; Story of William Penn. ...William Henry Giles Kings-

Trug-Gold. ...Rudolf Baumbach.

Truth ; or, Persis Clareton. ...Charles B. Taylor.

Two knights of Delany castle. ...Mary M. Sherwood.

Two swords; being a story of old Bristol. ...Emma Marshall.

Um den Kaiserstuhl. ...Wilhelm Jensen.

Die Vierhundert von Pforzheim. ...Carl A. F von Witzleben.

Waldemar. ...W. H. Harrison.

Waldner von Wildenstein. ...Joseph Schreyvogel.

Washingtons : a tale of a country parish in the seventeenth century. ...John N. Simpkinson.

Waldstein. ...Heinrich Laube.

Wallenstein. ...Ernst Willkomm.

Wallensteins erste Liebe. ...Georg C. R. Herlosssohn.

EUROPE, HISTORY OF MEDIÆVAL AND MODERN — *Continued.*

Wallensteins Letzte Tage.   ...Franz Lubojatzky.

Wearyholme ; or, Seedtime and harvest : a tale of the restoration of Charles II.   ...Emily Sarah Holt.

Whitefriars ; or, Times and days of Charles II.   ...Jane Robinson.

White gauntlet.   ...Mayne Reid.   [son.

Whitehall ; or, Days and times of Oliver Cromwell.   ...Jane Robin-

Willitoft ; or, Days of James I : a tale.   ...Anon.

Winchester meads in the time of Thomas Ken.   ...Emma Marshall.

Winifred ; or, English maiden in the seventeenth century.   ...Lucy Ellen Guernsey.

Woodstock ; or, Cavalier : a tale of the year 1651.   ...Sir Walter Scott.

Young carpenters of Freiberg : a tale of the thirty years' war. ...Anon.

Young deserter.   ...Anon.

Ein Zeitbild aus Braunschweig's Kirchen und Stadtgeschichte in den ersten Jahrh. des sieben Zehnten Jahrh.   ...Carl August Wildenhahn.

Eighteenth Century.

Adventures of Roderick Random.   ...George Tobias Smollett.

Adventures of Rob Roy.   ...James Grant.

Against the stream : the story of the heroic age in England. ...Elizabeth Charles.

Agathon.   ...Christoph M. Wieland.

Aid-de-camp.   ...James Grant.

Albert Lunel.   ...Henry Brougham.

Alide.   ...Emma Lazarus.

Der Alte Dessauer.   ...Franz Lubojatzky.

Der Alte Fritz und seine Zeit.   ...Clara M. Mundt. (Pseud., Louise Mühlbach.)

Un ami de la reine.   ...Paul Gaulot.   [trian.

L'an I. de la République.   ...Émile Erckmann and Alexandre Cha-

Ancient régime.   ...George Payne Rainsford James.

Andrée de Taverny.   ...Alexandre Dumas.

Angela Pisani.   ...Lord Strangford.

Anna Saint Ives.   ...Thomas Holcroft.

L'an '93.   ...Victor Hugo.

Ashcliffe Hall : a tale of the last century.   ...Emily Sarah Holt.

Atelier du Lys ; or, Art student in the reign of terror.   ...Margaret Roberts.

Auf Befehl des Königs.   ...Clarissa Lohde.

Aus den Tagen zweier Könige.   ...Friedrich Adami.

Aus drei Kaiserzeiten.   ...Johann Georg L. Hesekiel.

EUROPE, HISTORY OF MEDIÆVAL AND MODERN — *Continued.*

Count Mirabeau.   ...Theodor Mundt. English translation.
Countess de Charny.   ...Alexandre Dumas. English translation.
Creole.   ...J. B. Cobb.
Cyrus.   ...Christoph M. Wieland.
Dead Marquise.   ...Leonard Kip.
De Vane : a story of plebeians and patricians.   ...Henry W. Hilliard.
De Vere.   ...Robert P. Ward.
Devereux.   ...Edward George Earle Lytton Bulwer-Lytton.
Der Domherr.   ...Jodocus D. H. Temme.
Diary of Mrs. Kitty Trevelyan : a story of the times of Whitefield
    and the Wesleys.   ...Elizabeth Charles.
Dorothea Cappel.   ...Emilie Friedrich S. Lohmann.
Duchenier; or, Revolt of La Vendée.   ...John Mason Neale. (Pseud., `
    Aurelius Gratianus.)
Eighty years ago.   ...Harriet Morton.
Elizabeth of Guttenstein.   ...Caroline Pichler. English translation.
Elizabeth von Guttenstein.   ...Caroline Pichler.
Les États Généraux.   ...Émile Erckmann et Alexandre Chatrian.
Evalina.   ...Frances d'Arblay.
Expedition of Humphrey Clinker.   ...Tobias George Smollett.
Father Clement : a Roman Catholic story.   ...Grace Kennedy.
Le faubourg Saint Antoine.   ...Tony Révillon.
Fifteen years.   ...Thérèse Albertine Luise von Jacob Robinson.
    (Pseud., Talvj.) English translation.
Fool of quality; or, Henry Earl Moreland.   ...Henry Brooke.
For the king.   ...Charles Gibbon.
Four Georges.   ...William Makepeace Thackeray.
Frank Hilton ; or, Queen's own.   ...James Grant.
Frau von Staël : biographischer Roman.   ...Amalie C. E. M. Boelte.
Frederick the Great and his court.   ...Clara M. Mundt. (Pseud.,
    Louise Mühlbach.)
Die Freimaurer : eine Familien geschichte aus dem vorigen Jahr-
    hundert.   ...Ferdinand Gustav Kuehne.
French wine and politics.   ...Harriet Martineau.
Friedrich der Grosse und sein Hof.   ...Clara M. Mundt. (Pseud.,
    Louise Mühlbach.)
Friend of the queen.   ...Paul Gaulot. English translation.
Fünfzehn Jahre.   ...Thérèse Albertine Luise von Jacob Robinson.
    (Pseud., Talvj.)
Geschichte des weisen Danischmend.   ...Christoph M. Wieland.
Gipsy.   ...George Payne Rainsford James.
"God's providence " house : a story of 1791.   ...Isabel Banks.
Goethe and Schiller.   ...Clara M. Mundt. (Pseud., Louise Mühl-
    bach.) English translation.

EUROPE, HISTORY OF MEDIÆVAL AND MODERN — *Continued.*
Die Kaiserlichen in Sachsen.   ...Wilhelm R. Heller.
Kaunitz.   ...Leopold Sacher-Masoch.
King of Bath; or, Life at Spa.   ...Mary C. Ware.
King's own.   ...Frederick Marryat.
King's own borderers : a military romance.   ...James Grant.
König Friederich August III. von Sachsen und seine Zeit.   ...Franz
    Lubojatzky.
Lady Bell : a story of the last century.   ...Henrietta Keddie. (Pseud.,
    Sarah Tytler.)
Lady Grizel : an impression of a momentous epoch.   ...Lewis S.
    Wingfield.
Lady of the province.   ...Charlotte Tucker.   (Pseud., A. L. O. E.)
Lessing.   ...Alexandre Ungern-Sternberg.
Life, adventures, and piracies of the famous Captain Singleton.
    ...Daniel Defoe.
Limerick veteran.   ...Agnes M. Stewart.
Little schoolmaster Mark.   ...Joseph Henry Shorthouse.
Lord Harry Bellair : a tale of the last century.   ...Anne Manning.
Lord Mayor of London.   ...William Harrison Ainsworth.
Luck of Barry Lyndon : a romance of the last century.   ...William
    Makepeace Thackeray.
Lucy Arden; or, Hollywood Hall.   ...James Grant.
Madame De Staël ; an historical novel.   ...Amalie C. E. M. Boelte.
    English translation.   '
Madame Thérèse.   ...Émile Erckmann et Alexandre Chatrian.
Maiden's lodge : a tale of the time of Queen Anne.   ...Emily Sarah
    Holt.
Maria Theresa and her fireman.   ...Clara M. Mundt. (Pseud., Louise
    Mühlbach.)   English translation.
Maria Theresia und ihr Ofenheizer.   ...Clara M. Mundt.   (Pseud.,
    Louise Mühlbach.)
Maria Theresia und ihre Zeit.   ...Eduard Duller.
Maria Theresia und ihre Zeit.   ...Franz Lubojatzky.
Marston ; or, Soldier and statesman.   ...George Croly.
Maurice Tiernay, the soldier of fortune.   ...Charles Lever.
Les mémoires d'un médecin.   ...Alexandre Dumas.
Memoirs of an English officer.   ...Daniel Defoe.
Memoirs of a physician.   ...Alexandre Dumas.   English translation.
Merchant of Berlin.   ...Clara M. Mundt.   (Pseud., Louise Mühl-
    bach.)   English translation.
Miser's daughter.   ...William Harrison Ainsworth.
Miss Angel.   ...Anna Isabella Thackeray (Mrs. Richmond Ritchie).
Mit und ohne Vokation.   ...Elizabeth von Grotthuss.
Monsieur Jack : a tale of the old war-time.   ...Alfred H. Engelbach.

EUROPE, HISTORY OF MEDIÆVAL AND MODERN — *Continued.*

New voyage around the world. ...Daniel Defoe.

Nobleman of '89 : an episode of the French revolution. ...Abel Quinton.

Öfverste Stobée, ella holsteinska partiet under frihetstiden. ...Hermann Bjursten.

Old Chelsea Bun-house. ...Anne Manning.

Old Fritz and the new era. ...Clara M. Mundt. (Pseud., Louise Mühlbach.) English translation.

Old manor house. ...Charlotte Smith.

Oliver Ellis ; or, Fusiliers. ...James Grant.

On the edge of the storm. ...Margaret Roberts.

Ottilié. ...Violet Paget.

La patrie en danger. ...Émile Erckmann et Alexandre Chatrian.

Peasant and prince. ...Harriet Martineau.

Peg Woffington. ...Charles Reade.

Planter's daughter. ...Eliza Ann Dupuy.

Poor little Gaspard's drum. ...Alfred H. Engelbach.

Preston fight ; or, Insurrection of 1715. ...William Harrison Ainsworth.

Die Prinzessin von Wolfenbüttel. ...Johann Heinrich D. Zschokke.

Prisoner's daughter : a story of 1758. ...Esmé Stuart.

Ralf Skirlaugh, the Lancashire squire. ...Edward Peacock.

Reign of Terror. ...Catherine G. F. Gore.

Rival apprentices. ...Anon.

Rob Roy. ...Sir Walter Scott.

Ronald Morton ; or, Fireships : a story of the last naval war. ...William Henry Giles Kingston.

Die Rosenkreuzer in Wien. ...Eduard Breier.

Rosenkreuzer und Illuminaten. ...Max Ring.

Roseville family. ...Mrs. Alexander S. Orr.

Rückwirkungen ; oder wer regiert denn ? ...Johann Heinrich D. Zschokke.

Saint James ; or, Court of Queen Anne. ...William Harrison Ainsworth.

Saville House : an historical romance of the time of George the First. ...Mary M. Hardy. (Pseud., Addleston Hill.)

Schiller. Kultur historischer Roman. ...Johannes Scherr.

Schubart und seine Zeitgenossen. ...Albert E. Brachvogel.

Seats of the mighty. ...Gilbert Parker.

Shelley. Biographische Novelle. ...Wilhelm Ritter von Hamm.

Die Soldaten Friedrichs des Grossen. ...Julius von Wickede.

Sophie Charlotte, die philosophische Königin. ...Julius Bacher.

South-sea bubble : a tale of the year 1720. ...William Harrison Ainsworth.

EUROPE, HISTORY OF MEDIÆVAL AND MODERN — *Continued.*
Spiritual Don Quixote; or, Summer ramble of Mr. Geoffry Wild-
goose. ...Richard Graves.
States General. ...Émile Erckmann and Alexandre Chatrian. Eng-
lish translation.
Surgeon's daughter. ...Sir Walter Scott.
Tale of two cities. ...Charles Dickens.
Tapestried chamber. ...Sir Walter Scott.
Täuschungen. ...Heinrich J. Koenig.
Thomas Thyrnau. ...Henrietta von Paalzow.
Too strange not to be true. ...Georgiana C. L. G. Fullerton.
Treasure trove. ...Samuel Lover.
Tremaine; or, Man of refinement. ...Robert P. Ward.
Tuileries. ...Catherine G. F. Gore.
Two campaigns. ...Alfred H. Engelbach.
Die unsichtbare Loge. ...Jean Paul Richter.
La Vendée. ...Anthony Trollope.
Vera; or, War of the peasants. ...Hendrik Conscience. English
translation.
Verrath und Liebe. ...Adolf Muetzeburg.
Der Vetter im Consistorium. ...Philipp F. W. Oertel. [eray.
Virginians: a tale of the last century. ...William Makepeace Thack-
Von Fels zum Meer. ...Hans von Zollern.
Vor hundert Jahren. ...Carl W. Ritter von Martini.
Waverley. ...Sir Walter Scott.
When George III. was king. ...Anon.
White slave. ...Richard Hildreth.
Wien und Rom. ...Eduard Breier.
Wild fire. ...George Walter Thornbury.
Wilhelm Hogarth. Historischer Roman. ...Albert E. Brachvogel.
William Hogarth: a romance. ...Albert E. Brachvogel. English
translation. •
World went very well then. ...Walter Besant.
Year one of the Republic. ...Émile Erckmann and Alexandre
Chatrian. English translation.
Young Marmaduke. ...William D. H. Adams.
Youth and manhood of Cyril Thornton. ...Thomas Hamilton.
Zanoni. ...Edward George Earle Lytton Bulwer-Lytton.
Der Zigeuner-zogling, oder Schlangenwege des Verbrechens. ...Lud-
wig von Alvensleben.
Zrini a költő. ...Miklós Jósika.

Nineteenth Century.

Abigel Rowe: a chronicle of the regency. ...Lewis S. Wingfield.
Adventures of Brigadier Gerard. ...Arthur Conan Doyle.

EUROPE, HISTORY OF MEDIÆVAL AND MODERN — *Continued.*

Dangerfield's rest.   ...H. Sedley.
La débâcle.   ...Émile Zola.
Debit and credit.   ...Gustav Freytag.   English translation.
Detained in France.   ...Julia Kavanagh.
Ein deutsches Reiterleben.   ...Julius von Wickede.
Eine deutsche Revolution; oder, Der Karneval von 1848.   ...Eugen
    H. von Dedenroth.
Deutsche Wunden.   ...Louise Otto.
Deutschlands Kampf und Sieg.   ...Julius Woeniger.
Diana's crescent.   ...Anne Manning.
Der Domherr.   ...Jodocus D. H. Temme.
Downfall.   ...Émile Zola.   English translation.
Drei Tage in Mittenwalde im bayerischen Alpengebirge.   ...Philipp
    F. W. Oertel.
Durch Nacht zum Licht.   ...Friedrich Spielhagen.
Egy Magyar Nábob.   ...Mór Jókai.
1805; oder, die Franzosen zum erstenmal in Wien.   ...Eduard
    Breier.
1809.   Historischer Roman.   ...Eduard Breier.
1848; oder, Nacht und Licht.   ...Franz Lubojatzky.
1849; oder des Königs Maienblüte.   ...Franz Lubojatzky.
1858–89.   ...Bruno Rodwald.
1866; oder in Böhmen und am Main.   ...Franz Lubojatzky.
1870; oder die Heldin von Wörth.   ...Adolf Schirmer.
Elba und Waterloo : Forsetzung von 1813.   ...F. Stolle.
Elsie's dowry : a tale of the Franco-Prussian war.   ...Emma
    Leslie.
Empress Josephine.   ...Clara M. Mundt.   (Pseud., Louise Mühl-
    bach.)   English translation.
Enemy's friendship.   ...S. M. S. Clarke.
Erdély aranykora.   ...Mór Jókai.
Erlach coust.   ...Lolo Kirschner.   (Pseud., Ossip Schubin.)
Europäische Minen und Gegenminen.   ...Johann Ferdinand Martin
    Oskar Meding.   (Pseud., Gregor Samarow.)
Falu jegyzője.   ...József Eötvös.
Une famille pendant la guerre.   ...Mme. B. Boissonnas.
Der Feldmarschall Blücher und der Pfarrer Kretzschmar.   ...Philipp
    F. W. Oertel.
Fiddler of Lugau.   ...Margaret Roberts.
Fleur de lys : a story of the Franco-German war.   ...Grenville
    Murray.
Forest and the fortress.   ...Laura Jewry.
For the right.   ...Karl Emil Franzos.   English translation.
Franz Alzeyer.   ...Paul Heyse.

EUROPE, HISTORY OF MEDIÆVAL AND MODERN — *Continued.*
Invasion of France in 1814: comprising the night march of the Russian army past Phalsbourg. ...Émile Erckmann and Alexandre Chatrian. English translation.
L'Invasion; ou, Le fou Yégof. ...Émile Erckmann et Alexandre Chatrian.
Der Jäger von Königgratz. ...Eugen H. von Dedenroth.
Das Jahr 1866. ...Johann Ferdinand Martin Oskar Meding. (Pseud., Gregor Samarow.)
Jews of Barnow. ...Karl Emil Franzos. English translation.
Die Juden von Barnow. ...Karl Emil Franzos.
Kaiser Wilhelm und seine Zeitgenossen. ...Clara M. Mundt. (Pseud., Louise Mühlbach.)
Kaiserin Josephine. ...Clara M. Mundt. (Pseud., Louise Mühlbach.)
Ein Kampf um's Recht. ...Karl Emil Franzos.
Der Kaplan von Königgrätz. ...Franz Lubojatzky.
Kárpáthy Zoltán. ...Mór Jókai.
Kenneth; or, Rear-guard of the grand army. ...Charlotte Mary Yonge.
Kentucky's love; or, Roughing it about Paris. ...Edward King.
King of the mountain. ...Edmond About. English translation.
König Jérôme's Carneval. ...Heinrich J. Koenig.
Königin Hortense: ein Napoleonisches Lebensbild. ...Clara M. Mundt. (Pseud., Louise Mühlbach.)
Der Krieg am Rhein im Jahre 1870. ...Stanislaus S. A. Grabowski.
Kreuz und Schwert. ...Franz W. Ditfurth.
Kreuz und Schwert. ...Johann Ferdinand Martin Oskar Meding. (Pseud., Gregor Samarow.)
Krummensee. ...Johann Georg L. Hesekiel.
Lady Wedderburn's wish: a tale of the Crimean war. ...James Grant.
Der lange Isaack. ...Julius von Wickede.
Laura Everingham; or, Highlanders of Glen Ora. ...James Grant.
Leonie; or, Light out of darkness. ...Annie Lucas.
Lindau; oder, der unsichtbare Bund. ...Johann Weitzel.
Lord Hermitage. ...James Grant.
Lord of himself. ...F. H. Underwood.
Lothair. ...Benjamin Disraeli.
Lotta Schmidt and other stories. ...Anthony Trollope.
Louisa of Prussia and her time. ...Clara M. Mundt. (Pseud., Louise Mühlbach.) English translation.
Louisa von Preussen und ihre Zeit. ...Clara M. Mundt. (Pseud., Louise Mühlbach.)
Maid, wife, or widow? ...Annie F. Hector. (Pseud., Mrs. Alexander.)

EUROPE, HISTORY OF MEDIÆVAL AND MODERN — *Continued.*

Mainstone's housekeeper. ...Eliza Meteyard. (Pseud., Silver pen.)

Manchester strike. ...Harriet Martineau.

Man of the people. ...Émile Erckmann and Alexandre Chatrian. English translation.

Margaret Muller: a story of the late war in France. ...Eugenie Bersier.

Margaret von Ehrenberg, the artist's wife. ...William and Mary Howitt.

Marmone. ...Philip Gilbert Hamerton.

Martin der Stellmacher. ...L. Kreutzer.

Martyrs to circumstance. ...Maria Thérèse Longworth Yelverton.

Mary Barton: a tale of Manchester life. ...Elizabeth C. Gaskell.

Max Kroner: a story of the siege of Strasburg. ...Hannah Smith. (Pseud., Hesba Stretton.)

Member for Paris: a tale of the Second Empire. ...Grenville Murray.

Les Misérables. ...Victor Hugo.

Les Miserables. ...Victor Hugo. English translation.

Moderne Grössen. ...B. Aba.

Napoleon in Deutschland. ...Clara M. Mundt. (Pseud., Louise Mühlbach.)

Napoleon in Germany. ...Clara M. Mundt. (Pseud., Louise Mühlbach.) English translation.

Neudeutsch. ...Joseph E. C. Bischoff.

Der neue Hiob. ...Leopold Sacher-Masoch.

New Job. ...Leopold Sacher-Masoch. English translation.

North and South. ...Elizabeth C. Gaskell.

Old plantation. ...J. Hungerford.

One of the "Six hundred." ...James Grant.

Only half a hero: a tale of the Franco-German war. ...Alfred T. Story.

Other people's money. ...Émile Gaboriau. English translation.

Otto der Schutz. ...Johann Gottfried Kinkel.

Parisians. ...Edward George Earle Lytton Bulwer-Lytton.

Partisan leader. ...Beverly.

Perlen aus der Krone des letzten deutschen Kaisers. ...Franz Isidor Proschko.

Peter Mayr. ...Petri K. Rosegger.

Pförtner Jugend. ...Friedrich Kenner.

Phantom regiment; or, Stories of "Ours." ...James Grant.

Poet hero. ...Minny Bothmer.

Powder and gold: a story of the Franco-Prussian war. ...Levin Schuecking. English translation.

Prince Bismark: friend or foe ? ...Minny Bothmer.

EUROPE, HISTORY OF MEDIÆVAL AND MODERN — *Continued.*

Prinz Louis Ferdinand. ...Fanny Lewald Stahr.

Problematic characters. ...Friedrich Spielhagen. English translation.

Problematische naturen. ...Friedrich Spielhagen.

Prussian spy. ...V. Valmont.

Pulver und Gold. ...Levin Schuecking.

Queen Hortense. ...Clara M. Mundt. (Pseud., Louise Mühlbach.) English translation.

Ravenshoe. ...Henry Kingsley.

Red band: the adventures of a young girl during the siege of Paris. ...Fortuné Abraham Du Boisgobey. English translation.

Die Ritter vom Geiste. ...Carl F. Gutzhow.

Le roi des Montagnes. ...Edmond About.

Roman aus dem Berliner. ...Oskar Heller.

Ein Roman aus den Zeiten des schleswig-holsteineschen Krieges. ...Constantin M. Reichenbach.

Le roman d'un brave homme. ...Edmond About.

Romance of Vienna. ...Frances Trollope.

Romance of war. ...L. Bellstab.

Romance of war ; or, Highlanders in Spain. ...James Grant.

Die Römerfahrt der Epigonen. ...Johann Ferdinand Martin Oskar Meding. (Pseud., Gregor Samarow.)

Die Rose von Sadowa. ...Ludwig A. R. Kuerbis.

De Sadowa à Sedan. ...Johann Ferdinand Martin Oskar Meding. (Pseud., Gregor Samarow.)

Saly's Revolutionstage. ...Ulrich Hegner.

Scenes from the Ghetto. ...Leopold Kompert. English translation.

Schach-Bismark. ...J. G. Findel.

Schleswig-Holstein meerum schlugen. ...Carl von Kessel.

Shadow of the sword. ...Robert W. Buchanan.

She-wolves of Machecoul. ...Alexandre Dumas. English translation.

Shut up in Paris. ...Nathan Sheppard.

Sister Martha. ...Benjamin Wilson.

Six years ago. ...James Grant.

Soll und Haben. ...Gustav Freytag.

Der Spion. ...Franz F. Wangenheim.

Story of an honest man. ...Edmond About. English translation.

Story of the plébiscite. ...Émile Erckmann and Alexandre Chatrian. English translation.

Stories of Waterloo. ...William H. Maxwell.

Subaltern. ...George Robert Gleig.

Sybil ; or, Two nations. ...Benjamin Disraeli.

EUROPE, HISTORY OF MEDIÆVAL AND MODERN — *Continued.*

Thekla. ...William Armstrong.

Theodor Korner. ...Heribert Rau.

Through night to light. ...Friedrich Spielhagen. English translation.

Tithe-proctor. ...William Carlton.

Twins of Saint-Marcel: a tale of Paris incendié. ...Mrs. Alexander S. Orr.

Under the Red Dragon. ...James Grant.

Unter Preussens Fahnen: historischer Roman aus dem Jahre 1866. ...Stanislaus S. A. Grabowski.

Ut de Franzosentid. ...Fritz Reuter.

Valentin: a French boy's story of Sedan. ...Henry Kingsley.

Vanished Emperor. ...Percy Andreae.

Vera; or, Russian princess and English earl. ...Charlotte L. H. Dempster.

Village notary. ...József Eötvös. English translation.

Die Völkerschlacht bei Leipzig ...Joseph von Hinsberg

Die von Hohenstein. ...Friedrich Spielhagen.

Von Saalfeld bis Aspern. ...Heinrich J. Koenig.

Vor dem Sturm. ...Theodor Fontane.

Vor dem Sturm. ...Johann Ferdinand Martin Oskar Meding. (Pseud., Gregor Samarow.)

Waldfried. ...Berthold Auerbach.

Waterloo. ...Émile Erckmann et Alexandre Chatrian.

Waterloo. ...Émile Erckmann and Alexandre Chatrian. English translation.

Wenderholme: a story of Lancashire and Yorkshire. ...Philip Gilbert Hamerton.

Whitehall. ...William Maginn.

Whites and the blues. ...Alexandre Dumas. English translation.

Within iron walls: a tale of the siege of Paris. ...Annie Lucas.

Workman and soldier: a tale of Paris life during the siege and rule of the Commune. ...James F. Cobb.

Year nine: a tale of the Tyrol. ...Anne Manning.

Young buglers: a tale of the Peninsular war. ...George Alfred Henty.

Young Franc-Tireurs and their adventures in the Franco-Prussian war. ...George Alfred Henty.

Zu spät erkannt. ...Anon.

Zwei Kaiserkronen. ...Johann Ferdinand Martin Oskar Meding. (Pseud., Gregor Samarow.)

EVANGELINE. See **ACADIA.**

EXILE.

Le brigadier Frédéric.   ...Émile Erckmann et Alexandre Chatrian.
Brigadier Frederick.   ...Émile Erckmann and Alexandre Chatrian.
   English translation.
Driven into exile.   ...Charlotte Tucker.   (Pseud., A. L. O. E.)
Élisabeth; ou, Les Exilés de Sibérie.   ...Sophie Risteau Cottin.
Exiles.   ...Richard Harding Davis.
Exiles in Babylon.   ...Charlotte Tucker.   (Pseud., A. L. O. E.)
Exiles of Lucerne.   ...John R. Macduff.
Exile's trust.   ...Frances Browne.
Forest exiles.   ...Mayne Reid.
From exile.   ...James Payn.
Huguenot exiles; or, Times of Louis XIV.   ...Eliza Ann Dupuy.
Im exil.   ...Johann Ferdinand Martin Oskar Meding.   (Pseud.,
   Gregor Samarow.)
In the dwellings of silence.   ...Walter Kennedy.
In exile.   ...Mrs. H. H. Foote.
Jacobite exile: adventures in the service of Charles XII. of Sweden.
   ...George Alfred Henty.
Kings in exile.   ...Alphonse Daudet.   English translation.
Les rois en exil.   ...Alphonse Daudet.
Roman exile.   ...Guglielmo Gajani.
Czarina.   (Meuzikoff.)   ...Barjara Hofland.
Lichtenstein.   ...Wilhelm Hauff.
Lichtenstein; or, Swabian League.   ...Wilhelm Hauff.   English
   translation.
Vera Vorontzoff.   .:.Sonia Krukovsky Kovalevsky.

FAIRY TALES.   *A Selection.*

Adventures in Fairyland.   ...David H. Brewer.
Ahmed Ibn Hemden: Turkish evenings' entertainment.   ...Anon.
Baron Gravenstein in Fairy-land.   ...T. H. Chandler.
Billedbog uden Billeder.   ...Hans Christian Andersen.
Book of English fairy-tales from the north country.   ...A. C. Fryer.
Brownies and bogles.   ...Louise I. Guiney.
Bubbling teapot.   ...Elizabeth Williams Champney.
Celtic fairy tales.   ...J. Jacobs.
Céra una volta.   ...Luigi Capuana.
Cuckoo clock.   ...Mary Molesworth.   (Pseud., Ennis Graham.)
Danish fairy legends and tales.   ...Hans Christian Andersen.
   English translation.
Dealings with the fairies.   ...George MacDonald.
Deutsche Sagen.   ...Jacob L. Carl and Wilhelm Carl Grimm.
Echoes from Fairy-land.   ...Mary D. Brine.
Eventyr.   ...Hans Christian Andersen.

FAIRY TALES — *Continued.*
Once upon a time.   ...Luigi Capuana.   English translation.
Pansie's flour-bin.   ...Eliza Tabor (Mrs. Stephenson).
Popular tales from the Norse.   ...George W. Dasent.
Popular tales of the West Highlands.   ...J. F. Campbell.
Prince of Argolis.   ...J. Moyr Smith.
Rainbows for children.   ...Caroline Tappan.
Rose and the lily: how they became the emblems of England and
    France.   ...Octavian Blewitt.
Scenes in Fairy-land.   ...John C. Atkinson.
Something occurred.   ...Benjamin Leopold Fargeon.
Southern cross fairy tale.   ...Rebecca S. Clarke.
Tales of sixty mandarins.   ...P. V. Raju.
Tangled tale.   ...Charles Lutwidge.   (Pseud., Lewis Carroll.)
Turkish fairy tales and folk tales.   ...Ignācz Kunos.   English
    translation from the Hungarian.
Tuscan fairy tales.   ...Violet Paget.   (Pseud., Vernon Lee.)
Twilight land.   ...Howard Pyle.
Under the sunset.   ...Bram Stoker.
Wonder stories told for children.   ...Hans Christian Andersen.
    English translation.
See also **LEGENDS** and **FOLK-LORE.**

FAWKES, GUIDO (GUY).   *Lived 1570–1606.*
Coombe, Abbey.   ...Selina Bunbury.
Father Darcy: an historical romance.   ...Anne Marsh.
Guy Fawkes; or, Gunpowder treason.   ...William Harrison
    Ainsworth.

FEMGERICHTE.
Anne of Geierstein.   ...Sir Walter Scott.
Hermann von Unna.   ...Christiane Benedicte Eugenie Naubert.

FERGUSON, ROBERT.   *Lived 1750–1774.*
Rye House plot.   ...George W. M. Reynolds.
Traitor or patriot? a tale of the Rye House plot.   ...Mary C.
    Rowsell.

FIRES, FAMOUS.
Chicago.   1871.

Barriers burned away.   ...Edward Payson Roe.
Daniel Trentworthy.   ...J. McGovern.

London.   1666.

Andrew Golding: a tale of the great plague.   ...Annie E. Keeling.
Brave man of Eyam; or, Tale of the great plague.   ...Edward N.
    Hoare.

FIRES, FAMOUS — *Continued.*

Caleb Field : a tale of the Puritans. ...Margaret O. W. Oliphant.

Captain Jacques ; or, Romance of the time of the plague. ...Edward Fitzgibbon. (Pseud., Somerville Gibney.)

Carved cartoon. ...Austin Clare.

Cherry and violet : a tale of the great plague. ...Anne Manning.

History of the great plague in London in 1665. ...Daniel Defoe.

Jennett Cragg, the Quakeress. ...Maria Wright.

Old Saint Paul's : a tale of the plague and the fire. ...William Harrison Ainsworth.

Oliver Wyndham : a tale of the great plague. ...Mrs. J. B. Peploe (Webb).

Rosemary : a tale of the fire of London. ...Georgiana C. L. G. Fullerton.

Saint Dunstan's clock : a story of 1666. ...E. Ward.

When London burned : a story of restoration times and the great fire. ...George Alfred Henty.

When London was burned. ...J. Finnemore.

Rome. 64.

Beric the Briton. ...George Alfred Henty.

Burning of Rome ; or, Story of the days of Nero. ...Alfred John Church.

FLORENCE, ITALY.

Fast of Saint Magdalen. ...Anna Maria Porter.

Legends of Florence. ...Charles Godfrey Leland.

Romola. ...Marian Evans Lewes Cross. (Pseud., George Eliot.)

Di Vasari. ...Charles Edwards.

See also **ITALIAN HISTORY** and **ITALY**.

FLORIDA.

Bewitched. ...Louis Pendleton.

Canoe-mates : a tale of the Florida reef and Everglades. ...Kirk Munroe.

Coral ship. ...Kirk Munroe.

East Angels. ...Constance Fenimore Woolson.

Flamingo feather. ...Kirk Munroe.

Grahams. ...Jane G. Fuller.

Jack Tier ; or, Florida reef. ...James Fenimore Cooper.

In the wilds of Florida. ...William Henry Giles Kingston.

Lily and the totem. ...William Gilmore Simms.

Loss of the Swansea. ...William L. Alden.

Remember the Alamo. ...Amelia Edith Barr.

Ringwood the rover. ...Henry W. Herbert. (Pseud., Frank Forester.)

Vasconselos : a romance of the new world. ...William Gilmore Simms.

FOLK-LORE.

Afro-American folk-lore told on the sea islands of South Carolina.
...A. M. H. Christensen.

Baital-Pāchchīsī : Twenty-five tales of a sprite. ...English translation from the Hindi test of Duncan Forbes.

Beside the fire : a collection of Irish tales and Gaelic folk stories.
...Douglas Hyde.

Blackfoot Lodge tales. ...George Bird Grinnell.

Book of New England legends and folk-lore in prose and poetry.
...S. A. Drake.

Céra una volta. ...Luigi Capuana.

Collection of Irish Gaelic folk stories. ...Douglas Hyde.

Fairy and folk-tales of the Irish peasantry. ...W. B. Yeats.

Folk and fairy tales. ...Peter C. Asbjørnsen. English translation.

Folk-lore and legends. ...Lál Behári Day. (8 vols.)

Folk-lore and legends, Russian and Polish. ...Anon.

Folk-tales of Angola. ...Heli Chatelain.

Folk-tales of Bengal. ...Lál Behári Day.

Ghosts and family legends. ...Catherine Crowe.

Green fairy book. ...Andrew Lang.

In the land of marvels : folk-tales of Austria and Bohemia. ...Frie-
drich Theodor Vernaleken. English translation.

Italian popular tales. ...Thomas F. Crane.

Korean tales. ...H. N. Allen.

Legend of Montrose. ...Sir Walter Scott.

Legend of Thomas Didymus, the Jewish skeptic. ...James Freeman Clarke.

Légendes des plantes et des oiseaux. ...Xavier Marmier.

Legends and popular tales of the Basque people. ...Mariana Mon-
teiro.

Legends of Charlemagne. ...Edward Everett Hale.

Legends of fairyland. ...Anna Bache.

Legends of Florence. ...Charles Godfrey Leland.

Legends of the Middle Ages. ...H. A. Guerber.

Legends of the Wagner Drama. ...Jessie L. Weston.

Louisiana folk-tales. ...Alcée Fortier.

Mamma's black nurse stories : West Indian folk lore. ...Mary P.
M. Home.

Man who married the moon. ...Charles F. Lummis.

Mr. Rabbit at home. ...Joel Chandler Harris.

Myths and folk-lore of Ireland. ...Jeremiah Curtin.

Negro myths from the Georgia coast, told in the vernacular.
...Charles Colcock Jones, Jr.

Nights with Uncle Remus. ...Joel Chandler Harris.

Old Celtic romances. ...Patrick Weston Joyce.

FOLK-LORE — *Continued.*
Once upon a time. ...Luigi Capuana. English translation.
Österreichische Kinder- und Hausmärchen. ...Friedrich Theodor
Vernaleken.
Pawnee stories and folk lore. ...George Bird Grinnell.
Shadow-land on Ellan Vannin ; or, Folk tales of the Isle of Man.
...Mrs. J. W. Russell. (Pseud., I. H. Leney.)
Sixty folk-tales from exclusively Slavonic sources. ...A. H. Wratis-
law, translator.
Tales and legends of national origin, or widely current in England,
from the earliest times. ...William C. Hazlitt.
Tales of Punjab: told by the people. ...Flora Annie Steel.
Die Tellfrage.Versuch ihrer Geschichte und Lösung. ...Anton Gisler.
Through green glasses. ...F. M. Allen.
Turkish fairy tales and folk tales. ...Ignácz Kunos. English trans-
lation from the Hungarian.
Uncle Remus and his friends. ...Joel Chandler Harris.
Van Gelder papers. ...John T. Irving.
Voodoo tales as told among the negroes of the Southwest. ...Mary
Alicia Owen.
West Irish folk-tales and romances. ...William Larmine.
Winter evenings tales, collected among the cottages in the South
· of Scotland. ...James Hogg.
Wonder world stories from the Chinese, French, German, and
Hebrew. ...Marie Pabke and Margaret J. Pitman.
See also **FAIRY TALES** and **LEGENDS**.

FORECASTS. *A Selection.*
Cæsar's column: a story of the twentieth century. ...Ignatius
Donnelly. (Pseud., Edmund Boisgilbert.)
Dame Fortune smiled. ...Willis Barnes.
Fifty years hence ; or what may be in 1943. ...Robert Grimshaw.
Final war. ...Louis Tracy.
Great war in England in 1897. ...William Le Queux.
Hartmann the anarchist. ...E. Douglas Fawcett.
Ireland's dream. ...Edmund D. Lyon.
Journey in other worlds. ...John Jacob Astor.
Looking backwards. 2000–1887. ...Edward Bellamy.
Marshal Duke of Denver ; or, Revolution of 1920. ...Hugo Barnaby.
Mr. East's experiences in Mr. Bellamy's world. ...Conrad Wilbrandt.
New Centurion. ...James Eastwick.
News from Nowhere. ...William Norris.
1900 A. D. ...Marianne Farningham.
Pictures of the socialistic future. ...Eugene Richter. English
translation.

FORECASTS — *Continued.*

Sozialdemokratische Zukunftbilder.   ...Eugene Richter.

2894 ; or, Fossil man.   ...Walter Browne.

Traveller from Altruria.   ...William Dean Howells.

Utopia.   ...Sir Thomas More.

FRANCE.   *A Selection.*

Description, Manners, and Customs.

Before the dawn : a story of Paris and the Jacquerie.   ...George Dulac.

Beside the river.   ...Katharine S. Macquoid.

Conscript of 1813.   ...Émile Erckmann and Alexandre Chatrian. English translation.

Constable of France.   ...James Grant.

Cruise of the Mystery.   ...Louise Seymour Houghton.

From the memoirs of a minister of France.   ...Stanley J. Weyman.

Gowrie ; or, King's plot.   ...George Payne Rainsford James.

Her majesty the Queen.   ...John Esten Cooke.

Histoire d'un Conscrit de 1813.   ...Émile Erckmann et Alexandre Chatrian.

Historical tales ; or, Romance of reality.   ...Charles Morris.

Kenneth ; or, Rear-guard of the great army.   ...Charlotte Mary Yonge.

Red cockade.   ...Stanley J. Weyman.

Refugees : a tale of two continents.   ...Arthur Conan Doyle.

Romance of Prince Eugene.   ...Albert Pulitzer.   English translation.

Saint Katharine's by the tower.   ...Walter Besant.

Shield of the fleur de lis.   ...Constance Goddard DuBois.

La Vendée : an historical romance.   ...Anthony Trollope.

Young Breton volunteer.   ...Frances M. Wilbraham.

See also **FRENCH HISTORY** and **HUGUENOTS.**

FRANCIS I. OF FRANCE.   *Reigned 1515-1547.*

Anne de Montmorency, connétable de France : nouvelle historique· ...Pierre de Lesconvel.

Ascanio.   ...Alexandre Dumas.

Ascanio.   ...Alexandre Dumas.   English translation.

Constable de Bourbon.   ...William Harrison Ainsworth.

Darnley ; or, Field of the cloth of gold.   ...George Payne Rainsford James,

Saint Leon : a tale of the sixteenth century.   ...William Godwin.

Die Schlacht von Marignano.   ...Carl A. F. von Witzleben.

See also **FRENCH HISTORY.**

FRANCIS II. OF FRANCE. *Reigned 1559-1560.*
Le chevalier de la Renaudie: roman historique. ...Édouard Devicque.
See also **MARY, QUEEN OF SCOTS,** and **FRENCH HISTORY.**

FRANCO-GERMAN WAR. *1870-1871.*
American in Paris. ...Eugene Coleman Savidge.
L'argent des autres. ...Émile Gaboriau.
Aulnay Tower. ...Blanche Willis Howard.
La bande rouge. ...Fortuné Abraham Du Boisgobey.
Barackenleben. ...Lodovika Hesekiel.
Battle of Dorking: reminiscences of a volunteer. ...George T. Chesney.
Belfry of Saint Jude. ...Esmé Stuart.
Die Bluttaufe der Deutschen Einheit im Jahre 1870. ...Eugen H. von Dedenroth.
Die Braut aus Frankreich. ...Hilmar Juetbock. (Pseud., J. Hilmar.)
Le brigadier Frédéric. ...Émile Erckmann et Alexandre Chatrian.
Brigadier Frederick. ...Émile Erckmann and Alexandre Chatrian. English translation.
Change for France and other stories. ...John Heard, Jr.
Comedy of terrors. ...James De Mille.
La débâcle. ...Émile Zola.
Deutschlands Kampf und Sieg. ...Julius Woeniger.
Downfall. ...Émile Zola. English translation.
Dreams and reality: a tale of the siege of Paris. ...Anon.
Eagle and dove: a tale of the Franco-Prussian war. ...M. E. Clements.
Echoes of a famous year. ...Harriet Parr. (Pseud., Holme Lee.)
1870; oder die Heldin von Wörth. ...Adolf Schirmer.
Elsie's dowry: a tale of the Franco-German war. ...Emma Leslie.
Une famille pendant la guerre. ...Mme. B. Boissonnas.
Fleur de lys: a story of the Franco-German war. ...Grenville Murray.
French prisoners. ...Edward Bertz.
Fritz of the tower: a tale of the Franco-German war. ...L. Lobenhoffer.
Fünf Milliarden. ...Bernhard Hesslein.
Gate and the glory beyond it: a tale of the Franco-Prussian war. ...Onyx, pseud.
Geschichten aus der Pariser Belagerung. ...Paul d'Abrest.
Golden moment: a tale of the Franco-Prussian war. ...Marian F. Fernando.
Die Grandidiers. ...Julius Rodenberg.
Der Grenadier von Weissenburg. ...Robert Neumann.

FRANCO-GERMAN WAR — *Continued.*

Great mistake : a story of adventure.   ...Thomas S. Millington.

Hester.   ...Thomas Hardy.

Histoire du plébiscite.   ...Émile Erckmann et Alexandre Chatrian.

Une idylle pendant le siège.   ...François Coppée.

Im Exil.   ...Johann Ferdinand Martin Oskar Meding. (Pseud., Gregor Samarow.)

Kentucky's love ; or, Roughing it about Paris.   ...Edward King.

Der Krieg am Rhein im Jahre 1870.   ...Stanislaus S. A. Grabowski.

Kreuz und Schwert.   ...Franz W. Ditfurth.

Kreuz und Schwert.   ...Johann Ferdinand Martin Oskar Meding. (Pseud., Gregor Samarow.)

Leaf in the storm.   ...Louise De la Ramé. (Pseud., Ouida.)

Leonie ; or, Light out of darkness.   ...Annie Lucas.

Margaret Muller : a story of the late war in France.   ...Eugénie Bersier.

Marmorne.   ...Philip Gilbert Hamerton.

Max Kroner : a story of the siege of Strasburg.   ...Hannah Smith. (Pseud., Hesba Stretton.)

New Magdalen.   ...William Wilkie Collins.

Only half a hero : a tale of the Franco-German war.   ...Alfred T. Story.

Other people's money.   ...Émile Gaboriau.   English translation.

Paquette.   ...James Grant.

Parisians.   ...Edward George Earle Lytton Bulwer-Lytton.

Powder and gold : a story of the Franco-Prussian war.   ...Levin Schuecking.   English translation.

Princess Clarice : a story of 1871.   ...Mabel Collins Cook.

Prussian spy.   ...V. Valmont.

Pulver und Gold.   ...Levin Schuecking.

Red band : the adventures of a young girl during the siege of Paris.   ...Fortuné Abraham Du Boisgobey.   English translation.

Reverse of the shield ; or, Adventures of Gremille de Marchant during the Franco-Prussian war.   ...Augusta Marryat.

Robert Helmont : diary of a recluse, 1870–1871.   ...Alphonse Daudet.   English translation.

Robert Helmont.   ...Alphonse Daudet.

Le roman d'un brave homme.   ...Edmond About.

Secret of Fontaine-La Croix.   ...Margaret Field.

Shut up in Paris.   ...Nathan Sheppard.

Sinless secret : a romance of the Franco-Prussian war.   ...Eliza M. J. von Booth. (Pseud., Rita.)

Sister Martha : a tale of the Franco-Prussian war.   ...Benjamin Wilson.

Six years ago.   ...James Grant.

FRANCO-GERMAN WAR — *Continued.*
Story of an honest man. ...Edmond About. English translation.
Story of the plébiscite. ...Émile Erckmann and Alexandre Chatrian. English translation.
Sunbeam Susette. ...Emma Leslie.
Der Todesgruss der Legionen. ...Johann Ferdinand Martin Oskar Meding. (Pseud., Gregor Samarow.)
Trojan : the history of a sentimental young man. ...Henry F. Keenan.
Twins of Saint-Marcel : a tale of Paris incendié. ...Mrs. Alexander S. Orr.
Valentin : a French boy's story of Sedan. ...Henry Kingsley.
Waldfried. ...Berthold Auerbach.
White month. ...Frances M. Peard.
Within iron walls : a tale of the siege of Paris. ...Annie Lucas.
Workman and soldier : a tale of Paris life during the siege and rule of the commune. ...James F. Cobb.
Young Franc-Tireurs and their adventures in the Franco-Prussian war. ...George Alfred Henty.
Zu spät erkannt. ...Anon.

FREDERICK I. OF GERMANY. *Reigned 1152–1190.*
Barbarossa. Historischer Roman. ...Joseph E. C. Bischoff.
Friedrich's des Ersten letzte Lebenstage. ...Julius Bacher.
Kaiser Friedrich, genannt Barbarossa. ...Carl J. Simrock.
See also **GERMAN HISTORY.**

FREDERICK I. OF PRUSSIA. *Reigned 1701–1713.*
Öfverste Stobée, ella holsteinska partiet under frihetstiden. ...Herman Bjursten.
Sophia Charlotte die philosophische Königin. ...Julius Bacher.
See also **PRUSSIA.**

FREDERICK II. OF PRUSSIA. *Reigned 1740–1786.*
Der alte Fritz und seine Zeit. ...Clara M. Mundt. (Pseud., Louise Mühlbach.)
Auf Befehl des Königs. ...Clarissa Lohde.
Der Bachtanz zu Langenselbold. ...Philipp F. W. Oertel.
Die Bären von Augustusburg : eine Erzählung aus der sächsischen Geschichte des 18 Jahrhunderts. ...Carl Gustav Nieritz.
Bears of Augustusburg. ...Carl Gustav Nieritz. English translation.
Berlin and Sans-Souci. ...Clara M. Mundt. (Pseud., Louise Mühlbach.) English translation.
Bischof und König. ...Mariam Tenger.
Die Brautschau Friedrichs des Grossen. ...Julius Bacher.
Cabanis, Vaterländischer Roman. ...Georg Wilhelm H. Haering.

FREDERICK II. OF PRUSSIA — *Continued.*
Cyrus. ...Christoph M. Wieland. (The Cyrus of this story is the
Prussian disguise of Frederick II.)
Dorothea Cappel. ...Emilie F. S. Lohmann.
Elizabeth von Guttenstein. ...Caroline Pichler.
Fifteen years. ...Therese Albertine Luise von Jacob Robinson.
English translation.
Frederick the Great and his court. ...Clara M. Mundt. (Pseud.,
Louise Mühlbach.) English translation.
Friedrich der Grosse. ...Eugen H. von Dedenroth.
Friedrich der Grosse und sein Hof. ...Clara M. Mundt. (Pseud.,
Louise Mühlbach.)
Fünfzehn Jahre. ...Therese Albertine Luise von Jacob Robinson.
Historische novellen über Friedrich II. von Preussen und seine
Zeit. ...Joseph E. C. Bischoff.
Hölderlin. ...Heribert Rau.
Johann Gotzkowsky, der Kaufmann von Berlin. ...Clara M. Mundt.
(Pseud., Louise Mühlbach.)
Merchant of Berlin. ...Clara M. Mundt. (Pseud., Louise Mühlbach.)
English translation.
Old Fritz and the new era. ...Clara M. Mundt. (Pseud., Louise
Mühlbach.) English translation.
Die Soldaten Friedrichs des Grossen. ...Julius von Wickede.
Der Vetter im Consistorium. ...Philipp F. W. Oertel.
Von Fels zum Meer. ...Hans von Zollern.
Vor hundert Jahren. ...Carl W. Ritter von Martini.
See also **FRENCH HISTORY** and **GERMAN HISTORY.**

FREDERICK WILLIAM I. OF PRUSSIA. *Reigned 1713-1740.*
Der alte Dessauer. ...Franz Lubojatzky.
Aus den Tagen zweier Könige. ...Friedrich Adami.
Der Domherr. ...Jodocus D. H. Temme.
Gräfin Lichtenau. ...Robert Springer.
Die Prinzessin von Wolfenbüttel. ...Johann Heinrich D. Zschokke.

FREDERICK WILLIAM, ELECTOR OF BRANDENBURG. *Lived
1640-1688.*
Brandenburgische Geschichte. ...Gustav H. G. Putlitz.
Dorothe. ...Georg Wilhelm H. Haering.
Der Fels von Erz. ...Albert E. Brachvogel.
Der grosse Kurfürst in Preussen. ...Ernst A. A. G. Wichert.
Der grosse Kurfürst, und der Schöppenmeister. ...Max Ring.
Paul Gerhardt. ...Carl August Wildenhahn.

FRENCH AND INDIAN WARS, U. S. A. *1756-1775.*
Les anciens Canadiens. ...Philippe Aubert de Gaspé.
Braddock: a story of the French and Indian wars. ...John R.
Musick.

FRENCH AND INDIAN WARS, U. S. A. — *Continued.*

Brandon : a tale of the American colonies. ...O. Tiffany.

Canadians of old. ...Philippe Aubert de Gaspé. English translation.

Fort Braddock letters ; or, Tales of the French and Indian wars in America at the beginning of the eighteenth century. ...Anon.

Last of the Mohicans. ...James Fenimore Cooper.

Legend of the vision of Campbell of Inverawe. ...T. D. Lauder. (In his Legendary tales of the Highlands.)

Legends of the West. ...James Hall.

Marcus Blair. ...Caleb E. Wright.

Old Fort Duquesne : a tale of the early toils, struggles and adventures of the first settlers at the forks of the Ohio in 1754. ...Anon.

Old Fort Duquesne ; or, Captain Jack the scout. ...Charles McKnight.

Ticonderoga ; or, Black eagle. ...George Payne Rainsford James.

Twice taken : an historical romance of the maritime British provinces. ...Charles W. Hall.

Unfinished tale ; or, Daughter of the mill : romance of Lake George. ...F. Claiborne.

Wilderness ; or, Braddock's times. ...J. MacHenry.

William and Mary : a tale of the siege of Louisburg, 1745. ...Daniel Hickey.

With Wolfe in Canada. ...George Alfred Henty.

See also **UNITED STATES HISTORY.**

FRENCH HISTORY. *A Selection.*

Early and General.

La Avarchide. ...Luigi Alamanni.

Bathilde ; ou, L'héroisme de l'amour. ...François T. B. d'Arnaud.

Bathilde, reine des Francs. ...Amélie J. Candeille.

Bibliothèque bleue, réimpression des romans de chevalerie des XIIᵉ–XVᵉ siècles. ...Alfred Delvau.

La France chevalresque ; ou, Les origines héroïques de la nation française. ...Paul Renan.

Frédégonde et Brunehaut. ...Jacques M. Boutet de Monvel.

Histoire de Melusine princesse du Lusignan et de ses fils. ...Melusina Melusine.

Invasion. ...Gerald Griffin.

La Judith française ; ou, Elmonde et Clotilde. ...Jean E. Paccard.

Les mystères du peuple. ...Marie Joseph Eugene Sue.

Old chest ; or, Journal of a family of the French people from the Merovingian times to our own days. ...Anon.

Pre-historic world. ...Élie B. Berthet. English translation.

Rival races. ...Marie Joseph Eugene Sue. English translation.

FRENCH HISTORY — *Continued.*

Le Roi Flore et la belle Jehanne.   ...Anon.

Romance of history.   ...Leitch Ritchie.

Romans préhistoriques.   Le monde inconnu.   ...Élie B. Berthet.

Stories from French history.   ...Emma Leslie.

Tales of a grandfather.   ...Sir Walter Scott.

Fifth to the Eleventh Century.

Four sons of Aymon.   ...Anon.

Ierne of Armorica : a tale of the time of Chlovis.   ...J. C. Bateman.

Invasion.   ...Gerald Griffin.

Les mystères du peuple.   ...Marie Joseph Eugene Sue.

Ogine : chronique du Xe siècle.   ...Armand Garreau.

Raymond de Saint Gilles at the crusades, 1095–1099.   ...Henrietta de Witt.

Rival races.   ...Marie Joseph Eugene Sue.   English translation.

Twelfth and Thirteenth Centuries.

L'Abbaye de Typhaines.   ...Joseph Arthur, Comte de Gobineau.

Abbey of Typhaine.   ...Joseph Arthur, Comte de Gobineau.   English translation.

Abélard and Héloïse : a mediæval romance.   ...Pierre Abélard and Héloïse.

Agnes de France ; ou, Le XIIe siècle : roman historique.   ...Amélia J. Candeille.

Albigenses : a romance.   ...Charles R. Maturin.

Arthur of Brittany.   ...Peter Leicester.

Boy crusaders : a story of the days of Saint Louis.   ...John George Edgar.

Florine, princesse of Burgundy : a tale of the first crusade.   ...William B. MacCabe.

Franchise.   ...Joséphine B. Colomb.

Good Saint Louis and his times.   ...Anna Eliza Bray.

Guillaume Hubray : scènes de la vie féodale.   ...C. Guénot.

Histoire de Robert le Diable, duc de Normandie.   ...Anon.

L'histoire du châtelain de Coucy et de la Dame de Fayel.   ...Raoul de Coucy.

In His name : a story of the Waldenses seven hundred years ago.   ...Edward Everett Hale.

Julio : a tale of the Vaudois.   ...Mrs. J. B. Peploe (Webb).

Knight's ransom.   ...Laura Valentine.

Little duke ; or, Richard the fearless.   ...Charlotte Mary Yonge.

Marie de Brabant, reine de France.   ...A. P. F. Menegault.

Michel : chronique Normande du XIe siècle.   ...Felix Courty.

Page, squire, and knight : a romance of the days of chivalry.   ...William H. D. Adams.

FRENCH HISTORY — *Continued.*

Philip Augustus; or, Brothers in arms. ...George Payne Rainsford James.

Priest and man; or, Abélard and Heloïsa. ...William W. Newton.

Ransom: a tale of the thirteenth century. ...Laura Valentine.

Sir Raoul de Broc and his son Tristam. ...George A. Poole.

Story of Viteau. ...Frank Richard Stockton.

Summons to judgment. ...Henry C. Adams.

Typhaines: a tale of the twelfth century. ...Joseph Arthur, Comte de Gobineau. English translation.

Fourteenth Century.

Le bâtard de Mauléon. ...Alexandre Dumas.

Before the dawn: a story of Paris and the Jacquerie. ...George Dulac.

Die Belagerung von Calais. ...Carl Weichselbaumer.

Bertrand Du Guesclin. ...Céline Fallet.

Blanche d'Évereaux. ...Amélia J. Candeille.

Cesare Torneo: épisode de l'histoire du Quercy au XIV. siècle. ...Bertrandy.

Charles-le-Mauvais; ou, La cour de Navarre. ...Élisabeth Guénard.

Chronique de Bertrand Du Guesclin. ...Cuvelier.

Cressy and Poictiers; or, Story of the Black Prince's page. ...John George Edgar.

De Foix: a romance of Béarn of the fourteenth century. ...Anna Eliza Bray.

Eustache de Saint Pierre; or, Surrender of Calays. ...Henry W. Herbert. (Pseud., Frank Forester.)

Half-brothers; or, Head and the hand. ...Alexandre Dumas. English translation.

Hélêne de Poitiers. ...G. Touchard Lafosse.

Henry of Monmouth; or, Field of Agincourt. ...Major Michel.

Isabel de Bavière. ...Alexandre Dumas.

Isabel of Bavaria; or, Chronicles of France for the reign of Charles VI. ...Alexandre Dumas. English translation.

Jacquerie. ...George Payne Rainsford James.

La Jacquerie: scènes féodales suivies de la famille de La Carvajal. ...Prosper Mérimée.

Le jeune Du Guesclin. ...René de Mont-Louis.

Knight of Mauléon. ...Alexandre Dumas. English translation. (Same as " Half-brothers.")

Le Leeuw van Vlaanderen. ...Hendrik Conscience.

Lily of Paris. ...John P. Simpson.

Lion of Flanders. ...Hendrik Conscience. English translation.

Phélippa, souvenirs du règne de Charles VI. ...C. Guénot.

FRENCH HISTORY — *Continued.*

Provost of Paris : a tale.   ...William S. Browning.
Robert d'Artois.   ...John Humphrey Saint-Aubyn.
Le roman du Mont Saint Michel, 1364.   ...E. Meunier.
Le siège de Calais.   ...Claudine A. G. Tencin.
Sword of De Bardwell : a tale of Agincourt.   ...C. M. Katherine
    Phipps.
La tour de Nesle.   ...Frédéric Girard.
Vassel : a story of the old Normandy.   ...Laura Valentine.
White company.   ...Arthur Conan Doyle.

Fifteenth Century.

Agincourt.   ...George Payne Rainsford James.
Agnes Sorel.   ...George Payne Rainsford James.
Agnès Sorel ; ou, La cour de Charles VII.   ...Élisabeth Guénard.
L'amazone française ; ou, Jeanne d'Arc.   ...Maria Thérèse Peroux
    d'Abany.
Anne of Geierstein.   ...Sir Walter Scott.
Armourer of Paris.   ...Anon.
Le bâtard de Mauléon.   ...Alexandre Dumas.
Before the dawn : a story of Paris and the Jacquerie.   ...George
    Dulac.
Charles the Bold.   ...Alexandre Dumas.   English translation.
Charles le Téméraire.   ...Alexandre Dumas.
Le château de Lavardin.   ...Alexandre de Salies.
Le Chevalier Bayard.   ...Saint Albin Berville.
Cloister and the hearth.   ...Charles Reade.
Les écorcheurs ; ou, L'usurpation et la peste, fragmens hist. 1418.
    ...Victor, Vicomte d'Arlincourt.
Faithful but not famous : a historical tale.   ...Anon.
Le fratricide ; ou, Gilles de Bretagne.   ...Joseph A. Walsh.
Half-brothers ; or, Head and the hand.   ...Alexandre Dumas.   Eng-
    lish translation.
Histoire de Jean de, Paris, roi de France.   ...Anon.
Hunchback of Notre Dame.   ...Victor Hugo.   English translation.
In the city.   ...Deborah Alcock.
Isabel de Bavière.   ...Alexandre Dumas.
Isabel of Bavaria ; or, Chronicles of France for the reign of Charles
    VI.   ...Alexandre Dumas.   English translation.
Jeanne d'Arc.   ...Caroline d'Hautefeuille.
Jeanne d'Arc ; ou, La France au XVᵉ siècle.   ...Jules Foussette.
Jeanne d'Arc ; ou, L'héroïne de Vaucouleurs.   ...René de Mont-Louis.
Jeanne de France.   ...Stéphanie Félicité, Comtesse de Genlis.
Jehanne la Pucelle.   ...Alexandre Dumas.
Joan of Arc : an historical tale.   ...Anon.

FRENCH HISTORY — *Continued.*

Joan of Arc; or, Story of a noble life. ...Anon.

Joan of Arc, the maid of Orleans. ...Orlando W. Wight.

Joan of Arc, the maid of Orleans. ...Thomas J. Serle.

Joan the maid deliverer of France and England. ...Elizabeth Charles.

Die Jungfrau von Orleans. ...Stanislaus S. A. Grabowski.

King of a day. ...Florence Wilford.

Knight of Mauléon. ...Alexandre Dumas. English translation. (Same as Half-brothers.)

Legends of Feudal days. ...Henry W. Herbert. (Pseud., Frank Forester.)

Leonora D'Orco. ...George Payne Rainsford James.

Lodging for the night. ...Robert Louis Stevenson.

Louis de la Trémoille; ou, Les frères d'armes. ...Justin J. E. Roy.

Mademoiselle de Montdidier; ou, La cour de Louis XI. ...Marie A. Barthélemy-Hadot.

Maid of Orleans. ...William H. D. Adams.

Maid of Orleans. ...Jane Robinson.

Marie de Bourgogne. ...Mme. de Saint-Venant.

Mary of Burgundy. ...George Payne Rainsford James.

Notre Dame de Paris. ...Victor Hugo.

Notre Dame; or, Bellringer of Paris. ...Victor Hugo. English translation. (Same as Hunchback of Notre Dame.)

La paysanne de Domremy. ...Eugénie Foa.

Personal recollections of Joan of Arc. ...Samuel Langhorne Clemens. (Pseud., Mark Twain.)

Philippe-Monsieur. ...Charles Buet.

Le prince de Bretagne. ...François T. M. B. d'Arnaud.

La pucelle. ...Carl W. T. Frenzel.

Queen's favorite. ...Anon.

Quentin Durward. ...Sir Walter Scott.

Romans relatifs à l'histoire de France aux XVᵉ et XVIᵉ siècles. ...Paul Lacroix, ed.

Le siège de Calais. ...Claudine A. G. Tencin.

Le Sire d'Aubigny. ...Pierre de Lesconvel.

Spotless and fearless: the story of Chevalier Bayard. ...Saint Albin Berville.

Stormy life. ...Georgiana C. L. G. Fullerton.

Thomassine Spinola. ...Mme. d'Omatu.

Under Bayard's banner. ...Henry Frith.

Sixteenth Century.

L'Abbesse de Montmartre. ...Henri Augu.

Anne de Montmorency, connétable de France: nouvelle historique. ...Pierre de Lesconvel.

FRENCH HISTORY — *Continued.*
Arthur Blanc.   ...James Grant.
Ascanio.   ...Alexandre Dumas.
Ascanio.   ...Alexandre Dumas.   English translation.
Astrologer's daughter : an historical novel.   ...Rose E. Temple.
   (Pseud., R. E. Hendricks.)
Les beaux Messieurs de Bois-Doré.   ...Amantine L. A. D. Dudevant.
   (Pseud., George Sand.)
Black tulip.   ...Alexandre Dumas.   English translation.
Brigand; or, Corse de Leon.   ...George Payne Rainsford James.
Le Capitaine muet.   ...Adolphe Racot.
Catharine de Medicis ; or, Queen mother.   ...Louisa S. Costello.
Chaplet of pearls; or, White and black Ribaumont.   ...Charlotte
   Mary Yonge.
Le chevalier de la Renaudie : roman historique.   ...Édouard
   Devicque.
Chicot the jester.   ...Alexandre Dumas.   English translation.
Chronique du temps de Charles IX., 1572.   ...Prosper Mérimée.
Les Comtes de Montgomery : roman historique.   ...Lottin de Laval.
La Comtesse de Tende.   ...Comtesse de Lafayette.
Constable de Bourbon.   ...William Harrison Ainsworth.
Crichton.   ...William Harrison Ainsworth.
La dame d'entremont récit du temps de Charles IX.   ...Ernest
   d'Hervilly.
La dame de Monsoreau.   ...Alexandre Dumas.
La dame de Saint-Bris.   ...Alexandre F. Guesdon.
Darnley ; or, Field of the cloth of gold.   ...George Payne Rainsford
   James.
Les deux Dianes.   ...Alexandre Dumas.
Diane de Poitiers.   ...Jean B. H. R. Capefigue.
Diane de Poitiers.   ...Maurice A. Maurage.
Le duc de Guise.   ...Louis P. P. Legay.
Duplessis.   ...E. Robinson.
Faithful but not famous : a historical tale.   ...Anon.
La famille Prudhomme.   ...Victor Derode.
La fille d'honneur.   ...Alexandrine S. C. de C. Baronne de Bawr.
Forty-five guardsmen.   ...Alexandre Dumas.   English translation.
François de Guise.   ...Joseph Mathurin Brisset.
François le Balafré.   ...Charles Buet.
Gallant lords of Bois-Doré.   ...Amantine L. A. D. Dudevant.
   (Pseud., George Sand.)
Le gantier d'Orléans, histoire du XVIe siècle.   ...Jean B. P. Lafitte.
Gaston de Blondville ; or, Court òf Henry III. resting in Ardennes.
   ...Anne Radcliffe.
Gentleman of France.   ...Stanley J. Weyman.

FRENCH HISTORY — *Continued.*

Good old times : a tale of Auvergne. ...Anne Manning.

Le gueux de Bruges. ...Frédéric de Courcy and Ferdinand Langlé.

Hamilton of Bothwellhaugh : a dark scene in Paris. ...Henry W. Herbert. (Pseud., Frank Forester.)

Heinrich des Vierten Vermählung mit Bertha von Susa. ...Luise Zeller.

Helen of Tournon. ...Adélaïde M. É. F. Souza-Botelho. English translation.

Henri de Rohan; or, Huguenot refugee. ...Francisca I. Ouvry.

Henri Quatre; or, Days of the league. ...John H. Mancur.

Henry the Fourth. ...Luise Zeller. English translation.

Henry of Guise; or, States of Blois. ...George Payne Rainsford James.

Hermit of Livry : a story of the sixteenth century. ...M. R. Housekeeper.

Histoire de Marguerite de Valois, reine de Navarre, sœur de François I. ...Charlotte R. de Caumont de la Force.

History of Nicholas Muss. ...Charles DuBois-Melly.

House of the Wolf : a romance. ...Stanley J. Weyman.

In the vulture's nest; or, Huguenots at the court of France in 1572. ...Mildred Fairfax.

Louis Belat : a tale of the reformation in Savoy. ...Mrs. Alexander S. Orr.

Madaron; or, Artizan of Nismes : an historical romance of the sixteenth century. ...D'Aubigné White.

Mademoiselle de Tournon. ...Adélaïde M. É. F. Souza-Botelho.

Man at arms ; or, Henry de Cerons. ...George Payne Rainsford James.

Margaret de Valois : a historical romance. ...Alexandre Dumas. English translation.

Mountain patriots: a tale of the reformation in Savoy. ...Mrs. Alexander S. Orr.

Une nièce de Balafré : histoire du temps de la Ligue. ...Ernest Faligan.

One in a thousand ; or, Days of Henri Quatre. ...George Payne Rainsford James.

De l'Orme. ...George Payne Rainsford James.

Owen Tudor. ...Jane Robinson.

Le page du duc de Savoie. ...Alexandre Dumas.

Page of the Duke of Savoy. ...Alexandre Dumas. English translation.

Princess of Clèves. ...Comtesse de Lafayette. English translation.

La princesse de Clèves. ...Comtesse de Lafayette.

Provocations of Mme. Palissy. ...Anne Manning.

FRENCH HISTORY — *Continued.*

Les quarante-cinq.   ...Alexandre Dumas.
La reine Margot.   ...Alexandre Dumas.
Le roi Charlot : scènes de la Saint-Barthélemy.   ...Charles Buet.
Rose d'Albert ; or, Troublous times.   ...George Payne Rainsford James.
Saint Bartholomew's eve : a tale of the Huguenot wars.   ...George Alfred Henty.
Saint Leon ; a tale of the sixteenth century.   ...William Godwin.
Die Schlacht von Marignano.   ...Carl A. F. von Witzleben.
Secret of Narcisse.   ...Edmund W. Gosse.
Sister Rose ; or, Eve of Saint Bartholomew's eve.   ...Emily Sarah Holt.
Soldiers of the cross : a story of the Huguenots.   ...Edith S. Floyèr.
Spotless and fearless : the story of Chevalier Bayard.   ...Saint Albin Berville.
Two Dianas.   ...Alexandre Dumas.   English translation.
Under Bayard's banner.   ...Henry Frith.
Under the bells.   ...Leonard Kip.
See also **HUGUENOTS, SAINT BARTHOLOMEW'S DAY, 1572.**

Seventeenth Century.

Alamontade, der Galeeren-Sklave.   ...Johann Heinrich D. Zschokke.
Alamontade ; or, Galley slave.   ...Johann Heinrich D. Zschokke. English translation.
Antoine de Bonneval : a tale of Paris in the days of Saint Vincent de Paul.   ...William T. Anderdon.
Arnold Delahaize ; or, Huguenot pastor.   ...Francisca I. Ouvry.
Artamène.   ...Madeleine de Scudéry.
Arthur Blane ; or, Hundred cuirassiers.   ...James Grant.
Asylum Christi : a story of the dragonnades.   ...Edward Gilliat.
Der Aufruhr in den Cevennen.   ...Johann Ludwig Tieck.
Les beaux messieurs de Bois Doré.   ...Amantine L. A. D. Dudevant. (Pseud., George Sand.)
Die beiden Condé : historische novelle.   ...Eugen H. von Dedenroth.
Belle-Rose.   ...Louis Amédée Eugène Achard.
Black tulip.   ...Alexandre Dumas.   English translation.
Blanche Gamond : a heroine of the faith.   ...Anon.
Blue pavilions.   ...Arthur T. Q. Couch. (Pseud., "Q.")
Bourdaloue and Louis XIV. ; or, Preacher and king.   ...Laurence Louis Félix Bungener.   English translation.
Brothers : a tale of the Fronde.   ...Henry W. Herbert. (Pseud., Frank Forester.)
Camisard ; or, Protestants of Languedoc.   ...Anon.
Le Capitaine Fracasse.   ...Théophile Gautier.

FRENCH HISTORY — *Continued.*

Captain Fracasse. ...Théophile Gautier. English translation.

Cardillac, the jeweller. ...Ernst Theodor Wilhelm Hoffmann. (In his Strange stories.)

Château de Louard : a story of the Edict of Nantes. ...Henry C. Coape.

Cinq-Mars ; or, Conspiracy under Louis XIII. ...Alfred Victor, Comte de Vigny. English translation.

Cinq-Mars ; ou, Une conjuration sous Louis XIII. ...Alfred Victor, Comte de Vigny.

Cross and the crown : a tale of the revocation of the Edict of Nantes. ...Deborah Alcock.

Les dernieres maîtresses de Louis XIV. ...Félix de Servan.

Deux soirées à l'Hôtel de Rambouillet. ...Laurence Louis Félix Bungener.

Dorothy Arden : a story of England and France two hundred years ago. ...J. M. Callwell.

Driven into exile : a story of the Huguenots. ...Charlotte Tucker. (Pseud., A. L. O. E.)

Duchess of Burgundy ; or, Scenes in the court of Louis XIV. ...Anon.

Duchess of Mazarin : a tale of the times of Louis XIV. ...Anon,

Duchess of La Vallière. ...Stéphanie Félicité, Comtesse de Genlis. English translation.

La duchesse de La Vallière. ...Stéphanie Félicité, Comtesse de Genlis.

La duchesse de Montpensier. ...Pierre A., Vicomte de Ponson du Terrail.

Die eiserne Maske. ...Paul V. Wichmann.

Every man his own trumpeter. ...George Walter Thornbury.

Une fantaisie de Louis XIV. ...Anne Bignan.

Gabrielle. ...Alice M. C. Durand. (Pseud., Henry Gréville.)

Gabrielle ; or, House of Maurèze. ...Alice M. C. Durand. (Pseud., Henry Gréville.) English translation.

Gabrielle ; or, Pictures of a reign. ...Louisa S. Costello.

Gallant lords of Bois Doré. ...Amantine L. A. D. Dudevant. (Pseud., George Sand.) English translation.

Das Geheimniss der Frau von Nizza. ...Emil M. Vacano.

Geneviève ; or, Children of Port Royal. ...Deborah Alcock.

La guerre des femmes. ...Alexandre Dumas.

L'homme au masque de fer. ...Élisabeth Guénard.

L'homme au masque de fer. ...Chevalier de Mouhy.

Henri de Rohan ; or, Huguenot refugee. ...Francisca I. Ouvry.

How they kept the faith : a tale of Huguenots of Languedoc. ...Grace Raymond.

FRENCH HISTORY — *Continued.*

Hubert Montreuil; or, Huguenot and the dragoon.   ...Francisca I. Ouvry.

Huguenot : a tale of the French Protestants.   ...George Payne Rainsford James.

Huguenot exiles ; or, Times of Louis XIV.   ...Eliza Ann Dupuy.

Huguenot family.   ...Henrietta Keddie.   (Pseud., Sarah Tytler.)

In the days of adversity.   ...J. Bloundelle Burton.

Isabeau's hero : a story of the revolt of the Cevennes.   ...Esmé Stuart

Jacques Bonneval : or, Days of the dragonnades.   ...Anne Manning.

Jean Cavalier ; ou, Les fanatiques de Cévennes.   ...Marie Joseph Eugene Sue.

Jeannette : a story of the Huguenots.   ...Frances·M. Peard.

Letters writ by a Turkish spy.   ...Giovanni P. Marana.

Lauzun.   ...Paul E. de Musset.

Légendes de Fontainebleau.   ...Julie O. Lavergne.

Lettre de cachet.   ...Catherine G. F. Gore.   (In her Romances of real life.)

Lord Montagu's page : an historical romance of the seventeenth century.   ...George Payne Rainsford James.

Louis ; or, Doomed to the cloister : a tale of religious life in the time of Louis XIV.   ...M. J. Hope.

Mme. D'Aubigné la petite créole.   ...Eugénie Foa.

Mme. de Lafayette.   ...Stéphanie Félicité, Comtesse de Genlis.

Man in black.   ...Stanley J. Weyman.

Der Mann mit der eisernen Maske.   ...E. Koenigs.

Marchioness of Brinvilliers.   ...Albert Smith.

Marie's story : a tale of the days of Louis XIV.   ...Mary E. Bamford.

La marquise de Courcelles.   ...Charles J. B. Jacquot.

Minister of Andouse.   ...Heinrich Moewes.   English translation.

Les missionnaires bottés de Louis XIV.   ...Adolphe Michel.

Molière.   ...Alexander Ungern Sternberg.

Nameless nobleman.   ...Jane Goodwin Austin.

Nachtstücke.   ...Ernst Theodor Wilhelm Hoffmann.

Nanon ; or, Women's war.   ...Alexandre Dumas.   English translation.

Olympe de Clèves.   ...Alexandre Dumas.

De l'Orme.   ...George Payne Rainsford James.

Der Pfarrer von Andouse.   ...Heinrich Moewes.

Pignerol.   ...Paul La Croix.

Priest and the Huguenot ; or, Persecution in the age of Louis XV.   ...Laurence Louis Félix Bungener.   English translation.

Protestant leader : a novel.   ...Marie Joseph Eugene Sue.   English translation.

Refugees : a tale of two continents.   ...Arthur Conan Doyle.

FRENCH HISTORY — *Continued.*

Reparation : a story of the reign of Louis XIV. and other tales. ...Georgiana C. L. G. Fullerton.

Richelieu ; or, Broken heart : an historical tale. ...Catherine G. F. Gore.

Richelieu ; or, Conspiracy. ...Edward George Earle Lytton Bulwer-Lytton.

Richelieu : a tale of France. ...George Payne Rainsford James.

Le roi des raffinés sous Louis XIII. ...L. J. Emmanuel Gonzalés.

Royal captives. ...Ann Yearsley.

Royal hunt : a story of Huguenot emigration. ...Mrs. E. C. Wilson.

Le Roy de Polexandre. ...Marin Gomberville.

Saint Eustace ; or, Hundred and one. ...Vane I. Saint John.

School for husbands. ...Rosina W. L. Bulwer-Lytton.

Die Serapionsbrüder. ...Ernst Theodor Wilhelm Hoffmann.

Un sermon sous Louis XIV. ...Laurence Louis Félix Bungener.

Shoulder knot. ...Benjamin F. Tefft.

Le siège de La Rochelle. ...Stéphanie Félicité, Comtesse de Genlis.

Siege of La Rochelle. ...R. C. Dallas.

Siege of La Rochelle. ...Stéphanie Félicité, Comtesse de Genlis. English translation.

Sister Louise ; or, Story of a woman's repentance. ...George John Whyte Melville.

Six sisters of the Valley. ...William Bramley Moore.

Stray pearls. ...Charlotte Mary Yonge.

Three musketeers. ...Alexandre Dumas. English translation.

Les trois mousquetaires. ...Alexandre Dumas.

Trois sermons sous Louis XV. ...Laurence Louis Félix Bungener.

Turkish spy. ...Giovanni P. Marana.

Twenty years after. ...Alexandre Dumas. English translation.

Two evenings at the Hôtel de Rambouillet. ...Laurence Louis Félix Bungener. English translation.

Under the red robe. ...Stanley J. Weyman.

Vicomte de Bragelonne. ...Alexandre Dumas.

Vingt ans après. ...Alexandre Dumas.

Eighteenth Century.

Adèle de Sénanges. ...Adélaïde M. É. F. Souza-Botelho.

Albert Lunel ; or, Château of Languedoc. ...Henry Brougham.

Un ami de la reine. ...Paul Gaulot.

L'an '93. ...Victor Hugo.

L'an 1 de la république, 1793. ...Émile Erckmann et Alexandre Chatrian.

Ancien régime. ...George Payne Rainsford James.

FRENCH HISTORY — *Continued.*

Andrée de Taverny.   ...Alexandre Dumas.

Angela Pisani.   ...Lord Strangford.

Anna Saint Ives.   ...Thomas Holcroft.

Atelier du Lys ; or, Art student in the Reign of Terror.   ...Margaret Roberts.

Beaumarchais.   ...Albert E. Brachvogel.

Bellah : a tale of La Vendée.   ...Octave Feuillet.   English translation.

Birth and education.   ...Marie Sophie Schwartz.   English translation.

Blockade : an episode of the fall of the first French empire.   ...Émile Erckmann and Alexandre Chatrian.   English translation.

Le Blocus.   ...Émile Erckmann et Alexandre Chatrian.

Blue ribbons : a tale of the last century.   ...Anna H. Drury.

De Boerenkryg.   ...Hendrik Conscience.

Börd och Bildning.   ...Marie Sophie Schwartz.

By fire and sword : a story of the Huguenots.   ...Thomas Archer.

Camisard ; or, Protestants of Languedoc.   ...Anon.

Les catacombes de Paris.   ...Élie B. Berthet.

Cerise : a tale of the last century.   ...George John Whyte Melville.

Charlotte Corday.   ...Henri Alphonse Esquiros.

Charlotte Corday : an historical novel.   ...Henri Alphonse Esquiros.   English translation.

Le chevalier d'Harmental.   ...Alexandre Dumas.

Le chevalier de Maison-Rouge.   ...Alexandre Dumas.

Le chevalier de Maison-Rouge : a tale of the reign of terror.   ...Alexandre Dumas.   English translation.

Citizen Bonaparte.   ...Émile Erckmann and Alexandre Chatrian.   English translation.

Le citoyen Bonaparte.   ...Émile Erckmann et Alexandre Chatrian.

Citoyenne Jacqueline : a woman's lot in the great French revolution.   ...Henrietta Keddie.   (Pseud., Sarah Tytler.)

Le collier de la reine.   ...Alexandre Dumas.

La comtesse de Charny.   ...Alexandre Dumas.

La comtesse d'Egmont.   ...Sophie Gay.

Comtesse de Fargy.   ...Adélaïde M. É. F. Souza-Botelho.

Conspirators ; or, Chevalier d'Harmental.   ...Alexandre Dumas.   English translation.

Contes de la Montague.   ...Émile Erckmann et Alexandre Chatrian.

Count Mirabeau : an historical novel.   ...Theodor Mundt.   English translation.

Countess Eva.   ...Joseph Henry Shorthouse.

Country in danger, 1792 ; or, Episodes of the great French revolu-

FRENCH HISTORY — *Continued.*

Madame de Staël: an historical novel.   ...Amalie C. E. M. Boelte. English translation.

Madame Thérèse.   ...Émile Erckmann et Alexandre Chatrian.

Madame Thérèse; or, Volunteers of '92.   ...Émile Erckmann and Alexandre Chatrian.   English translation.

Maria Antoinette and her son.   ...Clara M. Mundt.   (Pseud., Louise Mühlbach.)   English translation.

Marie Antoinette, la reine de France décapitée.   ...Ernest Pitavall.

Marie Antoinette.   ...Eugen H. von Dedenroth.

Marie Antoinette.   ...Carl L. Haeberlein.

Marie Antoinette und ihr Sohn.   ...Clara M. Mundt.   (Pseud., Louise Mühlbach.)

Marquis Jeanne Hyacinthe de Saint Palaye.   ...Joseph Henry Shorthouse.

Marston; or, Soldier and statesman.   ...George Croly.

Maurice Tiernay, the soldier of fortune.   ...Charles Lever.

Les mémoires d'un médecin.   ...Alexandre Dumas.

Memoirs of a physician.   ...Alexandre Dumas.   English translation.

Monsieur Jack: a tale of the old war-time.   ...Alfred H. Engelbach.

Montesquieu à Marseille.   ...Louis S. Mercier.

La mort de Louis XVI.   ...Armand R. M. Du Chatellier.

Ninety-three.   ...Victor Hugo.   English translation.

Nobleman of '89: an episode of the French revolution.   ...Abel Quinton.

Noblesse oblige.   ...Margaret Roberts.

Olympe de Clèves.   ...Alexandre Dumas.

Olympe de Cleves.   ...Alexandre Dumas.   English translation.

On the edge of the storm.   ...Margaret Roberts.

Orphan of La Vendée.   ...Anna Eliza Bray.

La patrie en danger.   ...Émile Erckmann et Alexandre Chatrian.

Peasant and the prince.   ...Harriet Martineau.

Peer's daughter.   ...Rosina W. L. Bulwer-Lytton.

Die Pilger der Wildniss.   ...Johannes Scherr.

Poor little Gaspard's drum: a tale of the French revolution.   ...Alfred H. Engelbach.

Priest and the Huguenots; or, Persecution in the age of Louis XV.   ...Laurence Louis Félix Bungener.   English translation.

Quatrevingt-treize.   ...Victor Hugo.

Queen's necklace.   ...Alexandre Dumas.   English translation.

Red cap and blue jacket.   ...George Dunn.

Red republic.   ...Robert W. Chambers.

Reds of the Midi.   ...Felix Gras.

Regent's daughter.   ...Alexandre Dumas.   English translation.

FRENCH HISTORY — *Continued.*

Reign of terror. ...Catherine G. F. Gore. (In her Romances of real life.)

Roseville family : an historical tale of the eighteenth century. ...Mrs. Alexander S. Orr.

Sister Cora : a tale of the eighteenth century. ...Anon.

Six years later. ...Alexandre Dumas. English translation.

States General. ...Émile Erckmann and Alexandre Chatrian. English translation.

Tale of two cities. ...Charles Dickens.

Trois sermons sous Louis XV. ...Laurence Louis Félix Bungener.

Tuileries. ...Catherine G. F. Gore.

La Vendée : an historical romance. ...Anthony Trollope.

Vera ; or, War of the peasants. ...Hendrik Conscience. English translation.

Watteau. ...Carl W. T. Frenzel.

Wild fire. ...George Walter Thornbury.

Year one of the Republic. ...Émile Erckmann and Alexandre Chatrian. English translation.

Young Marmaduke. ...William Henry Davenport Adams.

Zanoni. ...Edward George Earle Lytton Bulwer-Lytton.

See also **REPUBLICS, France,** and **REVOLUTIONS, France.**

Nineteenth Century.

Adventures of Brigadier Gerard. ...Arthur Conan Doyle.

Aims and obstacles. ...George Payne Rainsford James.

Andreas Hofer. ...Clara M. Mundt. (Pseud., Louise Mühlbach.)

Andreas Hofer. ...Clara M. Mundt. (Pseud., Louise Mühlbach.) English translation.

Andreas Hofer. ...Maximilian von Schenkendorf.

L'argent des autres. ...Émile Gaboriau.

L'assommoir. ...Émile Zola.

Battle of Dorking : reminiscences of a volunteer. ...George T. Chesney.

Belfry of Saint Jude. ...Esmé Stuart.

Ben Brace, the last of Nelson's Agamemnons. ...Frederick Chamier.

Bivouac ; or, Stories of the Peninsular war. ...William H. Maxwell.

Les blancs et les bleus. ...Alexandre Dumas.

Die Bluttaufe der deutschen Einheit im Jahre 1870. ...Eugene H. von Dedenroth.

Boy of the first empire. ...Elbridge S. Brooks.

Brackenleben. ...Lodovika Hesekiel.

Die Braut aus Frankreich. ...Hilmar Juetbock. (Pseud., J. Hilmar.)

Cäsar und Napoleon III. ...Julius Gunding.

Château Morville ; or, Life in Touraine. ...Emily Reed.

FRENCH HISTORY — *Continued.*

Los cien mil hijos de San Luis. ...Benito Pérez Gladós.
Die Clubisten in Mainz. ...Heinrich J. Koenig.
Comedy of terrors. ...James De Mille.
Compagnons de Jéhu. ...Alexandre Dumas.
Companions of Jehu. ...Alexandre Dumas. English translation.
Conscript : an historical novel of the days of the first Napoleon.
 ...Émile Erckmann and Alexander Chatrian. ...English trans-
 lation.
Ein deutsches Reiterleben. ...Julius von Wickede.
Diana's crescent. ...Anne Manning.
Der Domherr. ...Jodocus D. H. Temme.
Doña Flor. ...Gustave Aimard.
Edmond Dantes. ...Alexandre Dumas.
Edmond Dantes. ...Alexandre Dumas. English translation.
1805; oder, Die Franzosen zum erstenmal in Wien. ...Eduard
 Breier.
1809: Historischer Roman. ...Eduard Breier.
Elba und Waterloo : Fortsetzung von 1813. ...F. Stolle.
Elsie's dowry : a tale of the Franco-German war. ...Emma Leslie.
Empress Josephine. ...Clara M. Mundt. (Pseud., Louise Mühl-
 bach.) English translation.
Enemy's friendship. ...S. M. S. Clarke.
Enriqueta Faber. ...Andres Clemente Vazqueq.
Der Feldmarschall Blücher und der Pfarrer Kretzschmar. ...Philipp
 F. W. Oertel.
Fiddler of Lugau. ...Margaret Roberts.
Franz Alzeyer. ...Paul Heyse.
Die Franzosen in Berlin; 1806–8. ...Friedrich H. Unger.
Fudge family in Paris. ...Thomas Moore.
Gate and the glory beyond it : a tale of the Franco-Prussian war.
 ...Onyx, pseud.
General Rapp und die Belagerung von Danig. ...Maria von Ros-
 kowska.
Gervaise : the natural and social life of a family under the second
 empire. ...Émile Zola. English translation.
Histoire d'un conscrit de 1813. ...Émile Erckmann et Alexandre
 Chatrian.
Histoire d'un crime. ...Victor Hugo.
Histoire d'un homme du peuple. ...Émile Erckmann et Alexandre
 Chatrian.
Histoire du plébiscite. ...Émile Erckmann et Alexandre Chatrian.
History of a crime. ...Victor Hugo. English translation.
History of the conscript of 1813. ...Émile Erckmann and Alexandre
 Chatrian. English translation.

FRENCH HISTORY — *Continued.*

Hofer. ...Bramley-Moore.

Hungarian brothers. ...Anna Maria Porter.

Im Herzen von Deutschland. ...Carl G. Berneck.

In the year '13. ...Fritz Reuter. English translation.

Invasion of France in 1814 : comprising the night march of the Russian army past Phalsburg. ...Émile Erckmann and Alexandre Chatrian. English translation.

L'Invasion; ou, Lefou Yégof. ...Émile Erckmann et Alexandre Chatrian.

Kaiserin Josephine. ...Clara M. Mundt. (Pseud., Louise Mühlbach.)

Kenneth ; or, Rear-guard of the grand army. ...Charlotte Mary Yonge.

Kentucky's love ; or, Roughing it about Paris. ...Edward King.

König Jérôme's Carneval. ...Heinrich J. Koenig.

Königin Hortense : ein Napoleonisches Lebensbild. ...Clara M. Mundt. (Pseud., Louise Mühlbach.)

Krummensee. ...Johann Georg L. Hesekiel.

Der lange Isaack. ...Julius von Wickede.

Leonie ; or, Light out of darkness. ...Annie Lucas.

Lindau; oder, Der unsichtbare Bund. ...Johann Weitzel.

Louis Napoleon : Roman und Geschichte. ...Eugen H. von Dedenroth.

Louis Napoleon. ...Julius Gundling.

Louisa of Prussia and her time. ...Clara M. Mundt. (Pseud., Louise Mühlbach) English translation.

Louise von Preussen und ihre Zeit. ...Clara M. Mundt. (Pseud., Louise Mühlbach.)

Les louves de Machecoul. ...Alexandre Dumas.

Malgrétout. ...Amantine L. A. D. Dudevant. (Pseud., George Sand.)

Margaret Muller : a story of the late war in France. ...Eugenie Bersier.

Max Cromer : a story of the siege of Strasburg. ...Hannah Smith. (Pseud., Hesba Stretton.)

Member for Paris : a tale of the second empire. ...Grenville Murray.

Les Misérables. ...Victor Hugo.

Les Miserables. ...Victor Hugo. English translation.

Monsieur de Camors. ...Octave Feuillet.

Napoleon III. : Fortsetzung von Louis Napoleon. ...Julius Gundling.

Napoleon in Deutschland. ...Clara M. Mundt. (Pseud., Louise Mühlbach.)

Napoleon in Germany. ...Clara M. Mundt. (Pseud., Louise Mühlbach.) English translation.

Nest of royalists. ...Esmé Stuart.

Other people's money. ...Émile Gaboriau. English translation.

FRENCH HISTORY— *Continued.*
White lies.   ...Charles Reade.
Whites and the blues.   ...Alexandre Dumas.   English transla-
tion.
Within iron walls : a tale of the siege of Paris.   ...Annie Lucas.
Workman and soldier : a tale of Paris life during the siege and the
rule of the commune.   ...James F. Cobb.
Year nine : a tale of the Tyrol.   ...Anne Manning.
Young Breton volunteer.   ...Frances M. Wilbraham.
Young buglers : a tale of the Peninsular war.   ...George Alfred
Henty.
Young Franco-tireurs and their adventures in the Franco-Prussian
war.   ...George Alfred Henty.
See also **FRANCO-GERMAN WAR ; NAPOLEON I. OF FRANCE ;
REPUBLICS, France : CHARLES X. OF FRANCE ; CRUSADES**
and **WALDENSES.**

FRENCH REVOLUTION.   See **REVOLUTION, France.**

FRONTIER AND PIONEER LIFE.   *A Selection.*
L'Aigle noir des Dacotahs.   ...Gustave Aimard.
Baby Rue : her friends and her enemies.   ...Charlotte M. Clark.
(Pseud., Charles M. Clay.)
Backwoodsman.   ...Frederick Lascelles.
Les bandits de l'Arizona.   ...Gustave Aimard.
Bandits of the Osage.   ...Emerson Bennett.
Betrothed.   ...Sir Walter Scott.
Big Cypress.   ...Kirk Munroe.
Border reminiscences.   ...Randolph Barnes Marcy.
Boy settlers.   ...Noah Brooks.
Camp notes : a story of the plains.   ...Kirk Munroe.
Le cher cheur de pistes.   ...Gustave Aimard.
Children of the New Forest.   ...Frederick Marryat.
Le cœur de pierre.   ...Gustave Aimard.
Ella Barnwell.   ...Emerson Bennett.
La fièvre d'or.   ...Gustave Aimard.
For freedom's sake.   ...Arthur Paterson.
Gold-seekers : a tale of California.   ...Gustave Aimard.   English
translation.
Jericho road.   ...John Habberton.
John Brent's field.   ...L. Bates.
Maple range : a frontier romance.   ...Edna A. Barnard.
Marked bullet.   ...George H. Prentice.
Nakoma : a story of frontier life.   ...George Huntington.
Picture of pioneer times in California.   ...William Grey.
Planting the wilderness.   ...James D. MacCabe, Jr.

FRONTIER AND PIONEER LIFE — *Continued.*

Saratoga: an Indian tale of frontier life: a true story of 1787. ...D. Shepherd.

Starlight Ranch and other stories of army life on the frontier. ...Charles King.

Stoneheart : a romance of Indian life. ...Gustave Aimard.   English translation.

Tales of the Border.   ...James Hall.

Talking leaves : an Indian story.   ...William Oliver Stoddard.

Thirteen stories of the far West.   ...Forbes Heermans.

Thirty years of army life on the Border.   ...Randolph Barnes Marcy.

Trail-hunters : a tale of the far West.   ...Gustave Aimard.   English translation.

Les trappeurs de l'Arkansas.   ...Gustave Aimard.

Western character.   ...John L. MacConnel.

Wyandotte; or, Hutted knoll.   ...James Fenimore Cooper.

FUTURE LIFE.

At the threshold.   ...Nina Pictou.   (Pseud., Laura Dearborn.)

Beauty for ashes.   ...Kate Clark Brown.

From whose bourne ?   ...Robert Barr.

Gates ajar.   ...Elizabeth Stuart Phelps Ward (Mrs. Herbert D. Ward).

John Ward, preacher.   ...Margaret Deland.

Little pilgrim.   ...Margaret O. W. Oliphant.

One traveller returns.   ...Daniel Christie and Herman Henry Murray.

Quick or the dead?   ...Amélie Rives Chanler.

Wedding garments: a tale of the life to come.   ...Louis Pendleton.

GALICIA, SPAIN.

For the right.   ...Karl Emil Franzos.   ...English translation.

Jews of Barnow.   ...Karl Emil Franzos.   English translation.

Die Juden von Barnow.   ...Karl Emil Franzos.

Ein Kampf um's Recht.   ...Karl Emil Franzos.

Der neue Hiob.   ...Leopold Sacher-Masoch.

New Job.   ...Leopold Sacher-Masoch.   English translation.

GENEVIEVE DE BRABANT.

Geschichte von der heiligen Pfalzangräfin Genoeva.   ... G. O. Marbach.

L'Innocence reconnue ; ou, La vie admirable de Geneviève, Princesse du Brabant.   ...Anon.

Stories of old renown.   ...Robert Hope Moncrieff.   (Pseud., Ascott R. Hope.)

GEORGE I. OF ENGLAND. *Reigned 1714–1737.*
Father Clement: a Roman Catholic story. ...Grace Kennedy.
For the king. ...Charles Gibbon.
Four Georges. ...William Makepeace Thackeray.
Hartland forest: a legend of North Devon. ...Anna Eliza Bray.
Henry Smeaton. ...George Payne Rainsford James.
Himself his worst enemy; or, Duke of Wharton's career. ...A. P. Brotherhead.
John Law, the projector. ...William Harrison Ainsworth.
Life, adventures, and piracies of the famous Captain Singleton. ...Daniel Defoe.
Lucy Arden; or, Hollywood Hall. ...James Grant.
New voyage around the world. ...Daniel Defoe.
Preston fight; or, Insurrection of 1715. ...William Harrison Ainsworth.
Saville House: an historical romance of the time of George the First. ...Mary M. Hardy.
South-sea bubble: a tale of the year 1720. ...William Harrison Ainsworth.
See also **ENGLISH HISTORY.**

GEORGE II. OF ENGLAND. *Reigned 1727–1760.*
De Vane: a tale of plebeians and patricians. ...Henry W. Hilliard.
Diary of Mrs. Kitty Trevelyan: a story of the times of Whitefield and the Wesleys. ...Elizabeth Charles.
Expedition of Humphrey Clinker. ...Tobias George Smollett.
For James or George: schoolboy's tale of 1745. ...Henry C. Adams.
Four Georges. ...William Makepeace Thackeray.
Gipsy. ...George Payne Rainsford James.
Good old times: a story of the Manchester rebels of '45. ...William Harrison Ainsworth.
Hetty Hyde's lovers; or, Household brigade. ...James Grant.
Lady Grizel: an impression of a momentous epoch. ...Lewis Wingfield.
Lord Harry Bellair: a tale of the last century. ...Anne Manning.
Lord Mayor of London. ...William Harrison Ainsworth.
Miser's daughter. ...William Harrison Ainsworth.
Old Chelsea Bun-house. ...Anne Manning.
Old manor-house. ...Charlotte Smith.
Peg Woffington. ...Charles Reade.
Ralf Skirlaugh, the Lancashire squire. ...Edward Peacock.
Spiritual Don Quixote; or, Summer ramble of Mr. Geoffry Wildgoose. ...Richard Graves.
Treasure trove. ...Samuel Lover.
Virginians: a tale of the last century. ...William Makepeace Thackeray.

GEORGE II. OF ENGLAND — *Continued.*
Waverley.   ...Sir Walter Scott.
Wilhelm Hogarth : historische Roman.   ...Albert E. Brachvogel.
William Hogarth : a romance.   ...Albert E. Brachvogel.   English
   translation.

GEORGE III. OF ENGLAND.   *Reigned 1760–1820.*
Against the stream : the story of the heroic age in England.   ...Elizabeth Charles.
Aims and obstacles.   ...George Payne Rainsford James.
Ashcliffe Hall : a tale of the last century.   ...Emily Sarah Holt.
Barnaby Rudge.   ...Charles Dickens.
Beau Nash ; or, Bath in the eighteenth century.   ...William Harrison Ainsworth.
Ben Brace, the last of Nelson's Agamemnons.   ...Frederick Chamier.
De Vere.   ...Robert P. Ward.
Diana's crescent.   ...Anne Manning.
Evelina.   ...Frances D'Arblay.
Fool of quality ; or, Henry, Earl Moreland.   ...Henry Brooke.
Four Georges.   ...William Makepeace Thackeray.
Frank Hilton ; or, Queen's own.   ...James Grant.
"God's providence " house : a story of 1791.   ...Isabel Banks.
King of Bath ; or, Life at Spa.   ...Mary C. Ware.
King's own.   ...Frederick Marryat.
King's own borderers : a military romance.   ...James Grant.
Lady Bell : a story of the last century.   ...Henrietta Keddie. (Pseud., Sarah Tytler.)
Miss Angel.   ...Anna Isabella Thackeray (Mrs. Richmond Ritchie).
Northanger Abbey.   ...Jane Austen.
Oliver Ellis ; or, Fusiliers.   ...James Grant.
Phantom regiment ; or, Stories of "Ours."   ...James Grant.
Pride and prejudice.   ...Jane Austen.
Rival apprentices.   ...Anon.
Romance of war ; or, Highlanders in Spain.   ...James Grant.
Ronald Morton ; or, Fireships : a story of the last naval war.   ...William Henry Giles Kingston.
Shadow of the sword.   ...Robert W. Buchanan.
Shelley : biographische Novelle.   ...Wilhelm Ritter von Hamm.
Strawberry Hill.   ...Robert Folk Williams.
Subaltern.   ...George Robert Gleig.
Surgeon's daughter.   ...Sir Walter Scott.
Tapestried chamber.   ...Sir Walter Scott.
Trascaden Hall : " When George III. was king."   ...William G. Hamley.
Tremaine ; or, Man of refinement.   ...Robert P. Ward.

GEORGE III. OF ENGLAND — *Continued.*
Vanity fair. ...William Makepeace Thackeray.
Vicar of Wakefield. ...Oliver Goldsmith.
Young buglers: a tale of the Peninsular war. ...George Alfred Henty.
Youth and manhood of Cyril Thornton. ...Thomas Hamilton.

GEORGE IV. OF ENGLAND. *Reigned 1820–1830.*
Abigel Rowe : a chronicle of the regency. ...Lewis S. Wingfield.
Four Georges. ...William Makepeace Thackeray.
Weird of the Wentworths: a tale of George IV.'s time. ...Johannes Scotus, Pseud.
Whitehall. ...William Maginn.

GEORGE V., OF HANOVER. *Lived 1819–1878.*
Im exil. ...Johann Ferdinand Martin Oskar Meding. (Pseud., Gregor Samarow.)
Die Römerfahrt der Epigonen. ...Johann Ferdinand Martin Oskar Meding. (Pseud., Gregor Samarow.)

GEORGIA.
Chronicles of Pineville. ...William T. Thompson.
Dady Jake, the runaway. ...Joel Chandler Harris.
Dukesborough tales: Chronicles of Mr. Bill Williams. ...Robert M. Johnston.
Free Joe and other Georgian sketches. ...Joel Chandler Harris.
Guy Rivers. ...William Gilmore Simms.
Her second campaign. ...Maurice Thompson.
Mr. Absalom Billingslea. ...Richard Malcolm Johnston.
Primes and their neighbors. ...Richard M. Johnston.
Sister Jane. ...Joel Chandler Harris.
White Marie. ...William N. Harben.

GERMAN HISTORY. *A Selection.*
Early and General.
Die Ahnen. ...Gustav Freytag.
Astaroth und Mentha. ...Wilhelm Jensen.
Bissula : historischen Roman aus der Völkerwanderung. ...Felix Dahn.
Charakterbilder aus der Geschichte und Sage. ...August W. Grube.
Chriemhildens Rache. ...Carl F. Eichhorn.
Deutsche Heldensagen für Schule und Volk. ...Wilhelm Waegner.
Didrik af Berns Saga. ...Carl C. Rafn.
Dietrich und seine Gesellen aus der Heidelberger Handschrift.
Dietrich von Bern. ...Adolf Wechsler.
Dietrichs Brautfahrt von Albrecht von Kemenat. ...Friedrich H. von Hagen.

GERMAN HISTORY — *Continued.*

Dietrichs erste Ausfahrt. ...Franz Stark, ed.

Felicitas. ...Felix Dahn.

German classics from the fourth to the nineteenth century. ...Friedrich Max Mueller.

Historical tales. ...Charles Morris.

Lays and legends of Germany. ...William J. Thomas.

Popular romances of the middle ages. ...George W. Cox and E. H. Jones.

Preussens Tausend und eine Nacht. ...Bernhard Hesslein.

Romances of the middle ages. ...Wilhelm Waegner. English translation.

Romans et Épopées chevaleresques de l'Allemagne au moyen âge. ...Gustav von Bonstetten.

Die Römer in Deutschland. ...Joseph Marius Babo.

Sämundis Fuhrungen. ...Johann Arnold Kanne.

Scenes and characters of the middle ages. ...Edward L. Cutts.

Stories of old renown. ...Robert Hope Moncrieff. (Pseud., Ascott R. Hope.)

Vor Zeiten. ...Theodor Storm.

Fourth to the Ninth Centuries.

Eudoxia, die Kaiserinn. ...Ida Hahn-Hahn.

Jetta; or, Heidelberg under the Romans. ...Adolf Hausrath. English translation.

Jetta: historische Roman aus der Zeit Völkerwanderung. ...Adolf Hausrath.

Ninth Century.

Albrecht. ...Arlo Bates.

Les chevaliers du Cygne; ou, La Cour de Charlemagne. ...Stéphanie Félicité Comtesse de Genlis.

Ekkhard. ...Joseph V. Scheffel.

Knights of the Swan; or, Court of Charlemagne. ...Stéphanie Félicité Comtesse de Genlis. English translation.

Legends of Charlemagne. ...Thomas Bulfinch.

Magic runes. ...Emma Leslie.

Passe Rose. ...Arthur S. Hardy.

Pepin and Charlemagne. ...Alexandre Dumas. English translation.

Pépin et Charlemagne. ...Alexandre Dumas.

Romances relating to Charlemagne. ...George Ellis.

Vikings of the Baltic. ...George W. Dasent.

Eleventh Century.

Bertha: historical romance of the time of Henry IV. of Germany. ...Joseph E. C. Bischoff. English translation.

Bertha; or, Pope and the emperor. ...William B. MacCabe.

GERMAN HISTORY — *Continued.*

Boy crusaders. ...John George Edgar.

Count Robert of Paris. ...Sir Walter Scott.

Die Fahrten Thiodolfs des Isländers. ...Friedrich H. C. La Motte Fouqué.

Fighting the Saracens; or, Boy knight. ...George Alfred Henty.

Florine, princess of Burgundy : a tale of the first crusade. ...William B. MacCabe.

Foure prentises of London, with the conquest of Jerusalem. ...Thomas Heywood.

Heroines of the crusades. ...C. A. Bloss.

Hildebrand and the excommunicated emperor. ...Joseph Sortain.

I Lombardi alla prima crociata. ...Tommaso Grossi.

In the brave days of old : the story of the crusades. ...Henry Frith.

Königin Bertha : historischer Roman. ...Joseph E. C. Bischoff.

Lady Sybil's choice : a tale of the crusades. ...Emily Sarah Holt.

Mathilde; ou, mémoires tirés de l'histoire des croisades. ...Sophie Risteau Cottin.

Parcival : Roman. ...Albert E. Brachvogel.

Prince and the page : the last crusade. ...Charlotte Mary Yonge.

Raymond de Saint Gilles at the crusades. 1095–1099. ...Henriette de Witt.

Saracen; or, Matilda and Malek Adhel : a crusade romance. ...Sophia Risteau Cottin. English translation.

Sir Walter's ward : a tale of the crusades. ...William Everard.

Stories of the crusades. ...William E. Dutton.

Sword and scimetar : the romance of the crusades. ...Alfred Trumble.

Tales of the early ages. ...Horace Smith. (Pseud., Paul Chatfield.)

Thiodolf the Icelander. ...Friedrich H. C. La Motte Fouqué. English translation.

Truce of God : a tale of the eleventh century. ...George H. Miles.

Twelfth Century.

Barbarossa : historischer Roman. ...Joseph E. C. Bischoff.

Friedrich's des Ersten letzte Lebenstage. ...Julius Bacher.

Heinrich der Löwe. ...Carl L. Haeberlin.

Heinrich der Löwe. ...Carl C. F. Reidmann.

Heinrich der Sechste teutscher Kaiser. ...Carl F. A. Buchner.

Kaiser Friedrich genannt Barbarossa. ...Carl J. Simrock.

Thirteenth Century.

Castle of Ehrenstein : its lords, spiritual and temporal : its inhabitants, earthly and unearthly. ...George Payne Rainsford James.

Crusaders and captives. ...George E. Morrill.

GERMAN HISTORY — *Continued.*

Earl Hubert's daughter; or, Polishing of the pearl.   ...Emily Sarah Holt.

Henry of Ofterdingen.   ...Friedrich von Hardenberg.   English translation.

Henry von Ofterdingen.   ..:.Friedrich L. von Hardenberg.

Knight of the lion : a romance of the thirteenth century.   ...Anon.

Lieder aus Heinrich von Ofterdingen Zeit.   ...Joseph V. von Scheffel.

Philip Augustus; or, Brothers in arms.   ...George Payne Rainsford James.

Rudolf von Habsbury.   ...Friedrich C. Schlenkert.

Versunkene Welten.   ...William Jessen.

With script and staff : a tale of the children's crusade.   ...Elia W. Peattie.

Wolfram von Eschenbach.   ...Ludwig Lang.

Fourteenth Century.

Der falsche Wöldemar.   ...Georg Wilhelm Heinrich Haering.

Der Klosterjäger.   ...Ludwig A. Ganghofer.

Tower of the hawk : some passages in the history of the House of Hapsburg.   ...Jane L. Willyams.

Fifteenth Century.

Die Abenteuer des Herzogs Christoph von Bayern genannt der Kämpfer.   ...Franz Trautmann.

Anne of Geierstein.   ...Sir Walter Scott.

Briefe eines Frauenzimmers aus dem XV. Jahrh. nach alten Urschriften.   ...Paul von Stetten.

Burgomaster of Berlin.   ...Georg Wilhelm Heinrich Haering. English translation.

Cloister and the hearth.   ...Charles Reade.

Dove in the eagle's nest.   ...Charlotte Mary Yonge.

Dürer in Venedig.   ...Adolf Stern.

Ernest and Albert; or, Story of the stolen princes.   ...Anon.

Franz Sternbald's Wanderungen.   ...Johann Ludwig Tieck.

Die Gred: Roman aus dem alten Nuremberg.   ...Georg Moritz Ebers.

Gutenberg.   ...Emily C. Pearson.   (Pseud., Ervie.)

Hermann von Unna.   ...Christiane Benedicte Eugenie Naubert.

Jew.   ...Carl Spindler.   English translation.

Johannes Gutenberg.   ...Adolf Stern.

Der Jude.   ...Carl Spindler.

Die Künstlerehe.   ...Leopold Schefer.

Kunz von Kauffungen.   ...Ludwig Storch.

GERMAN HISTORY — *Continued.*

Last knight : a romance-garland.   ...Anton A., Count von Auersperg.
  English translation.

Der letzte Ritter.   ...Anton A., Count von Auersperg.

Lia : a tale of Nuremberg.   ...Esmé Stuart.

Margery : a tale of old Nuremberg.   ...Georg Moritz Ebers.
  English translation.

Marie de Bourgogne.   ...Mme. de Saint-Venant.

Mary of Burgundy ; or, Revolt of Ghent.   ...George Payne Rainsford James.

Norica, das sind Nürnbergische Novellen aus alter Zeit.   ...August
  Hagen.

Norica ; or, Tales of Nürnberg from the olden times.   ...August
  Hagen.   English translation.

Nüremberg tand, eine Geschichte aus dem fünfzehnten Jahrhundert.
  ...Ludovika Hesekiel.

Prinz Sigmund von Sachsen und seine Brüder.   ...Auguste Wilhelm
  Lorenz.

Der Roland von Berlin.   ...Georg Wilhelm Heinrich Haering.

Salt-master.   ...Julius Wolff.   English translation.

Scenes in feudal times.   ...R. H. Wilmot.

Der Sülfmeister.   ...Julius Wolff.

Der Weiss Kunig.   ...Maximilian I. arr. by Maria Treitzsaurwein.

Wenzel's inheritance.   ...Annie Lucas.

Sixteenth Century.

Agnes de Mansfeldt.   ...Thomas C. Grattan.

Allerlie von Luther.   ...Joachim L. Kaupt.

Aus dem sechzehnten Jahrhundert.   ...Wilhelm Jensen.

Banished.   ...William Hauff.

Barbara Ittenhausen.   ...Emma Wuttke-Biller.

Baron de Hertz.   ...Albert de Labadye.

Baron of Hertz.   ...Albert de Labadye.   English translation.

Die Belagerung von Gotha.   ...Augusta W. Lorenz.

Die Belagerung von Leipzig in den Jahren 1546-7.   ...F. Lohmann.

Die Belagerung von Madgeburg.   ...Friedrich L. Schmidt.

Blind girl of Wittenberg.   ...Carl August Wildenhahn.   English
  translation.

Burgomaster's wife.   ...Georg Moritz Ebers.   English translation.

Das Buch vom Doktor Luther.   ...Hermann O. Nietschmann.

Christian or Romanist ?   ...Eduard Jost.   English translation.

Christlich oder Päpstlich ?   ...Eduard Jost.

Chronicles of the Schönberg-Cotta family.   ...Elizabeth Charles.

Clytia : a romance of the sixteenth century.   ...Adolf Hausrath.
  English translation.

Lukas Cranach. ...Hermann Klencke.
Luther and the Cardinal. ...Hermann O. Nietschmann.
Luther: historisches Gedicht. ...Gerhard Friedrich. (Pseud., Gustav Waller.)
Luther in Rom. ...Levin Schuecking.
Luther in Rome; or, Corradina, the last of the Hohenstaufen. ...Levin Schuecking. English translation.
Luther und die seinen. ...Franz Lubojatzky.
Luther und seine Zeit. ...Theodor Koenig.
Luther's Brautfahrt. ...Joseph E. C. Bischoff.
Margarethe · a tale of the sixteenth century. ...Emma Leslie.
Martin Luther: historischer Roman. ...Heinrich Eisenlohr.
Martin Luther und Graf Erbach. ...Hermann O. Nietschmann.
Maurice the elector of Saxony. ...Katharine Colquhoun.
Michael Kohlhaas. ...Heinrich von Kleist.
Monk and knight. ...Frank W. Gunsaulus.
Moritz von Sachsen. ...Friedrich C. Schlenkert.
Philippine Welser; oder, Vor dreihundert Jahren. ...Adelbert Baudissin.
Professor of Alchemy. ...Percy Ross.
Der Prophet. ...Theodor Muegge.
Saint Leon: a tale of the sixteenth century. ...Caleb Williams Godwin.
Die Schlacht bei Drakenburg. ...Werner Bergman.
Die Schlacht bei Hammingstedt. ...Amalie Schoppe.
Die Schlacht von Marignano. ...Carl A. F. von Witzleben.
Sibylle von Cleve. ...Julius Bacher.
Sidonia the Sorceress. ...Johann Wilhelm Meinhold.
Der Sohn des Kaisers. ...Adolf Muetzelburg.
Tabithe von Geyersberg. ...Amalie Schoppe.
Tales from Alsace. ...Anne Manning.
Thomas Müntzer: Roman. ...Theodor Mundt.
Trübe Tage. ...Wilhelm Koch.
True as steel. ...George Walter Thornbury.
Ulric; or, Voices. ...Theodor S. Fay.
Die Wiedertäufer. ...Adolf Stern.
Die Wiellinger. ...Friedrich Wilhelm Arming.
Wolfgang, Prince of Anhalt. ...Franz Hoffmann.

Seventeenth Century.

Der abenteuerliche Simplicissimus. ...Hans J. C. von Grimmelshausen.
Agnes de Mansfeldt. ...Thomas C. Grattan.
Allwin: ein Roman. ...Friedrich H. C. La Motte Fouqué.

GERMAN HISTORY — *Continued.*

Allzeit getreu. ...H. Brand.

Axel.   ...Carl F. van der Velde.

Axel: a tale of the thirty years' war.   ...Carl F. van der Velde. English translation.

Baron and squire: a story of the thirty years' war.   ...Wilhelm Noeldechen. English translation.

Bernhard. Herzog zu Sachsen-Weimar.   ...Friedrich G. Schlenkert.

Bílá hora, Aneb; Tři léta z třiceti.   ...Ludwig Rellstab.

Blameless knights; or, Lützen and La Vendée.   ...Alice H. F. Byng, Viscountess Enfield.

Blue pavilions.   ...Arthur T. Q. Couch.   (Pseud., " Q.")

Brave resolve ; or, Siege of Stralsund.   ...John B. de Liefde.

Breughel brothers.   ...Alexander Ungern-Sternberg. English translation.

Die Brüder; oder, Das Geheimniss.   ...Alexander Ungern-Sternberg.

Der Buchführer von Lemgo.   ...Johann Georg L. Hesekiel.

Der Bürgemeister von Neisse.   ...Georg Hartwig.

Das Burgerweib von Weimar.   ...Julius W. Grosse.

Cajus Rungholt.   ...Lucian Buerger.

Der Christlichen Königlichen Fürsten Herculiscus und Herculadisla Wundergeschichte.   ...Andreas H. Buchholtz.

Des Christlichen teutschen Grossfürsten Hercules und der bohmischen königlichen Fräulein Valisca Wundergeschichte.   ...Andreas H. Buchholtz.

La conspiration de Walstein.   ...Jean François Sarrasin.

Der deutsche Krieg: historischer Roman in drei Büchern.   ...Heinrich Laube.

Der deutsche Michel.   ...Johann Georg L. Hesekiel.

Deutschlands Kassandra.   ...Heribert Rau.

Dorothe.   ...Georg Wilhelm Heinrich Haering.

Die drei Wünsche.   ...Carl F. von Witzleben.

Der dreissigjahrige Krieg.   ...Clara M. Mundt. (Pseud., Louise Mühlbach.)

Duellists: a tale of the thirty years' war.   ...Anon.

Ein deutsches Schneiderlein.   ...Franz Isidor Proschko.

Exiles of Lucerna.   ...John R. Macduff.

Fair Else, Duke Ulrich, and other stories.   ...Margaret Roberts.

Fältskarns Berättelser.   ...Zacharias Topelius.

Der Fels von Erz.   ...Albert E. Brachvogel.

Fides: eine Erzählung aus der Zeit des dreissigjährigen Krieges ...F. Jordan.

Der Findling.   ...Carl A. F. von Witzleben.

Frau Schatz Regime.   ...Johann Georg L. Hesekiel.

GERMAN HISTORY — *Continued.*
Gabriel : a story of the Jews in Prague. ...Salomon Kohn. English translation.
Die Gallerinn auf der Rieggersburg. ...Joseph F. von Hammer-Purgstall.
Gawriel : historische Erzählung aus dem dreissigjährigen Krieges. ...Salomon Kohn.
Geglänzt und erloschen. ...Ferdinand Pflug.
Geschichte der Gräfin Thekla von Thurn. ...Marc Anton Niendorf.
Der Graf von der Liegnitz. ...Carl G. von Berneck.
Der grosse Kurfürst in Preussen. ...Ernst A. C. C. Wichard.
Gustaf och 30 Åriga kriget. ...Zacharias Topelius.
Gustave Adolf and the thirty years' war. ...Zacharias Topelius. English translation.
Heidelberg. ...George Payne Rainsford James.
Herzog Bernhard. ...Hans Blum.
Herzog Bernhard. ...Heinrich Laube.
Die historisch-politischen Volkslieder des dreissigjährigen Krieges. ...Franz W. Ditfurth.
Im Schloss zu Heidelberg. ...Eva Hartner.
Ingerstein Hall and Chadwick Rise. ...James Routledge.
Johann Georg I. Von Sachsen. ...Franz Lubojatzky.
King's service. ...Anon.
Klosterheim. ...Thomas De Quincey.
Das Kräuterweible von Wimpfen. ...Conrad Fron.
Kruitzner ; or, German's tale. ...Harriet Lee. (In her Canterbury tales.)
Lichtenstein. ...William Hauff.
Lichtenstein ; or, Swabian league. ...William Hauff. English translation.
Die Lichtensteiner. ...Carl F. van der Velde.
Lichtensteins : a tale of the thirty years' war. ...Carl F. Van der Velde. English translation.
Die Lieder des dreissigjährigen Krieges. ...Eduard Weller.
Lion of the North : a tale of the times of Gustavus Adolphus and wars of religion. ...George Alfred Henty.
Maid of Stralsund : a story of the thirty years' war. John B. de Liefde.
Maria Schweidler die Bernsteinhexe. ...Johann Wilhelm Meinhold.
Maria Schweidler the amber witch. ...Johann Wilhelm Meinhold. English translation.
Memoirs of a cavalier ; or, Military journal of the wars in Germany and England, from 1632-'48. ...Daniel Defoe.
Memoirs of the honourable Colonel Andrew Newport. ...Daniel Defoe.

GERMAN HISTORY — *Continued.*

Mönch und Gräfin.   ...Anton J. Gross-Hoffinger.

Die Mörder Wallensteins.   ...George C. R. Herlosssohn.

Odolan Pĕptipeský.   ...Beneš-Třebizský.

Der Page des Herzogs von Friedland.   ...Carl A. T. von Witzleben.

Paul Gerhardt.   ...Carl August Wildenhahn.

Phantom ship.   ...Frederick Marryat.

Philip Rollo; or, Scottish musketeers.   ...James Grant.

Princess Vasa.   ...Zacharias Topelius.   English translation.

Prinsessan ab Vasa.   ...Zacharias Topelius.

Prinz Eugen und seine Zeit.   ...Clara M. Mundt.   (Pseud., Louise
    Mühlbach.)

Prinz Eugen unter Kaiser Leopold.   ...Johann Georg L. Hesekiel.

Prinz Eugen von Savoyen.   ...Carl T. Griesinger.

Prince Eugene and his Times.   ...Clara M. Mundt.   (Pseud., Louise
    Mühlbach.)   English translation.

Prinz Eugenius von Savoyen.   ...Emil von Boxberger.

Der Raub Strassburgs im Jahre 1681.   ...Heribert Rau.

Die Retter Niederwesels.   ...Philipp F. W. Oertel.

Der Ring.   ...Carl A. F. von Witzleben.

Ritterlicher Sinn.   ...Carl A. F. von Witzleben.

Der Rittmeister von Alt-Rosen.   ...Gustav Freytag.

Die Rose von Heidelberg.   ...Louisa M. Robiano.

Schoolmaster and his son.   ...Carl H. Caspari.   English translation.

Der Schulmeister und sein Sohn: eine Erzählung aus dem dreissig-
    jährigen Kriege.   ...Carl H. Caspari.

Die Schweden in Prag.   ...Caroline Pichler.

Story of an abduction in the seventeenth century   ....Jacob van
    Lennep.

Tempest tossed; the story of Seejungfer.   ...Margaret Roberts.

Thorn fortress: a tale of the thirty years' war.   ...Mary Bramston.

Times of Gustav Adolf.   ...Zacharias Topelius.   English translation.

Die Tochter des Piccolomini.   ...Georg C. R. Herlosssohn.

Trnová koruna.   Historícký obraz z třicetilete války.   ...Beneš Tře-
    bizský.

Trug-Gold.   ...Rudolf Baumbach.

Um den Kaiserstuhl.   ...Wilhelm Jensen.

Under Gustaf III.'s första regeringsår.   ...Zacharias Topelius.

Die Vierhundert von Pforzheim.   ...Carl A. F. von Witzleben.

Waldemar.   ...W. H. Harrison.

Waldner von Wildenstein.   ...Joseph Schreyvogel.

Waldstein.   ...Heinrich Laube.

Wallenstein.   ...Ernest Wilkomm.

Wallensteins erste Liebe.   ...Georg C. R. Herlosssohn.

Wallensteins Letzte Tage.   ...Franz Lubojatzky.

GERMAN HISTORY — *Continued.*

Young carpenters of Freiberg : a tale of the thirty years' war. ...Anon.

Young deserter. ...Anon.

Eighteenth Century.

Agathon. ...Christóph M. Wieland.

Alide. ...Emma Lazarus.

Der alte Dessauer. ...Franz Lubojatzky.

Der alte Fritz und seine Zeit. ...Clara M. Mundt. (Pseud., Louise Mühlbach.)

Auf Befehl des Königs. ...Clarissa Lohde.

Aus den Tagen zweier Könige. ...Friedrich Adami.

Aus drei Kaiserzeiten. ...Johann Georg L. Hesekiel.

Der Bachtanz zu Langenselbold. ...Philipp F. W. Oertel.

Die Bären von Augustusburg : eine Erzählung aus der Sächsischen Geschichte des 18 Jahrhunderts. ...Carl Gustav Nieritz.

Bears of Augustusburg. ...Carl Gustav Nieritz. English translation.

Der Berggeist im Riesengebirge. ...Franz Isidor Proschko.

Berlin and Sans-Souci. ...Clara M. Mundt. (Pseud., Louise Mühlbach.) English translation.

Bischof und König. ...Mariam Tenger.

Bishop and the king. ...Mariam Tenger. English translation.

Die Brautschau Friedrich des Grossen. ...Julius Bacher.

Cabanis Vaterlandischer Roman. ...Georg Wilhelm Heinrich Haering.

Charlotte Ackermann ; ein Hamburger Roman aus dem vorigen Jahrhundert. ...Otto Mueller.

Charlotte Ackermann : a theatrical romance. ...Otto Mueller. English translation.

Citizen of Prague. ...Henrietta von Paalzow. English translation.

Claude the Colporteur. ...Anne Manning. •

Cyrus. ...Christoph M. Wieland. (Cyrus was the Persian disguise for Frederick II.)

Dorothea Cappel. ...Emilie Friedrich S. Lohmann.

Elizabeth of Guttenstein. ...Caroline Pichler. English translation.

Elizabeth von Guttenstein. ...Caroline Pichler.

Fifteen years. ...Thérèse Albertine Luise von Jacob Robinson. English translation.

Frederick the Great and his court. ...Clara M. Mundt. (Pseud., Louise Mühlbach.) English translation.

Die Freimaurer : eine Familiengeschichte aus dem vorigen Jahrhundert. ...Ferdinand Gustav Kuehne.

Friedrich der Grosse und sein Hof. ...Clara M. Mundt. (Pseud., Louise Mühlbach.)

Friedrich der Grosse. ...Eugen H. von Dedenroth.

GERMAN HISTORY — *Continued.*

Fünfzehn Jahre. ...Thérèse Albertine Luise von Jacob Robinson.

Geschichte des weisen Danischmend. ...Christoph M. Wieland.

Goethe and Schiller. ...Clara M. Mundt. (Pseud., Louise Mühlbach.) English translation.

Der goldne Spiegel ; oder die Könige von Scheschian. ...Christoph M. Wieland.

Göthe und Schiller. ...Clara M. Mundt. (Pseud., Louise Mühlbach.)

Gute Zeit im Lande. ...H. Brand.

Historische Novellen über Friedrich II. von Preussen und seine Zeit. ...Joseph E. C. Bischoff.

Die historischen Volkslieder des siebenjährigen Krieges. ...Franz W. Ditfurth.

Hülderlin. ...Heribert Rau.

Im Banne des Schwarzen Adlers. ...Carl Rudolf Gottschall.

Die Insel Felsenbury. ...Ludwig Tieck.

Jean Paul: Kultur-historisch biographischer Roman. ...Heribert Rau.

Johan Gotzkowsky, der Kaufmann von Berlin. ...Clara M. Mundt. (Pseud., Louise Mühlbach.)

Josef Kaiser. ...Eduard Breier.

Joseph II. and his court. ...Clara M. Mundt. (Pseud., Louise Mühlbach.) English translation.

Joseph der Zweite. ...Arthur Korn.

Kaiser Josef II. ...Eduard Ille.

Kaiser Josef II. und das Geheimniss des Freihauses. ...Carl T. Fockt.

Kaiser Josef und der Sekretär. ...Adolf Muetzelburg.

Die Kaiserlichen in Sachsen. ...Wilhelm R. Heller.

Kaiser Joseph der Zweite und sein Hof. ...Clara M. Mundt. (Pseud., Louise Mühlbach.)

Kaiser Leopold II. und seine Zeit. ...Clara M. Mundt. (Pseud., Louise Mühlbach.)

Kaunitz. ...Leopold Sacher-Masoch.

König Friedrich August III. von Sachsen und seine Zeit. ...Franz Lubojatzky.

Lessing. ...Alexandre Ungern-Sternberg.

Little schoolmaster Mark. ...Joseph Henry Shorthouse.

Luck of Barry Lyndon: a romance of the last century. ...William Makepeace Thackeray.

Maria Theresa and her fireman. ...Clara M. Mundt. (Pseud., Louise Mühlbach.) English translation.

Maria Thérèsia und ihr Ofenheizer. ...Clara M. Mundt. (Pseud., Louise Mühlbach.)

GERMAN HISTORY — *Continued.*

Maria Theresia und ihre Zeit. ...Franz Lubojatzky.
Maria Theresia und ihre Zeit. ...Edouard Duller.
Merchant of Berlin. ...Clara M. Mundt. (Pseud., Louise Mühlbach.) English translation.
Mit und ohne Vakation. ...Elizabeth von Grotthuss.
Öfverste Stobée, ella holsteinska patriet under frihetstiden. ...Herman Bjursten.
Old Fritz and the new era. ...Clara M. Mundt. (Pseud., Louise Mühlbach.) English translation.
Ottilié. ...Violet Paget.
Die Prinzessin von Wolfenbüttel. ...Johann Heinrich D. Zschokke.
Prisoner's daughter : a story of 1758. ...Esmé Stuart.
Die Rosenkreuzer in Wien. ...Eduard Breier.
Rückwirkungen, oder wer regiert denn ? ...Johann Heinrich D. Zschokke.
Schiller : Kultur-historischer Roman. ...Johannes Scherr.
Schubart und seine Zeitgenossen. ...Albert E. Brachvogel.
Die Soldaten Friedrichs des Grossen. ...Julius von Wickede.
Sophie Charlotte, die philosophische Königin. ...Julius Bacher.
Täuschungen. ...Heinrich J. Koenig.
Thomas Thyrnau. ...Henrietta von Paalzow.
Too strange not to be true. ...Georgiana C. L. G. Fullerton.
Two campaigns. ...Alfred H. Engelbach.
Die unsichtbare Loge. ...Jean Paul F. Richter.
Verrath und Liebe. ...Adolf Muetzelburg.
Der Vetter im Consistorium. ...Philipp F. W. Oertel.
Von Fels zum Meer. ...Hans von Zollern.
Vor hundert Jahren. ...Carl W. Ritter von Martini.
Wien und Rom. ...Eduard Breier.
Der Zigeuner-zögling, oder Schlangenwege des Verbrechens. ...Ludwig von Alvensleben.

Nineteenth Century.

Alexander von Humboldt. ...Heribert Rau.
Andreas Burns und seine Familie. ...Philipp Lange. (Pseud., Philipp Galen.)
Andreas Hofer. ...Clara M. Mundt. (Pseud., Louise Mühlbach.)
Andreas Hofer. ...Clara M. Mundt. (Pseud., Louise Mühlbach.) English translation.
Andreas Hofer. ...Maximilian von Schenkendorf.
Aus dem Berlin Kaiser Wilhelm's I. ...Paul Lindenberg.
Der Baigneur von Ostende. ...Philipp F. W. Oertel.
Barackenleben. ...Ludovika Hesekiel.
Bilder und Geschichten aus Schwaben. ...Ottilie Wildermuth.

GERMAN HISTORY — *Continued.*

Bis zum Kaiserthron. ...Bruno Garlepp.

Die Bluttaufe der deutschen Einheit im Jahre 1870. ...Eugen H. von Dedenroth.

Bombardier H. and Corporal Dose. ...Friedrich W. Hacklaender.

Der böse Blick. ...Ludwig Schneider.

Die Clubisten in Mainz. ...Heinrich J. Koenig.

Die Czechin. ...Eugen H. von Dedenroth.

Debit and credit. ...Gustav Freytag. English translation.

Eine deutsche Revolution; oder, Der Karneval von 1848. ...Eugen H. Dedenroth.

Deutsche Wunden. ...Louise Otto.

Ein deutsches Reiterleben. ...Julius von Wickede.

Deutschlands Kampf und Sieg. ...Julius Woeniger.

Der Domherr. ...Jodocus D. II. Temme.

Drei Tage in Mittenwalde im bayerischen Alpengebirge. ...Philipp F. W. Oertel.

Durch Nacht zum Licht. ...Friedrich Spielhagen.

1805; oder die Franzosen zum erstenmal in Wien. ...Eduard Breier.

1848; oder Nacht und Licht. ...Franz Lubojatzky.

1849; oder des Königs Maienblüte. ...Franz Lubojatzky.

1866; oder in Böhmen und am Main. ...Franz Lubojatzky.

1870; oder die Heldin von Wörth. ...Adolf Schirmer.

Europäische Minen und Gegenminen. ...Johann Ferdinand Martin Oskar Meding. (Pseud., Gregor Samarow.)

Der Feldmarschall Blücher und der Pfarrer Kretzschmar. ...Philipp F. W. Oertel.

Fiddler of Lugau. ...Margaret Roberts.

For sceptre and crown. ...George Samaron.

Franz Alzeyer. ...Paul Heyse.

Die Franzosen in Berlin. 1806–8. ...Friedrich H. Unger.

Fritz of the tower: a tale of the Franco-German war. ...L. Lobenhoffer.

Fünf Milliarden. ...Bernhard Hesslein.

General Rapps und die Belagerung von Danzig. ...Maria von Roskowska.

Gold und Blut. ...Johann Ferdinand Martin Oskar Meding. (Pseud., Gregor Samarow.)

Die Grandidiers. ...Julius Rodenburg.

Der Grenadier von Weissenberg. ...Robert Neumann.

Ein Hamburger Kauffburger. ...W. Christern.

Haus Hohenzollern. ...Stanislaus S. A. Grabowski.

Heinrich Heine's erste Liebe. ...Katherine Diez.

Held und Kaiser. ...Johann Ferdinand Martin Oskar Meding. (Pseud., Gregor Samarow.)

GERMAN HISTORY — *Continued.*

Hohensteins. ...Friedrich Spielhagen. English translation.

Hungarian brothers. ...Anna Maria Porter.

Im Exil. ...Johann Ferdinand Martin Oskar Meding. (Pseud., Gregor Samarow.)

Im Herzen von Deutschland. ...Carl G. Berneck.

In the year '13. ...Fritz Reuter. English translation.

Der Jäger von Königgratz. ...Eugen H. von Dedenroth.

Das Jahr 1866. ...Johann Ferdinand Martin Oskar Meding. (Pseud., Gregor Samarow.)

Der Kaplan von Königgrätz. ...Franz Lubojatzky.

Kaiser Wilhelm und seine Zeitgenossen. ...Clara M. Mundt. (Pseud., Louise Mühlbach.)

König Jérôme's carnival. ...Heinrich J. Koenig.

Der Krieg am Rhein im Jahre 1870. ...Stanislaus S. A. Grabowski.

Kreuz und Schwert. ...Franz W. Ditfurth.

Kreuz und Schwert. ...Johann Ferdinand Martin Oskar Meding. (Pseud., Gregor Samarow.)

Krummensee. ...Johann Georg L. Hesekiel.

Der lange Isaack. ...Julius von Wickede.

Lindau ; oder der unsichtbare Bund. ...Johann Weitzel.

Louisa of Prussia and her times. ...Clara M. Mundt. (Pseud., Louise Mühlbach.) English translation.

Louise von Preussen und ihre Zeit. ...Clara M. Mundt. (Pseud., Louise Mühlbach.)

Maid, wife, or widow ? ...Annie F. Hector. (Pseud., Mrs. Alexander.)

Margaret von Ehrenberg, the artist's wife. ...William and Mary Howitt.

Martin der Stellmacher. ...L. Kreutzer.

Max Kromer : a story of the siege of Strasburg. ...Hannah Smith. (Pseud., Hesba Stretton.)

Napoleon and Blucher. ...Clara M. Mundt. (Pseud., Louise Mühlbach.) English translation.

Napoleon and the Queen of Prussia. ...Clara M. Mundt. (Pseud., Louise Mühlbach.) English translation.

Napoleon in Deutschland. ...Clara M. Mundt. (Pseud., Louise Mühlbach.)

Napoleon in Germany. ...Clara M. Mundt. (Pseud., Louise Mühlbach.) English translation.

Neudeutsch. ..Joseph F. C. Bischoff.

Otto der Schütz. ...Johann Gottfried Kinkel.

Pförtner Jugend. ...Friedrich Kenner.

Poet hero. ...Minny Bothmer.

GERMAN HISTORY — *Continued.*

Powder and gold : a story of the Franco-Prussian war.   ...Levin Schuecking.   English translation.

Prince Bismarck, friend or foe?   ...Minny Bothmer.

Prinz Louis Ferdinand.   ...Fanny Lewald Stahr.

Problematic characters.   ...Friedrich Spielhagen.   English translation.

Problematische Naturen.   ...Friedrich Spielhagen.

Prussian spy.   ...V. Valmont.

Pulver und Gold.   ...Levin Schuecking.

Die Ritter vom Geiste.   ...Carl F. Gutzhow.

Ein Roman aus den Zeiten des schlesvig-holsteinischen Krieges.   ...Constantin M. Reichenbach.

Die Römerfahrt der Epigonen.   ...Johann Ferdinand Martin Oskar Meding.   (Pseud., Gregor Samarow.)

Roman aus dem Berliner.   ...Oskar Heller.

Die Rose von Sadowa.   ...Ludwig A. R. Kuerbis.

Saly's Revolutionstage.   ...Ulrich Hegner.

Schach-Bismark.   ...J. G. Findel.

Schleswig-Holstein Meerumschlugen.   ...Carl von Kessel.

Sister Martha.   ...Benjamin Wilson.

Soll und Haben.   ...Gustav Freytag.

Der Spion.   ...Franz T. Wangenheim.

Tagebuch und Kriegslieder aus dem Jahre 1813.   ...Theodor Körner.

Theodor Körner.   ...Heribert Rau.

Through night to light.   ...Friedrich Spielhagen.   English translation.

Der Todesgruss der Legionen.   ...Johann Ferdinand Martin Oskar Meding.   (Pseud., Gregor Samarow.)

Um Szepter und Kronen.   ...Johann Ferdinand Martin Oskar Meding.   (Pseud., Gregor Samarow.)

Unter dem letzten Welfenkönige.   ...F. Klinck.

Unter Preussens Fahnen.   ...Stanislaus E. A. Grabowski.

Ut de Franzosentid.   ...Fritz Reuter.

Die von Hohenstein.   ...Friedrich Spielhagen.

Vanished emperor.   ...Percy Andreae.

Vaterländische Romane.   ...Johann Georg L. Hesekiel.

Verschollen.   ...Johann Ferdinand Martin Oskar Meding.   (Pseud., Gregor Samarow.)

Die Völkerschlacht bei Leipzig.   ...Joseph von Hinsberg.

Von Saalfeld bis Aspern.   ...Heinrich J. Koenig.

Vor dem Sturm.   ...Theodor Fontane.

Vor dem Sturm.   ...Johann Ferdinand Martin Oskar Meding. (Pseud., Gregor Samarow.)

GERMAN HISTORY — *Continued.*
Waldfried. ...Berthold Auerbach.
Zu spät erkannt. ...Anon.
Zwei Kaiserkronena. ...Johann Ferdinand Martin Oskar Meding.
(Pseud., Gregor Samarow.)

GERMANY.

Description, Manners, and Customs.

Adé: a story of German life. ...Esmé Stuart.
Die Ahnen. ...Gustav Freytag.
Albrecht. ...Arlo Bates.
Aus drei Kaiserzeiten. ...Johann Georg L. Hesekiel.
Bissula. ...Felix Dahn.
Buch der Sachsen. ...Adolf Boettger.
Der Buchführer von Lemgo. ...Johann Georg L. Hesekiel.
Castle on the Rhine. ...Johannes Carsten von Hauch. English
translation.
Chriemhildens Rache. ...Carl F. Eichhorn.
Contes des fords du Rhin. ...Émile Erckmann et Alexandre
Chatrian.
Didrik af Berns Saga. ...Carl C. Rafn.
Ekkhard. ...Joseph V. Scheffel.
Felicitas. ...Felix Dahn.
Gold und Blut. ...Johann Ferdinand Martin Oskar Meding.
(Pseud., Gregor Samarow.)
Historical tales ; or, Romance of reality. ...Charles Morris.
In the year '13. ...Fritz Reuter. English translation.
Ingo. ...Gustav Freytag.
Johann Kepler. ...Jo. Burow.
Leibnitz und die beiden Kurfürstinnen. ...Hermann Klencke.
Lost manuscript. ...Gustav Freytag. English translation.
Lutaniste of St. Jacob's. ...Catherine Drew.
Mit und Ohne Vakation. ...E. von Grotthuss.
Philipp Jacob Spencer. ...Carl August Wildenhahn.
Popular romances of the middle ages. ...George William Cox and
E. H. Jones.
Red knights of Germany. ...Peter Boyle.
Rent in a cloud. ...Charles James Lever. (Pseud., Cornelius
O'Dowd ; Harry Lorrequer.)
Rolandseck : a romance of the Rhine. ...Gerald Holcroft.
Romans et épopées chevaleresques de l'Allemagne au moyen âge.
...Gustav von Bonstetten.
Die Römer in Deutschland. ...Joseph Marius Babo.
Rosenkreuzer und Illuminaten. ...Max Ring.
Scenes and characters of the middle ages. ...Edward L. Cutts.

GERMANY — *Continued.*

Slottet ved Rhinen.   ...Johannes Carsten von Hauch.

Spinoza.   ...Berthold Auerbach.   English translation.

Spinoza : ein Denkerleben.   ...Berthold Auerbach.

Stories of old renown.   ...Robert Hope Moncrieff.   (Pseud., Ascott R. Hope.)

Stories of the Rhine.   ...Émile Erckmann and Alexandre Chatrian. English translation.

Vanished Emperor.   ...Percy Andreae.

Verlorene Handschrift.   ...Gustav Freytag.

Verrath und Liebe.   ...Adolf Muetzelburg.

Vikings of the Baltic.   ...George W. Dasent.

Von Saalfeld bis Aspern.   ...Heinrich J. Koenig.

See also **GERMAN HISTORY**.

GHETTO.

Aus dem Ghetto.   ...Leopold Kompert.

Christian and Leah.   ...Leopold Kompert.   English translation.

Vorhang-Purin.   ...Math. Kisch.

GHOST-STORIES AND THE SUPERNATURAL.   *A Selection.*

Archibald Malmaison.   ...Julian Hawthorne.

Cecilia de Noel.   ...Lanoe Falconer.

Der Chaldäische Zauberer.   ...Ernst Eckstein.

Chaldean magician.   ...Ernst Eckstein.   English translation.

Damen's ghost.   ...Edwin L. Bynner.

Dead man's story.   ...H. Herman.

Fantôme d'Orient.   ...Louis Marie Julien Viaud.   (Pseud., Pierre Loti.)

Frankenstein.   ...Mary Wollstonecraft Shelley.

Der Geisterseher : eine Geschichte aus den Memoires des Grafen von O.   ...J. C. F. von Schiller.

Ghost-house ; or, Story of Rose Lichen.   ...E. W. Mildred.

Ghost-hunter and his family.   ...Michael Banim.

Ghost of Redbrook.   ...J. G. A. Coulson.

Ghost-seer ; or, Apparitionist.   ...J. C. F. von Schiller.   English translation.

Ghost-seer.   ...Johann C. Friedrich von Schiller.   ...English translation.

Ghost stories and presentiments.   ...Anon.

Ghosts and family legends.   ...Catherine Crowe.

Guy Neville's ghost.   ...Percy Greg.

Haunted and the haunters ; or, House and the brain.   ...Edward George Earle Lytton Bulwer-Lytton.

Haunted hotel.   ...Wilkie Collins.

Haunted wood.   ...E. E. Ellis.

GHOST STORIES AND THE SUPERNATURAL — *Continued.*
Italian ghost story.   ...John Temple Leader.
Last tenant.   ...Benjamin Leopold Farjeon.
Lost Stradivarius.   ...J. Meade Falkner.
Lourdes.   ...Émile Zola.
Love is a spirit.   ...Julian Hawthorne.
Magic ink and other stories.   ...William Black.
Magic runes.   ...Emma Leslie.
Magic skin.   ...Honoré de Balzac. English translation.
Man in black.   ...Stanley J. Weyman.
Maurice ; or, Red jar.   ...Frances Villiers, Countess of Jersey.
Mrs. Lord's moonstone.   ...C. Stokes Wayne.
Necromancer.   ...George W. M. Reynolds.
Nephele.   ...Francis William Bourdillon.
Out of the past   ...E. Anson More, Jr.
La peau de chagrin.   ...Honoré de Balzac.
Phantom city : a volcanic romance.   ...William Westall
Phantom from the East.   ...Louis Marie Julien Viaud. (Pseud.,
   Pierre Loti.) English translation.
Phantom regiment ; or, Stories of " Ours."   ...James Grant.
Phantom Rickshaw and other tales.   ...Rudyard Kipling.
Phantom ship.   ...Frederick Marryat.
Rival ghosts.   ...James Brander Matthews. (In his Tales of fantasy
   and fact.)
Second opportunity of Mr. Staplehurst.   ...Anon.
Some Chinese ghosts.   ...Chita Lafcadio Hearn.
Spectre of Milaggio.   ...Andrew Wilson.
Stable for nightmares.   ...James Le Fanu, Sheridan, Young, and
   others.
Strange stories of coincidence and ghostly adventures.   ...Anon.
Tales for a stormy night.   ...Eugene F. Bliss, ed.
Tales of fantasy and fact.   ...James Brander Matthews.
Unlaid ghost : a story in metempsychosis.   ...Joseph Vila Prichard.
Water ghosts and others.   ...John Kendrick Bangs.

GÖTHE, JOHANN WOLFGANG VON. *Lived 1749-1832.*
Alide : an episode of Göthe's life.   ...Emma Lazarus.
Goethe and Schiller.   ...Clara M. Mundt. (Pseud., Louise
   Mühlbach.) English translation.
Göthe und Schiller.   ...Clara M. Mundt. (Pseud., Louise
   Mühlbach.)

GOLD AND SILVER SEEKING.
Adventures in the Comanche country.   ...Alice Webber.
Around the gold deep : a romance of the Sierras.   ...A. P.
   Reeder.

13

GOLD AND SILVER SEEKING — *Continued.*

Captain Bayley's heir : a tale of the gold fields of California. ...George Alfred Henty.

Digging for gold.   ...Robert Michael Ballantyne.

La fièvre d'or 1860.   ...Gustave Aimard.

'49 : the gold-seekers of the Sierras.   ...Cincinnatus H. Miller. (Pseud., Joaquin Miller.)

Gold : a Dutch-Indian story.   ...Annie Linden.

Golden days of '49.   ...Kirk Munroe.

Golden dream.   ...Robert Michael Ballantyne.   [translation,

Gold seekers : a tale of California   ...Gustave Aimard.   English

Land of gold.   ...George G. Spurr.

Lord John ; or, Search for gold.   ...George Manville Fenn.

Nelly's silver mine.   ...Helen Hunt Jackson.

Off to California.   ...James F. Cobb.

Painted desert.   ...Kirk Munroe.

Silver Cañon.   ...George Manville Fenn.

Silver caves : a mining story.   ...Ernest Ingersoll.

Story of the mine.   ...Charles Howard Shinn.

Twice bought : a tale of the Oregon gold fields.   ...Robert Michael Ballantyne.

Young silver seekers.   ...Samuel W. Cozzens.

See also **MINES AND MINING.**

GOLDSCHMIDT, MADAME.   See **LIND, JENNY.**

GONZALVO DE CORDOVA.   *Lived 1453–1515.*

Gonzalvo de Cordova.   ...Jean P. C. de Florian.

Great Captain.   ...Ulrick R. Burke.

GORDON, CHARLES GEORGE.   *Lived 1833–1885.*

For honor, not honors : being the story of Gordon of Khartoum ...William Gordon Stables.

GORDON, PATRICK.   *Lived 1635–1699.*

Mynheer Joe.   ...St. George Rathborne.

GOVERNESSES.

Curb of honor.   ...Matilda B. Edwards.

Good French governess.   ...Maria Edgeworth.

Janet.   ...Margaret O. W. Oliphant.

John Ward's governess.   ...Annie L. MacGregor.

Lady Betty's governess.   ...Lucy Ellen Guernsey.

Man of to-day.   ...Helen B. Mathers.

Miss Canary.   ...Clara B. Conant.

Not "a fool's errand."   ...Joseph Holt Ingraham.

Only the governess.   ...Rosa N. Carey.

Roy's repentance.   ...Adeline Sergeant.

GRANADA.

Albambra. ...Washington Irving.
Conquest of Granada. ...Washington Irving.
Leila. ...Edward George Earle Lytton Bulwer-Lytton.
Tales from Spanish history. ...Elizabeth J. Barbazoo.
See also **SPANISH HISTORY.**

GREECE.

Description, History, Manners, and Customs.

Amygdal : a tale of the Greek revolution. ...Mrs. Edmonds.
Anastasius. ...Thomas Hope.
Anthea : true history of the war of Greek independence. ...Cecile Cassavetti.
Apelles and his contemporaries. ...Henry Greenough.
Aspasia : a romance of art and love in ancient Hellas. ...Robert Hamerling.
Callias. ...Alfred John Church.
Charikles. ...Wilhelm Adolph Becker.
Charmione : a tale of the great Athenian revolution. ...Edward A. Leatham.
Constantine : a tale of Greece under King Otho. ...George Horton.
Demigod. ...Edward P. Jackson.
Fair Athens. ...Edmonds.
Few days at Athens. ...Frances Wright D'Arusmont.
Fountain of Arethusa. ...Robert Landor.
Glaucia. ...Emma Leslie.
Heroes of ancient Greece : a story of the days of Socrates the Athenian. ...Ellen Palmer.
Hypatia. ...Charles Kingsley.
King of the mountains. ...Edmond About. English translation.
Last Athenian. ...Abraham Viktor Rydberg. English translation.
Pausanias the Spartan. ...Edward George Earle Lytton Bulwer-Lytton.
Pericles. ...Cornwallis.
Philothea. ...Lydia Maria Child.
Prince of Argolis. ...J. Moyr Smith.
Le roi des montagnes. ...Edmond About.
Saint Paul in Greece. ...G. S. Davis.
Den siste Athenaren. ...Abraham Viktor Rydberg.
Stories from the Greek comedians. ...Alfred John Church.
Thoth : a romance. ...Joseph Shield Nicholson.
Three Greek children. ...Alfred John Church.
Voyage du jeune Anacharsis en Grèce, vers le milieu du quatrième siècle avant Jésus Christ. ...J. J. Barthélemy.
Young Anacharsis. ...J. J. Barthélemy. English translation.

GREEK REVOLUTION. See **REVOLUTIONS, Greek.**

GREGORY VII. POPE OF ROME. *Lived 1015–1085.*
Bertha; or, Pope and emperor.   ...William B. MacCabe.
Hildebrand and the excommunicated emperor.   ...Joseph Sortain.
Truce of God : a tale of the eleventh century.   ...George H. Miles.
See also **CATHOLICISM.**

GREY, LADY JANE. *Lived 1537–1554.*
Lady Jane Grey.   ...Thomas Miller.
Tower of London.   ...William Harrison Ainsworth.
See also **ENGLISH HISTORY.**

GUSTAVUS I. (VASA) OF SWEDEN. *Reigned 1523–1560.*
Adventures of Gustavus Vasa.   ...L. S. Griffith.
Gustav Wasa.   ...Louisa M. Gräfin von Robiano.
Gustavus Vasa and his stirring times.   ...Albert Alberg.
Hero of the North.   ...William Diamond.
Princess Vasa.   ...Zacharias Topelius.   English translation.
Prinsessan af Vasa.   ...Zacharias Topelius.
Die Söhne Gustav Wasas.   ...Carl Berkow.
See also **SWEDEN.**

GUSTAVUS II. (ADOLPHUS) OF SWEDEN. *Reigned 1611–1632.*
Blameless knights; or, Lützen and Vendée.   ...Alice H. F. Byng,
    Viscountess Enfield.
Gustaf Adolph.   ...Joseph E. C. Bischoff.
Gustaf och 30 åriga kriget.   ...Zacharias Topelius.
Gustavus Adolphus.   ...Zacharias Topelius.   English translation.
Lion of the North : a tale of the times of Gustavus Adolphus and
    the wars of religion.   ...George Alfred Henty.
Maid of Stralsund : a story of the thirty years' war.   ...John B. de
    Liefde.
Maria Schweidler die Bernsteinhexe.   ...Johann Wilhelm Meimhold.
Maria Schweidler the amber witch.   ...Johann Wilhelm Meimhold.
    English translation.
Memoirs of a cavalier.   ...Daniel Defoe.
My Lady Rotha.   ...Stanley J. Weyman.
Times of Gustaf Adolf.   ...Zacharias Topelius.   English translation.
See also **THIRTY YEARS' WAR.**

GUSTAVUS III. OF SWEDEN. *Reigned 1771–1792.*
Gustav den Tredje och hans hof.   ...Carl A. Kullberg.
Gustavus III.   ...Zacharias Topelius.   English translation.
Gustav IIIs testamente eller 1792 och 1815.   ...Henrik af Trolle.
Seton.   ...Carl S. F. von Zeipel.
Under Gustaf III's första regerings år.   ...Zacharias Topelius.
See also **SWEDEN.**

GUTENBERG, JOHANN. *Lived 1410-1468.*
Gutenberg. ...E. C. Pearson.
Johannes Gutenberg. ...Adolf Stern.
Noble printer and his adopted daughter. ...Anon.

HANDEL, GEORG FRIEDRICH. *Lived 1685-1759.*
Master of the musicians : a story of Handel's days. ...Emma
   Marshall.
Tone-masters. ...Charles Barnard. Vol. 2.

HAYDN, JOSEPH. *Lived 1732-1809.*
Consuelo. ...Amantine L. A. D. Dudevant. (Pseud., George Sand.)
Consuelo. ...Amantine L. A. D. Dudevant. (Pseud., George Sand.)
   English translation.
Seppi. ...Franz Hoffmann.
Tone masters. ...Charles Barnard. Vol. 2.

HEIDELBERG.
Heidelberg. ...George Payne Rainsford James.
Im schloss zu Heidelberg. ...Eva Hartner.
Jetta ; or, Heidelberg under the Romans. ...Adolf Hausrath.
   English translation.
Jetta : historischer Roman aus der Zeit der Völkerwanderung.
   ...Adolf Hausrath.
Die Rose von Heidelberg. ...Louise M. Gräfin von Robiano.

HEINE, HEINRICH. *Lived 1799-1856.*
Heinrich Heine's erste Liebe. ...Katharine Diez.

HENRY, DUKE OF BAVARIA. *Lived 1156-1180.*
Heinrich der Löwe. ...Carl L. Haeberlin.
Heinrich der Löwe. ...Carl C. F. Riedmann.

HENRY I. OF LORRAINE, THIRD DUKE OF GUISE. *Lived 1550-
1588.*
Henry of Guise ; or, States of Blois. ...George Payne Rainsford
   James.

HENRY I. OF ENGLAND. *Reigned 1100-1135.*
Leper-house of Janval. ...Catherine G. F. Gore.
Thomas of Reading ; or, Sixe worthie yeomen of the West.
   ...Thomas Deloney.

HENRY II. OF ENGLAND. *Reigned 1154-1189.*
Betrothed. ...Sir Walter Scott.
Court Life under the Plantagenets. ...Hubert Hall.
Eva. ...Edward Maturin.
Fair Rosamond ; or, Days of Henry II. ...Thomas Miller.

HENRY VIII. OF ENGLAND. *Reigned 1509–1547.*

Agnes Martin; or, Fall of Cardinal Wolsey. ...Anon.

Alice Sherwin: a tale of the days of Sir Thomas More. ...C. J. M.

Anne Boleyn. ...Mrs. K. Thomson.

Anne Boleyn; or, Suppression of the religious houses. ...Anon.

Armourer's prentices. ...Charlotte Mary Yonge.

Bolsover Castle: a tale from Protestant history of the sixteenth century. ...Anon.

Captain Cobbler; or, Lincolnshire rebellion. ...Thomas Cooper.

Cardinal's daughter. ...Robert M. Daniel.

Chained book. ...Emma Leslie.

Chronicles of Camber Castle: a tale of the reformation. ...Anon.

Church and the king. ...Evelyn E. Green.

Cloister and the hearth. ...Charles Reade.

Constable of the Tower. ...William Harrison Ainsworth.

Darnley; or, Field of the cloth of gold. ...George Payne Rainsford James.

England's daybreak: narratives of the reformation. ...E. Bickersteth.

Forest of Arden. ...William Gresley.

Freston Tower: a tale of the times of Wolsey. ...Richard Cobbold.

Friar Hildebrand's cross. ...Margaret Agnes Paul.

Heiress of Ravensby. ...J. L. Watson.

Household of Sir Thomas More. ...Anne Manning.

In Editha's days. ...Mary E. Bamford.

Isoult Barry of Wynscote: a tale of Tudor times. ...Emily Sarah Holt.

King and the cloister. ...Elizabeth M. Stewart.

King Henry and his court. ...Clara M. Mundt. (Pseud., Louise Mühlbach.) English translation.

König Heinrich und sein Hof ....Clara M. Mundt. (Pseud., Louise Mühlbach.)

Lady Rosamond's book. ...Lucy Ellen Guernsey.

Lancashire witches: a romance of Pendle forest. ...William Harrison Ainsworth.

Last abbot of Glastonbury. ...Anon.

Last of the abbots. ...Arthur Brown.

Lettice Eden: a tale of the last days of Henry the Eighth. ...Emily Sarah Holt.

Life and wonderful adventures of "Trotty Testudo." ...Flora F. Wylde.

Lincolnshire tragedy. ...Anne Manning.

Die Lollharden. ...Hans G. Lotz.

Maid and the monk. ...Walter Stanhope.

Margaret Roper. ...Agnes M. Stewart.

HENRY VIII. OF ENGLAND — *Continued.*

Memoirs of Henry VIII. of England. ...Henry W. Herbert. (Pseud., Frank Forester.)

Necromancer. ...George W. M. Reynolds.

New year's day. 1518. ...Elbridge S. Brooks. (In his Storied holidays.)

Noble wife. ...John Saunders.

Passages in the life of the fair gospeller Mistress Anne Askew. ...Anne Manning.

Pendower : a story of Cornwall in the time of Henry the Eighth. ...Marianne Filleul.

Pilgrimage of grace. ...M. Emery.

Pilgrims of Walsingham. ...Agnes and Susana Strickland.

Richard Hunne. ...George E. Sargent.

Stanfield Hall. ...J. Frederick Smith.

Story of John Heywood. ...Charles Bruce.

Tor Hill. ...Horace Smith. (Pseud., Paul Chatfield.)

Tower Hill. ...William Harrison Ainsworth.

Tragic coronation. ...Herbert V. Mills.

True story of Catherine Parr. ...Elsa D. E. Keeling.

Uline's escape ; or, Hid with the nuns. ...Mrs. Alexander S. Orr.

Westminster Abbey. ...Emma Robinson.

When you see me you know me. ...Samuel Rowley.

Windsor Castle. ...William Harrison Ainsworth.

Wooing of Catherine Parr. ...Anon.

HENRY II. OF FRANCE. *Reigned 1547–1559.*

Brigand ; or, Corse de Leon. ...George Payne Rainsford James.

Les comtes de Montgomery : roman historique. ...Lottin de Laval.

Les deux Dianes. ...Alexandre Dumas.

Diane de Poitiers. ...Jean B. H. R. Capefigue.

Diane de Poitiers. ...Maurice A. Maurage.

Helen of Tournon. ...Adélaïde M. E. F. Souza-Botelho. English translation.

Mademoiselle de Tournon. ...Adélaïde M. E. F. Souza-Botelho.

Le page du duc de Savoie. ...Alexandre Dumas.

Page of the Duke of Savoy. ...Alexandre Dumas. English translation.

Princess of Cleves. ...Comtesse de Lafayette. English translation.

La Princesse de Clèves. ...Comtesse de Lafayette.

Two Dianas. ...Alexandre Dumas. English translation.

HENRY III. OF FRANCE. *Reigned 1574–1589.*

Chicot the jester. ...Alexandre Dumas. English translation.

La dame de Monsoreau. ...Alexandre Dumas.

La dame de Saint-Bris ...Alexandre F. Guesdon.

HENRY III. OF FRANCE — *Continued.*

Le duc de Guise. ...Louis P. P. Legay.

Forty-five guardsmen. ...Alexandre Dumas. English translation.

Gaston de Blondville ; or, Court of Henry III. resting in Ardennes. ...Anne Radcliffe.

Henry of Guise ; or, States of Blois. ...George Payne Rainsford James.

Histoire de Marguerite de Valois, reine de Navarre, sœur de François I. ...Charlotte R. de Caumont de La Force.

Marguerite de Valois: a historical romance. ...Alexandre Dumas. English translation.

Une nièce de Balafré : histoire du temps de la Ligue. ...Ernest Faligan.

De l'Orco. ...George Payne Rainsford James.

Owen Tudor. ...Jane Robinson.

Les quarante-cinq. ...Alexandre Dumas.

La reine Margot. ...Alexandre Dumas.

Under the bells. ...Leonard Kip.

HENRY IV. OF FRANCE. *Reigned 1589–1610.*

L'Abbesse de Montmertre. ...Henri Augu.

Heinrich des Vierten Vermählung mit Bertha von Susa. ...Luise Zeller.

Henri Quatre. ...John II. Mancur.

Henry IV. ...Luise Zeller. English translation.

Henry IV. ...John H. Mancur. English translation.

One in a thousand ; or, Days of Henry Quatre. ...George Payne Rainsford James.

Rose d'Albret ; or, Troublous times. ...George Payne Rainsford James.

HENRY IV. OF GERMANY. *Reigned 1056–1106.*

Bertha ; historical romance of the time of Henry IV. of Germany. ...Joseph E. C. Bishoff. English translation.

Bertha ; or, Pope and the emperor. ...William B. MacCabe.

Boy crusaders. ...John George Edgar.

Die Fahrten Thiodolfs des Isländers. ...Friedrich H. C. La Motte Fouqué.

Fighting the Saracens ; or, Boy knight. ...George Alfred Henty.

Florine princess of Burgundy: a tale of the first crusade. ...William B. MacCabe.

Foure prentises of London, with the conquest of Jerusalem. ...Thomas Heywood.

Heroines of the crusades. ...C. A. Bloss.

Hildebrand and the excommunicated emperor. ...Joseph Sortain.

Lombardi alla prima crociata. ...Tommaso Grossi.

HENRY IV. OF GERMANY — *Continued.*

In the brave days of old : the story of the crusades.   ...Henry Frith.

Königin Bertha : historischer Roman.   ...Joseph E. C. Bischoff.

Lady Sybil's choice : a tale of the crusades.   ...Emily Sarah Holt.

Mathilde ; or, Mémoires tirés de l'histoire des croisades.   ...Sophia Risteau Cottin.

Parcival : Roman.   ...Albert E. Brachvogel.

Saracen ; or, Matilda and Malek Adhel : a crusade romance.   ...Sophia Risteau Cottin.   English translation.

Sir Walter's ward : a tale of the crusades.   ...William Everard.

Stories of the crusades.   ...William E. Dutton.

Truce of God : a tale of the eleventh century.   ...George H. Miles.

HENRY THE LION.   See **HENRY DUKE OF BAVARIA**.

HEREDITY.

Blessed Saint Certainty.   ...William Mumford Baker.   (Pseud., George Harrington.)

Captain Macdonald's daughter.   ...Archibald Campbell.

Counterparts.   ...Elizabeth Sheppard.   (Pseud., Beatrice Reynolds.)

Curb of honor.   ...Matilda B. Edwards.

Demigod.   ...Edward P. Jackson.

Elsie Venner : a romance of destiny.   ...Oliver Wendell Holmes.

From one generation to another.   ...H. Seton Merriman.

Heavenly twins.   ...Sarah Grand, pseud.

History of David Grieve.   ...Mary Augusta Ward (Mrs. Humphry Ward).

John Applegate, surgeon.   ...Mary Harriett Norris.

Our manifold nature.   ...Sarah Grand, pseud.

Out of step.   ...Maria Louise Pool.   (Sequel to Two Salomes.)

Pushed by unseen hands.   ...Helen H. Gardener.

Romance of a châlet.   ...Rose Murray Prior Praed (Mrs. Campbell Praed).

Story of an enthusiast.   ...Mrs. C. V. Jamison.

Superfluous woman.   ...Anon.

Two Salomes.   ...Maria Louise Pool.

HEREWARD.

Hereward the Saxon.   ...Charles Knight.

Hereward the Wake.   ...Charles Kingsley.

HERMANN.

Arminius oder Hermann, nebst seiner durchlauchtigsten Thusnelda in einer sinnreichen Staats-Liebes- und Heldengeschichte.   ...Daniel C. von Lohenstein.

Hermann.   ...E. Buerstenbinder.

HEROES AND HEROISM.

Bathilde; ou, L'héroïsme de l'amour. ...François T. B. d'Arnaud.
Crucifixion of Philip Strong. ...Charles Sheldon.
Decatur and Somers. ...Molly Elliot Seawell.
Friedrich de Algeroy, the hero of Camden Plains. ...Giles Gazer, pseud.
Hebrew heroes. ...Charlotte Tucker. (Pseud., A. L. O. E.)
Hero girl and how she became a captain in the army. ...Ellen T. H. Putnam.
Hero in the strife. ...Louisa C. Silke.
Hero of Cowpens. ...Rebecca MacConkey.
Hero of the north. ...Diamond.
Hero tales from American history. ...Henry Cabot Lodge and Theodore Roosevelt.
Heroes of ancient Greece. ...Ellen Palmer.
Heroes of the North. ...F. Scarlett Potter.
Heroes of Young America. ...Robert H. Moncrieff. (Pseud., Ascott R. Hope.)
Heroine of the confederacy. ...Florence J. O'Connor.
Heroines of the crusades. ...Charles A. Bloss.
Die heroischen Epopëen. Gustav Wasa und Columbus. ...F. M. Franzen.
Isabeau's hero: a story of the revolt of the Cevennes. ...Esmé Stuart.
Knight of the nineteenth century. ...Edward Payson Roe.
Poet hero. ...Minny Bothmer.
Triumphs of the cross. ...John Mason Neale. (Pseud., Aurelius Gratianus.)
True hero; or, Story of William Penn. ...William Henry Giles Kingston.
Twin heroes. ...F. A. Reed.

HIGHLANDERS OF SCOTLAND.

Altavona. ...John S. Blackie.
Away on the moorland. ...A. C. Chambers.
Chief of Glen-Orchay. ...William Bennett.
Clan Albin: a national tale. ...Christina Jane Johnstone (Mrs. John Johnstone).
Donald Ross of Heimra. ...William Black.
Duke of Albany's own Highlanders. ...James Grant.
Elizabeth de Bruce. ...Christina Jane Johnstone (Mrs. John Johnstone).
Glenmorwen; or, Child life in the Highlands. ...M. M. B.
Glennair; or, Life in the Highlands. ...Helen Hazlett. (Pseud., H. M. Tatem.)

HIGHLANDERS OF SCOTLAND — *Continued.*
Heather belles.   ...Sigma, pseud.
Highland legends.   ...Sir Thomas D. Lauder.
Highland nurse: a tale.   ...George Douglas Campbell.
Highland rambles and long legends to shorten the way.   ...Sir
Thomas D. Lauder.
In far Lochaber.   ...William Black.
Jessie Cameron: a Highland story.   ...Rachel Butler.
Laura Everingham; or, Highlanders of Glen Ora.   ...James Grant.
Legendary tales of the Highlands.   ...Sir Thomas D. Lauder.
Legends of the Black Watch; or, Forty-second Highlanders.
...James Grant.
Legends of the Isles and Highland gatherings.   ...Charles Mackay.
Macleod of Dare.   ...William Black.
Mr. Pisistratus Brown, M. P., in the Highlands.   ...William Black.
Morag: a tale of Highland life.   ...Mrs. Milne Rae.
Popular tales of the West Highlands.   ...John F. Campbell, ed.
Princess of Thule.   ...William Black.
Raffans folk: a story of a Highland parish.   ...Mary E. Gellie.
Le régiment des géants.   ...Paul Féval.
Reminiscences of a Highland parish.   ...Norman Macleod.
Rob Roy.   ...Sir Walter Scott.
Romance of war; or, Highlanders in Spain.   ...James Grant.
Roua Pass; or, Englishmen in the Highlands.   ...Erick Mackenzie.
Royal Highlanders; or, Black Watch in Egypt.   ...James Grant.
Startling.   ...Norman Macleod.
Stronbuy; or, Hanks of Highland yarn.   ...James C. Lees.
Tales of the Highlands.   ...Sir Thomas D. Lauder.
See also **SCOTLAND.**

HILDEBRAND.   See **GREGORY VII., POPE OF ROME.**

HOFER, ANDREAS.   *Lived 1767–1810.*
Andreas Hofer.   ...Maximilian von Schenkendorf.
Andreas Hofer.   ...Clara M. Mundt.   (Pseud., Louise Mühlbach.)
Andreas Hofer.   ...Clara M. Mundt.   (Pseud., Louise Mühlbach.)
English translation.
At odds.   ...Jemima Montgomery, Baroness von Tautphœus.
1809: historischer Roman.   ...Eduard Breier.
Year nine: a tale of the Tyrol.   ...Anne Manning.
See also **TYROLESE.**

HOGARTH, WILLIAM.   *Lived 1697–1764.*
Wilhelm Hogarth: historischer Roman.   ...Albert E. Brachvogel.
William Hogarth: a romance.   ...Albert E. Brachvogel.   English
translation.

HOME RULE FOR IRELAND. See **POLITICS,** Irish.

HOMILETICS.
Crucifixion of Philip Strong. ...Charles Sheldon.
His majesty myself. ...William Mumford Baker.
Jarousseau, le pasteur du désert. ...Pierre Clément Eugène Pelletan.
Jean Jarousseau, pastor of the desert. ...Pierre Clement Eugene
Pelletan. English translation.
Little minister. ...James Mathew Barrie.
Mark Rutherford, dissenting minister. ...William Hale White.
(Pseud., Mark Rutherford.)
1900: a forecast and a story. ...Marianne Farningham.
Parson's proxy. ...Kate W. Hamilton.
Prophet of the Great Smoky Mountains. ...Mary N. Murfree.
(Pseud., Charles Egbert Craddock.)
Supply at Saint Agatha's. ...Elizabeth Stuart Phelps Ward
(Mrs. Herbert D. Ward).

HOOD, ROBIN. *Lived 1160–?*
Boy foresters: a tale of the days of Robin Hood. ...Anne Bowman.
Famous exploits of Robin Hood. ...Allan Cunningham.
Forest days: a romance of old times. ...George Payne Rainsford
James.
Forest outlaws; or, Saint Hugh and the king. ...Edward Gilliat.
Maid Marian. ...Thomas L. Peacock.
Maid Marian and Robin Hood. ....J. E. Muddock.
Merry adventures of Robin Hood. ...Howard Pyle.
Noble birth and gallant achievements of that remarkable outlaw
Robin Hood. ...Anon.
Le prince des voleurs. ...Alexandre Dumas.
Robin Hood. ...Pierce Egan, Jr.
Royston Gower. ...Thomas Miller.
Sad shepherd. ...Benjamin Jonson.

HUGUENOTS.
Arnold Delahaize; or, Huguenot pastor. ...Francisca I. Ouvry.
Asylum Christi: a story of the dragonnades. ...Edward Gilliat.
Der Aufruhr in den Cevennen. ...Johann Ludwig Tieck.
Blanche Gamond: a heroine of the faith. ...Anon.
Blanche the Huguenot. ...William Anderson.
Bourdaloue and Louis XIV. ...Laurence Louis Félix Bungener.
English translation.
By sword and fire. ...Thomas Archer.
Camisard; or, Protestants of Languedoc. ...Anon.
Chansonnier Huguenot du XVIe siècle. ...Anon.
Château de Louard: a story of the Edict of Nantes. ...Henry C.
Coape.

HUGUENOTS — *Continued.*

HUGUENOTS — *Continued.*
Protestant leader. ...Marie Joseph Eugène Sue. English translation.
Provocations of Madame Palissy. ...Anne Manning.
Refugees : a tale of two continents. ...Arthur Conan Doyle.
Royal hunt : a story of Huguenot emigration. ...Mrs. E. C. Wilson.
Saint Augustine : a story of the Huguenots in America. ...John R. Musick.
Soldiers of the cross. ...Edith S. Floyer.
Sunset in the province. ...Deborah Alcock.
Suzanne de l'Orme. ...H. G.
Through stress and strain. ...Emma Leslie.
Under the red robe. ...Stanley J. Weyman.
Villegagnon : a tale of the Huguenot persecution. ...William Henry Giles Kingston.
See also **PERSECUTIONS** and **SAINT BARTHOLOMEW'S MASSACRE.**

HUMBOLDT, FRIEDRICH HEINRICH ALEXANDRE VON. *Lived 1769–1859.*
Alexandre von Humboldt. ...Heribert Rau.

HUNGARY.
      Description, History, Manners, and Customs.
Abafi. ...Miklós Jósika.
Az utolso Bátori. ...Miklós Jósika.
Csehek Magyarorszagban. ...Miklós Jósika.
Dr. Dumany's wife. ...Mór Jókai. English translation.
Drei Schlosser. ...Eduard Breier.
Egy Magyar nábob. ...Mór Jókai.
Erdély aranykora. ...Mór Jókai.
Falu Jegyzője ...József Etövös.
Forest and foresters. ...Laura Jewry.
Fürst Georg Rákócsy 1 in Reps. ...Theobald Wolf.
Good people of Pawlicz. ...Colomann Mikszath.
Hungarian brothers. ...Anna Maria Porter.
Hungarian castle. ...Julia Pardoe.
Hungarian girl. ...Mariana Tenger.
Hungarian tales. ...Catherine G. F. Gore.
Interpreter : a tale of the Crimean war. ...George John Whyte Melville.
Interrupted wedding. ...Anne Manning.
Irma. ...Charles Vetter Du Lys.
Kárpáthy Zoltán. ...Mór Jókai.
Kossuths Braut. ...Theodor Scheibe.
Der letzte König der Magyaren. ...Leopold Sacher-Masoch.

HUNGARY — *Continued.*
Life's discipline: a tale of the annals of Hungary.   ...Therese
    Albertine Luise von Jacob Robinson.   (Pseud., Talvj.)
Ludwig Kossuth und Clemens Metternich.   ...Sigmund Kolisch.
Modern Midas.   ...Mór Jókai.   English translation.
Pandour and his princess.   ...Anon.   (In Tales from Blackwood.)
Der Pascha von Buda.   ...Johann Heinrich Zschokke.
Politikai divatok.   ...Mór Jókai.
Rab Ráby.   ...Mór Jókai.
Rauschgold.   ...Stephanie Wohl.
Sham gold.   ...Stephanie Wohl.   English translation.
Siege of Buda.   ...Anon.
Story of the faith in Hungary.   ...Anon.   (Sundays at Home, vol. 14.)
Tales and traditions of Hungary.   ...Terencz A. and Terezia
    Pulszky.   English translation.
Village notary.   ...József Eötvös.   English translation.
Zord Idő.   ...Zsigmund Kerény.
Zriny.   ...Theodor Koerner.
Zrinyi a költő.   ...Miklós Jósika.
Zrinyi a költő.   ...Károly Szasz.

HUSS, JOHN.   *Lived 1373-1415.*
Crushed yet conquering: a story of Constance and Bohemia.
    ...Deborah Alcock.
Gisela.   ...M. Lehmann.
Jean Zyska.   ...Amantine L. A. D. Dudevant.   (Pseud., George Sand.)
Johann Ziska.   ...Franz T. Wangenheim.
Ottokar von Falkenburg.   ...L. Lehnert.
Der Rabbi von Leignitz.   ...A. Sammter.
Upálam Jana Husa Čili.   ...Felik Deriége.

HYPNOTISM.
David Elginbrod..   ...George MacDonald.
Herr Paulus, his rise, his greatness, and his fall.   ...Walter Besant.
Hypnotic tales and other tales.   ...James L. Ford.
Modern wizard.   ...Rodrigues Ottolingui.
Parasite.   ...Arthur Conan Doyle.
Queen of Ecuador.   ...R. M. Manley.
Six cent Sam's.   ...Julian Hawthorne.
Soul's underwood.   ...Charles Kelsey Gaines.
Study in hypnotism.   ...Sidney Flower.
Suggestion.   ...Mabel Collins.
Trilby.   ...George Louis Palmella Busson Du Maurier.
Trilby.   ...George Louis Palmella Busson Du Maurier.   English
    translation.
Witch of Prague.   ...Francis Marion Crawford.

ICELAND.

Description, History, Manners, and Customs.

Curate of Steinhollt. ...James Flamank.
Eric Brighteyes. ...H. Rider Haggard.
Grettie the outlaw : a story of Iceland. ....Sabine Baring-Gould.
Han d'Islande. ...Victor Hugo.
Iceland fisherman. ...Louis Marie Julien Viaud. (Pseud., Pierre
  Loti.) English translation.
Jüngling und Mädchen. ...Jóu Thoroddsen.
Lad and lass. ...Jóu Thoroddsen. English translation.
Northern lights. ...Anon.
Off the geysers. ...Charles A. Stephens.
Pêcheur d'Islande. ...Louis Marie Julien Viaud. (Pseud., Pierre
  Loti.)

IDENTITY.

Corsican brothers. ...Alexandre Dumas. English translation.
Doctor Jekyll and Mr. Hyde. ...Robert Louis Stevenson.
Les frères corses. ...Alexandre Dumas.
Guy Tresillian's fate. ...Harriet Lewis.
L'homme au masque de fer. ...Chevalier de Mouhy.
In the day of adversity. ...J. Bloundelle Burton.
Jane Field. ...Mary E. Wilkins.
Literary courtship under the auspices of Pike's Peak. ...Anna
  Fuller.
Miss Mordeck's father. ...Fani Pusey Gooch.
My double and how he undid me. ...Edward Everett Hale.
Prisoner of Zenda, being the history of three months in the life of
  an Englishman. ...Anthony Hope.
Question of identity. ...Louisa P. Dodge.
Ralph Rider of Brent. ...Florence Warden.
Soldier and a gentleman. ...J. M. Cobban.
Track of a storm. ...Owen Hall.
Tragic blunder. ...Mrs. H. Lovett Camoron.
Tresilian Court. ...Harriet Lewis.
Two Salomes. ...Maria Louise Pool.
Unbidden quest. ...E. W. Hornung.
Unlaid ghost : a story in metempsychosis. ...Joseph Vila Prichard.
Wedded to fate. ...Mrs. George Sheldon.

ILLINOIS.

Barriers burned away. ...Edward Payson Roe.
Cliff-dwellers. ...Henry B. Fuller.
Foiled by a lawyer. ...Anon.
George's mother. ...Stephen Crane.

14

ILLINOIS — *Continued.*

Graysons.　...Edward Eggleston.

Hardscrabble; or, Fall of Chicago.　...John Richardson.

Hoosier schoolmaster.　...Edward Eggleston.

Lucky number: a book of stories of the Chicago slums.　...J. K. Friedmann.

McVeys.　...Joseph Kirkland.

Two circuits.　...J. L. Crane.

Wau-nan-gee; or, Massacre at Chicago.　...John Richardson.

With the procession.　...Henry B. Fuller.

Zury, the meanest man in Spring Co.　...Joseph Kirkland.

IMAGINARY LANDS, CITIES, AND INSTITUTIONS.

Account of an extraordinary living hidden city in Central Africa. ...W. H. Middleton.

Around the world in eighty days.　...Jules Verne.　English translation.

Baron Munchausen's narrative of his marvellous travels and campaigns in Russia.　...Rudolf Erich.

Baron Raspe Trump's marvellous underground journey.　...Ingersoll Lockwood.

Battle of Dorking: reminiscences of a volunteer.　...George Tomkyns Chesney.

Chronicles of Count Antonio.　...Anthony Hope.

Cinq semaines en ballon.　...Jules Verne.

Daybreak: a romance of an old world.　...James Cowan.

Le désert de glace.　...Jules Verne.

Field of ice.　...Jules Verne.　English translation.

Finding of Lot's wife.　...Alfred Clark.

Five weeks in a balloon.　...Jules Verne.　English translation.

Flight of a Tartar tribe.　...Thomas De Quincey.

From the earth to the moon and a trip around it.　...Jules Verne. English translation.

Fur country.　...Jules Verne.　English translation.

Gulliver's travels.　...Dean Swift.

Hartmann the anarchist.　...E. Douglas Fawcett.

Hector Servadac.　...Jules Verne.

Hector Servadac.　...Jules Verne.　English translation.

Involuntary voyage.　...Lucien Biart.　English translation.

L'Ile mystérieuse.　...Jules Verne.

Journey in other worlds.　...John Jacob Astor.

Journey to the centre of the earth.　...Jules Verne.　English translation.

Looking backwards.　2000–1887.　...Edward Bellamy.

Man of mark.　...Anthony Hope.

IMAGINARY LANDS, CITIES, AND INSTITUTIONS — *Continued.*
- McVeys. ...Joseph Kirkland.
Mr. East's experience in Mr. Bellamy's world. ...Conrad Wilbrandt.
Mysterious Island. ...Jules Verne. English translation.
New Academe. ...Edward Hartington.
⊢New Eden. ...C. J. Cutcliffe Hyne.
1900: a forecast and a story. ...Marianne Farningham.
Le pays des fourrures. ...Jules Verne.
Purchase of the North Pole. ...Jules Verne. English translation.
Queer race. ...William Westall.
Le rayon vert. ...Jules Verne.
San Salvador. ...Mary A. Tucker.
ʻSenator at sea: a story of mine and thine. ...G. F. Duysters.
Stephen Remarx. ...James Adderley.
De la terre à la lune. ...Jules Verne.
Traveller from Altruria. ...William Dean Howells.
Twenty thousand leagues under the sea. ...Jules Verne. English translation.
Uranie. ...Camille Flammarion.
- Venus and Cupid. ...Anon.
Vingt mille lieues sous les mers. ...Jules Verne.
Visit to the asylum for aged and decayed punsters. ...Oliver Wendell Holmes.
Voyage au centre de la terre. ...Jules Verne.
Voyage autour du monde en quatre-vingts jours. ...Jules Verne.
Voyage involontaire. ...Lucien Biart.
Voyage round the world. ...Jules Verne.

INCAS.
L'Araucan. ...Gustave Aimard.
Coitlan: a tale of the Inca world. ...Anson Uriel Hancock.
Golden magnet, the land of the Incas. ...George Manville Fenn.
Inca queen; or, Lost in Peru. ...J. Evelyn. [montel.
Les Incas, ou la destruction de l'empire du Pérou. ...J. F. Mar-
Last of the Incas: a romance of the Pampas. ...Gustave Aimard. English translation.
Lost Incas. ...P. Ozollo. ＼
Puebla, oder die Franzosen in Mexico. ...H. O. F. Goedsche.
Der Schatz der Ynkas. ...H. O. F. Goedsche.
Der Schatz des Inka. ...Franz Hoffmann. ·
Treasure of the Inca. ...Franz Hoffmann. ...English translation.

INDIA.
Description, History, Manners, and Customs.
Adventures in India. ...William Henry Giles Kingston.
Axbar. ...A. S. Van Limburg-Bronner.

INDIA — *Continued.*

INDIA — *Continued.*

On to the rescue. ...William Gordon Stables.

Pandurang Hari. ...William Brown Hockley.

Phantom 'rickshaw. ....Rudyard Kipling.

Phantom ship. ...Frederick Marryat.

Plain tales from the hills. ...Rudyard Kipling.

Potter's thumb. ...Flora Annie Steel.

Prasanna and Kamini. ...Mrs. Mullens and others.

Queen of the regiment. ...Katharine King.

Ralph Darnell. ...Meadows Taylor.

Randall Davenant : a tale of the Mahrattas. ...Claude Bray.

Redemption of the Brahman. ...Richard Garbe.

Seeta. ...Meadows Taylor.

Simple adventures of a memsahib. ...Sarah Jeannette Duncan.

Soldier born. ...John Percy Groves.

Story of a Dacoit and the Lolapur week. ...G. K. Betham.

Stretton. ...Henry Kingsley.

Surgeon's daughter. ...Sir Walter Scott.

Tera : a Mahratta tale. ...Meadows Taylor.

Through the Sikh war : a tale of the conquest of the Punjaub. ...George Alfred Henty.

Tiger of Mysore : a story of the war with Tippoo Saib. ...George Alfred Henty.

Tippoo Sultaun : a tale of the Mysore war. ...Meadows Taylor.

War and peace. ...Charlotte Tucker. (Pseud., A. L. O. E.)

With Clive in India ; or, Beginnings of an empire. ...George Alfred Henty.

Young Rajah : a story of Indian life and adventure. ...William Henry Giles Kingston.

INDIANA.

From dark to daylight. ...Eunice W. Beecher.

John Thorn's folks. ...Angeline Teal, pseud.

INDIANS OF NORTH AMERICA.

Les anciens Canadiens. ...Philippe Aubert de Gaspé.

L'Araucan. ...Gustave Aimard.

Baby Rue : her friends and her enemies. ...Charlotte M. Clark. (Pseud., Charles M. Clay.)

Big brother. ...George C. Eggleston.

Blackfoot Lodge tales. ...George Bird Grinnell.

Braddock : a story of the French and Indian wars. ...John R. Musick.

Bridge of the gods. ...F. H. Balch.

Camp-fire and wigwam. ...Edward S. Ellis.

Camp-fires of the red men. ...J. R. Orton.

INDIANS OF NORTH AMERICA — *Continued.*

Canadians of old. ...Philippe Aubert de Gaspé. English translation.

Captain Smith and the Princess Pocahontas. ...J. Davis.

Cassique of Kiawah. ...William Gilmore Simms.

Christian Indian. ...Anon.

Crossing the quicksands. ...Samuel W. Cozzens.

Deerslayer. ...James Fenimore Cooper.

Doom of Mamelons : a legend of the Saguenay. ...William H. Murray.

Enchanted moccasins and other legends of the American Indians. ...Cornelius Mathews.

Father Laval; or, Jesuit missionary. ...J. M'Sherry.

First settlers of New England. ...Lydia Maria Child.

Forest tragedy and other tales. ...Sara J. Lippincott. (Pseud., Grace Greenwood.)

Hardscrabble ; or, Fall of Chicago. ...John Richardson.

Hiawatha and other legends of the Red American Indians. ...Cornelius Mathews.

Hidden power. ...T. N. Tibbles.

Hope Leslie ; or, Early times in Massachusetts. ...Catherine M. Sedgwick.

Howling Wolf and his trick pony. ...Elizabeth W. Champney.

Ish-noo-ju-sche ; or, Eagle of the Mohawks. ...J. L. Shecut.

Kabaosa; or, Warriors of the West. ...Anna L. Snelling.

Kin-da-shon's wife : an Alaskan story. ...Mrs. Eugene S. Willard.

King of the Hurons. ...P. H. Meyers.

" Laramie ;" or, Queen of Bedham. ...Charles King.

Last of the Mohicans. ...James Fenimore Cooper.

Little Smoke : a tale of the Sioux. ...William Osborn Stoddard.

Maid of Wyoming. ...J. L. Bowen.

Man who married the moon. ...Charles F. Lummis.

Mary Derwent. ...Ann S. W. Stephens.

Merry-mount : a romance of the Massachusetts colony. ...John Lothrop Motley.

Me-won-i-toc. ...S. Robinson.

Mila ; or, Last wigwam of the Pawnees. ...T. Bonin.

Mohegan maiden. ...J. L. Bowen.

Moravian Indian boy. ...Anon.

Nick of the woods. ...Robert Montgomery Bird.

Oak openings. ...James Fenimore Cooper.

Onaqua. ...Francis C. Sparhawk.

Osceola. ...Mayne Reid.

Osceola; or, Fact and fiction : a tale of the Seminole war. ...Seymour R. Duke.

INDIANS OF NORTH AMERICA — *Continued.*

Out of the past. ...E. Anson More, Jr.

Pathfinder. ...James Fenimore Cooper.

Pawnee stories and folk-lore. ...George Bird Grinnell.

Piokee and her people. ...Theodora R. Jenness.

Pioneers. ...James Fenimore Cooper.

Ploughed under. ...Judge Harsha.

Prairie. ...James Fenimore Cooper.

Prairie Bird. ...C. Murray.

Quabi; or, Virtues of nature. ...Sarah Wentworth A. Morton.

Ramona. ...Helen Hunt Jackson.

Red beauty: a story of the Pawnee trail. ...William Osborn Stoddard.

Red Cloud, the solitary Sioux. ...William F. Butler.

Red man and white. ...Owen Wister.

Redskin and cowboy: a tale of the Western plains. ...George Alfred Henty.

Romance of King Philip's war. ...Fanny B. Workman.

Saratoga: an Indian tale of frontier life: a true story of 1787. ...D. Shepherd.

Scarlet feather; or, Young chief of the Abenaquies. ...Joseph Holt Ingraham.

Senator Intrigue and Inspector Noseby. ...Molly Elliot Seawell.

Shoshonee valley. ...Timothy Flint.

Stories of the prairie. ...James Fenimore Cooper.

Storm mountain. ...Edward S. Ellis.

Story of the Indian. ...George Bird Grinnell.

Tadeuskund, the last king of the Lenape: an historical tale. ...Anon.

Tales of an Indian camp. ...J. A. Jones.

Tales of Indian life. ...Gustave Aimard. English translation.

Tales of the North American Indian. ...Barbara Hawes.

Talking leaves: an Indian story. ...William Osborn Stoddard.

Therese; or, Iroquois maiden. ...O. Bradbury.

Too strange not to be true. ...Georgiana C. L. G. Fullerton.

Totem tales. ...W. S. Phillips.

Two arrows: a story of red and white. ...William Osborn Stoddard.

Wacousta; or, Prophecy. ...W. Richardson.

Wanneta the Sioux. ...Warren K. Moorehead.

Wau-nan-gee; or, Massacre at Chicago. ...John Richardson.

Wa-wa-wanda: a legend of old Orange. ...Anon.

Wept of Wish-ton-wish ...James Fenimore Cooper.

White chief among the red men. ...J. T. Adams.

White Islanders. ...Mary Hartwell Catherwood.

Wyandotte. ...James Fenimore Cooper.

INDIANS OF NORTH AMERICA — *Continued.*
Wyoming.   ...Edward S. Ellis.
Yemasee.   ...William Gilmore Simms.
Young silver-seekers.   ...Samuel W. Cozzens.
Young trail-hunters.   ...Samuel W. Cozzens.
See also **FRENCH AND INDIAN WARS.   1756–1775.**

INSANITY.
Bella; or, Cradle of liberty.   ...Martha E. Berry.
Conspiracy of silence.   ...Mrs. Colmore Dunn.
Doctor Ben.   ...Anon.
Hard cash.   ...Charles Reade.
Lily of Paris.   ...John P. Simpson.
Mad Sir Uchtred of the hills.   ...Samuel Rutherford Crockett.
Man from Nowhere.   ...Flora Haines Loughead.
Maniac of Brussels.   ...F. W. N. Bayley.
Maris Stella.   ...Marie Clothilde Balfour.
Romance of a châlet.   ...Rose Murray Prior Praed (Mrs. Campbell Praed).

INSURRECTIONS, FAMOUS.

England. Ker. 1708.
Mistress Hazelwode.   ...Frederick H. More.

England. Wat Tyler. 1381.
Alice of Fobbing.   ...Anon. (In Tales illustrating Church History.)
Bondman: a story of the times of Wat Tyler.   ...Anon.
Dick Delver: a story of the peasant revolt of the fourteenth century.
    ...Hariette E. Burch.
Idol of the clownes.   ...John Cleaveland.
John Standish; or, Harrowing of London.   ...Edward Gilliat.
Mediation of Ralph Hardelot.   ...W. Minto.
Merry England; or, Nobles and serfs.   ...William Harrison Ains-
    worth. ·
Vox clamantis.   ...John Gower.
Wardship of Steepcoombe.   ...Charlotte Mary Yonge.
Wat Tyler.   ...Pierce Egan, Jr.

Ireland. 1798.
Croppy: a tale of 1798.   ...Michael Benim.
Donal Dun O'Byrne : a tale of the rebellion of '98.   ...Denis Holland.
Family of Glencarra: a tale of the Irish rebellion.   ...Sidney O.
    Moore.
Forge of Clohogue : a story of the rebellion of '98.   ...James Murphy.
Die Heideschenke.   ...Ludwig Storch.
Irish widow's son; or, Pikemen of '98.   ...Con. O'Leary.

INSURRECTIONS, FAMOUS — *Continued.*

Land of the Kelt : a tale of the days of '98.   ...Anon.
Lloyd Pennant : a tale of the West.   ...Ralph Neville.
Michael Dwyer ; or, Insurgent captain of the Wicklow Mountains and reminiscences of '98.   ...John T. Campion.
My Lords of Strogue : a chronicle of Ireland from the convention till the union.   ...Lewis S. Wingfield.
O'Halloran ; or, Insurgent chief.   ...James McHenry.
O'Hara ; or, 1798 : a historical novel.   ...William Hamilton Maxwell.
Olive Lacy : a tale of the Irish rebellion of 1798.   ...Anna Argyle.
O'Mahoney, chief of the Commeraghs : a tale of the rebellion of 1798.   ...David P. Conyngham.
Peggy : a tale of the Irish rebellion.   ...Mary Damant.
Die Rebellen in Irland.   ...Ferdinand G. Kuehne.
Rose Parnell, the flower of Avondale : a tale of the rebellion of '98.   ...David P. Conyngham.

INTEMPERANCE.

Amid the shadows.   ...Mary F. Martin.
Baby Rue, her friends and her enemies.   ...Charlotte M. Clark. (Pseud., Charles M. Clay.)
Barclay's daughter.   ...Jean Kate Ludlum.
Bar-rooms at Brantley.   ...Timothy Shay Arthur.
Devil's dream.   ...Evangeline B. Blanchard.
Digging a grave with a wine-glass.   ..Anna Maria Hall.
Elsie Magoon ; or, Old still-house in the hollow.   ...Francis D. Gage
First glass of wine.   ...Mrs. A. K. Dunning.
John Bremm.   ...Alphonso A. Hopkins.
Lady of quality.   ...Frances Hodgson Burnett.
Lost estate.   ...Mrs. J. P. Ballard.
Man's will.   ...Edgar Fawcett.
Man who became a savage.   ...W. T. Hornaday.
"Nothing but truth" : a picture of the effects of intemperance.   ...William Gilbert.
Our new crusade.   ...Edward Everett Hale.
Outcast of the Islands.   ...Joseph Conrad.
Pitiless passion.   ...Ella MacMahon.
Present problem.   ...Sarah K. Bolton.
Red bridge : a temperance story.   ...Ellen Putnam. (Pseud., Thrace Talmon).
Sinner and saint.   ...Alphonso A. Hopkins.
Temperance stories for the young.   ...Timothy Shay Arthur.
Ten nights in a bar-room.   ...Timothy Shay Arthur.
Willie Welsh.   ...Mrs. M. M. B. Goodwin.
Wormwood : a drama of Paris.   ...Maria Corelli.

IRELAND. *A Selection.*

Description, Manners, and Customs.

SUBJECT INDEX TO FICTION. 219

IRELAND — *Continued.*

Heiress of Kilorgan. ...Mary R. Sadlier (Mrs. James Sadlier).
Hero-tales of Ireland. ...Jeremiah Curtin.
Hogan, M. P. ...May Hartley.
Hon. Miss Ferrard. ...May Hartley.
Man's foes. ...E. H. Strain.
Miss Honoria. ...F. Langbridge.
Hurrish. ...Emily Lawless.
Ierne. ...William Stewart Trench.
Irish idylls. ...Jane Barlow.
Irish stories. ...W. B. Yeats, ed.
Irish widow's son : or, Pikemen of '98. ...Con. O'Leary.
Irrelagh ; or, Last of the chiefs. ...Miss Colthurst.
Ismay's child. ...May Hartley.
John Sherman and Dhoya. ..." Ganconagh," pseud.
Kerrigan's quality. ...Jane Barlow.
Kilgorman. ...Talbot Baines Reed.
Kilgroom : a story of Ireland. ...John A. Stuart.
King and viking. ...P. G. Smyth.
Knight of Gwynne. ...Charles J. Lever.
Knockagow. ...Charles Joseph Kickham.
Lloyd Pennant : a tale of the West. ...Ralph Neville.
Marcella Grace. ...Rosa Mulholland.
Mike. ...Anon.
Molly Bawn. ...Margaret Hungerford.
Moonlight by the Shannon shore. ...Norris Paul.
My Lords of Strogue : a chronicle of Ireland from the convention
    till the union. ...Lewis S. Wingfield.
Mystery of Killard. ...Richard Dowling.
Nevilles of Garretstown : a tale of 1870. ...Anne Marsh.
Old house by the Boyne. ...Mary A. Sadlier (Mrs. James Sadlier).
Olive Lacy : a tale of the Irish rebellion of 1798. ...Anna Argyle.
Parish province. ...E. M. Lynch.
Reddy the Rover. ...William Carleton.
Red Hugh's captivity : Ireland in the reign of Elizabeth. ...Stan-
    dish O'Grady.
Redmond, Count O'Hanlon, the Irish Rapparee. ...William
    Carleton.
Ridgeway : an historical romance of the Fenian invasion of Canada.
    ...Scian Dubh.
Rose de Blaquière ; or, Lake of Killarney. ...Anna Maria Porter.
Ruined race. ...Hester Sigerson.
Sally Cavanagh; or, Untenanted graves. ...Charles Joseph Kickham.
Sarsfield ; or, Last great struggle for Ireland. ...D. P. Conyngham
Shandon bells. ...William Black.

IRELAND — *Continued.*

Shilrich the drummer.   ...Julia Agnes Fraser.
Silk of the kine.   ...L. McManus.
Stella and Vanessa.   ...Lady Duff Gordon.
Tithe-proctor : a novel. Being a tale of the tithe rebellion in Ireland.
   ...William Carleton.
Ulrick the Ready.   ...Standish O'Grady.
Valentine McClutchy, the Irish Agent; or, Chronicles of Castle
   Cumber.   ...William Carleton.
Wearing of the green.   ...Richard Ashe King.
Where the Atlantic meets the land.   ...Caldwell Lipsett.
White boy : a story of Ireland in 1822.   ...Anna Maria Hall.
Wild birds of Killeevy.   ...Rose Mulholland.
Wild Irish boy.   ...Charles Robert Maturin.
Willy Reilly and his dear Colcen bawn.   ...William Carleton.
With Essex in Ireland.   ...Emily Lawless.
See also **IRISH HISTORY**, **POLITICS, Irish**, and **SOCIETY, Irish**.

IRISH HISTORY. *A Selection.*

Fifth to the Sixteenth Centuries.

Eva ; or, Tales of life and death.   ...Edward Maturin.
Geraldine of Desmond ; or, Ireland in the reign of Elizabeth.
   ...Miss Crumpe.
Juverna.   ...Harmer D. Spratt.
Kathleen : a tale of the fifth century.   ...E. A.
Knights of the Pale ; or, Ireland four hundred years ago.   ...C. M.
   O'Keefe.
Last earl of Desmond : a historical romance of 1559–1603.
   ...Charles B. Gibson.
Last king of Ulster.   ...Anon.
Last monarch of Tara : a tale of Ireland in the sixth century.
   ...Eblana, Pseud.
MacCarthy More ; or, Fortunes of an Irish chief in the reign of
   Queen Elizabeth.   ...Mary A. Sadlier (Mrs. James Sadlier).
Old Irish knight : a Milesian tale of the fifth century.   ...Anon.
Pirates' fort : a tale of the sixteenth century.   ...Louise MacNally.
Red Hugh's captivity : Ireland in the reign of Elizabeth.   ...Standish
   O'Grady.
Saint Patrick's cathedral.   ...L. O. Donnell.

Seventeenth Century.

Arrah Neil.   ...George Payne Rainsford James.
Baldeary O'Donnell : a tale of 1690–91.   ...Albert S. G. Canning.
Boyne-water : a tale.   ...John Banim. (Pseud., The O'Hara family.)
Chances of war : an Irish tale.   ...A. Whitelock.

Nineteenth Century.

IRISH HISTORY — *Continued.*

Falcon family; or, Young Ireland. ...Marmion W. Savage.

Golden hills : a tale of the Irish famine. ...Anon.

Grace Cassidy; or, Repealers. ...Marguerite Gardiner.

Ierne. ...William Steuart Trench.

Ireland's dream : a romance of the future. ...Edmund D. Lyon.

John Doe. ...John Banim. (Pseud., The O'Hara family.)

Kellys and O'Kellys. ...Anthony Trollope.

Kilgroom : a story of Ireland. ...John A. Steuart.

Knight of Gwynne : a tale of the time of the union. ...Charles James Lever.

Light and shade. ...Charlotte O'Brian.

My lords of Strogue : a chronicle of Ireland from the convention till the union. ...Lewis S. Wingfield.

O'Donnells of Glen Cottage : a tale of the famine year in Ireland. ...David P. Conyngham.

Old love and the new. ...Maurice Wilton.

O'Sullivan : épisode de la dernière insurrection d'Irlande. ...L. V. Denancé.

Our radicals : a tale of love and politics. ...Frederick G. Burnaby.

Red route; or, Saving a nation. ...William Sime.

Ridgeway : an historical romance of the Fenian invasion of Canada. ...Scian Dubh.

Rody the Rover; or, Ribbonman. ...William Carleton.

Steadfast unto death : a tale of the Irish famine of to-day. ...Mrs. E. M. Berens.

Tithe-proctor : a novel. Being a tale of the tithe rebellion in Ireland. ...William Carleton.

Valentine M'Clutchy, the Irish agent; or, Chronicles of Castle Cumber. ...William Carleton.

Viceroy. ...John Fisher Murray.

Vultures of Erin : a tale of the penal laws. ...N. J. Dunn.

Whiteboy : a story of Ireland in 1822. ...Anna Maria Hall.

Willy Reilly and his dear Coleen bawn. ...William Carleton.

See also **POLITICS, Irish,** and **SOCIETY, Irish.**

ISLE OF MAN.

Captain Davy's honeymoon. ...Thomas Henry Hall Caine.

Deemster : a romance of the Isle of Man. ...Thomas Henry Hall Caine.

Green hills of the sea : a Manx story. ...Hugh C. Davidson.

Manxman. ...Thomas Henry Hall Caine.

Peveril of the Peak. ...Sir Walter Scott.

Shadow-land on Ellan Vannin; or, Folk-tales of the Isle of Man. ...Mrs. J. W. Russell. (Pseud., I. H. Leney.)

ITALIAN HISTORY.

Eighth Century Before Christ.

Numa Pompilius.   ...Jean Pierre Claris de Florian.

First Century Before Christ.

Fawn of Sertorius.   ...Robert Landor.
Gallus: historischer Roman.   ...Wilhelm Adolf Becker.
Gallus; or, Roman scenes in the time of Augustus.   ...Wilhelm
    Adolf Becker.   English translation.

First Century Anno Domini.

Agathocles.   ...Caroline Pichler.
Aurelia; or, Jews of Capernagate.   ...Abel Quinton.
Burning of Rome; or, Story of the days of Nero.   ...Alfred John
    Church.
Dion and Sybils.   ...M. G. Keon.
Gaudentius.   ...Gerald S. Davies.
Gladiators: a tale of Rome and Judæa.   ...George John Whyte
    Melville.
Greek maid at the court of Nero.   ...Franz Hoffmann.   English
    translation.
Helena's household: a tale of Rome in the first century.   ...James
    De Mille.
Julia of Baioe.   ...J. W. Brown.
Lapsed, not lost.   ...Elizabeth Charles.
Last days of Pompeii.   ...Edward George Earle Lytton Bulwer-Lytton.
Light from the catacombs.   ...E. L. N.
Naomi; or, Last days of Jerusalem.   ...Mrs. J. B. Peploe (Webb).
Neither Rome nor Judæa.   ...E. Hoven.
Pomponia; or, Gospel in Cæsar's household.   ...Mrs. J. B. Peploe
    (Webb).
Quo Vadis.   ...Henryk Sienkiewicz.
Quo Vadis: a narrative of the time of Nero.   ...Henryk Sienkiewicz.
    English translation.
Roman traitor.   ...Henry W. Herbert.   (Pseud., Frank Forester.)
Triumphs of the cross.   ...John Mason Neale.   (Pseud., Aurelius
    Gratianus.)
Victory of the vanquished.   ...Elizabeth Charles.

Second Century.

Flavia.   ...Emma Leslie.
Letters from Rome.   ...Eustace Wace.
Roman exile.   ...Guglielmo Gajani.
Three Bernices.   ...Amanda M. Bright.
Valerius: a Roman story.   ...John Gibson Lockhart.

ITALIAN HISTORY — *Continued.*
Third Century.
Æmilius. ...Augustine D. Crake.
Callista : a sketch of the third century. ...John Henry Newman.
Child martyr and early Christians at Rome. ...Anon.
Diotima : eine culturhistorische Novelle aus der Zeit der diocle·
tianischen Verfolgung. ...Wilhelm Tangermann.
Farm of Aptonga : a story for children, of the time of Saint
Cyprian. ...John Mason Neale. (Pseud., Aurelius Gratianus.)
Martyrs of Carthage. ...Mrs. J. B. Peploe (Webb).
Money god ; or, Empire and the papacy. ...Abel Quinton.
Probus ; or, Rome in the third century. ...William Ware.
Theban legion : story of the times of Diocletian. ...William Max-
well Blackburn.
Fourth Century.
Claudius. ...Mrs. R. K. Causton.
Conquering and to conquer : story of Rome in the days of Saint
Jerome. ...Elizabeth Charles.
Constantine. ...Edmund Spencer.
Egyptian wanderers. ...John Mason Neale. (Pseud., Aurelius
Gratianus.)
Evanus : a tale of the days of Constantine the Great. ...Augus-
tine D. Crake.
Jovinian ; or, Early days of papal Rome. ...William Henry Giles
Kingston.
Julian's dream. ...Gerald S. Davies.
Norma. ...Ellen Palmer.
Out of the mouth of the lion. ...Anon.
Parthenia ; or, Last days of Paganism. ...Eliza Buckminster Lee.
Quadratus : a tale of the world in the church. ...Emma Leslie.
Fifth Century.
Alypius of Tagaste : a tale of the early church. ...Mrs. J. B. Pep-
loe (Webb).
Attila. ...George Payne Rainsford James.
Conquering and to conquer : story of Rome in the days of Saint
Jerome. ...Elizabeth Charles. [man.
Fabiola ; or, Church of the Catacombs. ...Cardinal Nicholas Wise-
Quadratus : a tale of the world in the church. ...Emma Leslie.
Sixth Century.
Antonina ; or, Fall of Rome. ...Wilkie Collins.
Bélisaire. ...Jean François Marmontel.
Belisarius. ...Jean François Marmontel. English translation.
Struggle for Rome. ...Felix Dahn.
15

ITALIAN HISTORY — *Continued.*

Seventh Century.

Last Athenian.   ...Abraham Viktor Rydberg.   English translation.
Martyrs of the Catacombs.   ...Anon.
Den siste Athenaren.   ...Abraham Viktor Rydberg.

Twelfth Century.

Sir Guy de Lusignan.   ...E. Cornelia Knight.

Fourteenth Century.

Foster brother.   ...Thornton Hunt.
Marco Visconti.   ...Tommaso Grossi.
Marco Visconti.   ...Tommaso Grossi.   English translation.
Rienzi, the last of the Roman Tribunes.   ...Edward George Earle
    Lytton Bulwer-Lytton.
Valperga.   ...Mary Wollstonecraft Shelley.

Fifteenth Century.

Agnes of Sorrento.   ...Harriet Beecher Stowe.
Ettore Fieramosca; or, Challenge of Barletta.   ...Massimo Tapa-
    relli, Marchese d'Azeglio.   English translation.
Ettore Fieramosca, ossia, La disfida di Barletta.   ...Massimo Tapa-
    relli, Marchese d'Azeglio.
House of Fiesole.   ...Catherine Shaw.
Isabella Orsini.   ...Francesco Domenico Guerrazzi.
Melanthe.   ...R. C. Maberly.
De l'Orco.   ...George Payne Rainsford James.
Romola.   ...Marian Evans Lewes Cross.   (Pseud., George Eliot.)
Villa Verocchio.   ...D. L. MacDonald.

Sixteenth Century.

Beatrice Cenci.   ...Francesco Domenico Guerrazzi.
Cæsar Borgia.   ...E. Robinson.
Caravaggio.   ...Grosse.
Catherine de' Medici.   ...Thomas Adolphus Trollope.
Fast of Saint Magdalen.   ...Anna Maria Porter.
Florence betrayed; or, Last days of the republic.   ...Massimo
    Taparelli, Marchese d'Azeglio.   English translation.
Flower of the Ticino.   ...Mrs. Alexander S. Orr.
Knight of St. John.   ...Anna Maria Porter.
Loyola.   ...Eduard Duller.
Niccolò de Lapi; ovvero I Palleschi e i Piagnoni.   ...Massimo
    Taparelli, Marchese d'Azeglio.
Pupil of Raphael.   ...Anon.
Die Schlacht von Marignano.   ...Karl A. F. von Witzleben.

ITALIAN HISTORY — *Continued.*

Struggle in Ferrara : a tale of the reformation in Italy. ...William Gilbert.

Tasso and Leonora. ...Anne Manning.

Valentius : an historical romance of the sixteenth century in Italy. ...William Waldorf Astor.

Seventeenth Century.

Betrothed. ...Alessandro Manzoni. English translation.

Castle of the three mysteries. ...Anon.

Daughter of Galileo. ...Anne Manning.

Idyl of the Alps. ...Anne Manning.

I promessi sposi. ...Alessandro Manzoni.

Masaniello : a nine days' wonder. ...F. Bayford Harrison.

Masque at Ludlow. ...Anne Manning.

Paul the pope and Paul the friar. ...Thomas Adolphus Trollope.

Nineteenth Century.

Aid-de-camp. ...James Grant.

Babes in the woods. ...James De Mille.

Courtship and a campaign. ...Talmage Dalin, pseud.

Doctor Antonio. ...Giovanni Dominico Ruffini.

Last days of a king. ...Moritz Hartmann. English translation.

Mademoiselle Mori : a tale of modern Rome. ...Margaret Roberts.

Modern society in Rome. ...John B. Beste.

Rule of the monk. ...Garibaldi.

Strife. ...Mrs. E. D. Wallace.

Tolla. ...Edmund About.

True stories from Italian history. ...F. Bayford Harrison.

Vittoria coronna. ...Anon.

Who breaks pays. ...Jenkins.

Wing and wing ; or, Le feu-follet. ...James Fenimore Cooper.

ITALY.

Description, Manners, and Customs.

Antonina ; or, Fall of Rome. ...Wilkie Collins.

Attila. ...George Payne Rainsford James.

La baraonda. ...Girolamo Rovetta.

Betrothed. ...Alessandro Manzoni. English translation.

Bianca. ...William E. Norris.

Bravo. ...James Fenimore Cooper.

Brigand's bride. ...Laurence Oliphant.

Buccaneer. ...Anna Maria Hall.

Burning of Rome ; or, Story of the days of Nero. ...Alfred John Church.

ITALY — *Continued.*
Casa Braccio.   ...Francis Marion Crawford.
Castellamonte.   ...Antonio Gallenga.
Chevalier of Pensieri Vani.   ...Henry B. Fuller.
Children of the king.   ...Francis Marion Crawford.
Conquering and to conquer: story of Rome in the days of Saint
    Jerome.   ...Elizabeth Charles.
Daughter of Galileo.   ...Anne Manning.
Le docteur Pascal.   ...Émile Zola.
Doctor Pascal.   ...Émile Zola.   English translation.
Don Orsino.   ...Francis Marion Crawford.
Dorothy and other Italian stories.   ...Constance Fenimore Woolson.
Ettore Fieramosca; or, Challenge of Barletta.   ...Massimo Tapa-
    relli, Marchese d'Azeglio.   English translation.
Ettore Fieramosca; ossia, La disfida di Barletta.   ...Massimo Tapa-
    relli, Marchese d'Azeglio.
Le fils du Titiens.   ...Alfred de Musset.
Giacomo.   ...W. C. Bamburgh.
Gladiators: a tale of Rome and Judæa.   ...George John Whyte
    Melville.
Goneril.   ...Agnes M. F. Robinson.
Haunted hotel.   ...Wilkie Collins.
Holmby House.   ...George John Whyte Melville.
Honour of Savelli.   ...S. Leavett Yeats.
Improvisatore; or, Life in Italy.   ...Hans Christian Andersen.
    English translation.
Improvisatoren.   ...Hans Christian Andersen.
Iola, the senator's daughter: a story of ancient Rome.   ...Mans-
    field L. Hillhouse.
Italian child life.   ...Marietta Ambrosi.
John Inglesant.   ...Joseph Henry Shorthouse.
Jovinian; or, Early days of papal Rome.   ...William Henry Giles
    Kingston.
Knight of Saint John.   ...Anna Maria Porter.
Lapsed but not lost.   ...Elizabeth Charles.
Legends of Florence.   ...Charles Godfrey Leland.
Life and opinions of Tristram Shandy.   ...Laurence Sterne.
Lion of Saint Mark: a tale of Venice.   ...George Alfred Henty.
Marion Darche.   ...Francis Marion Crawford.
Masque of Ludlow.   ...Anne Manning.
Maudeville.   ...William Godwin.
Mrs. General Talboys.   ...Anthony Trollope.
One that wins: story of a holiday in Italy.   ...Anon.
Ovingdean Grange.   ...William Harrison Ainsworth.
Pictures from Italy.   ...Charles Dickens.

ITALY — *Continued.*

JACOBITES.

JACOBITES — *Continued.*

Old manor house. ...Charlotte Smith.
Ralf Shirlaugh the Lancashire squire. ...Edward Peacock.
Redgauntlet. ...Sir Walter Scott.
Tale of two cities. ...Charles Dickens.
Thorndyke manor : a tale of Jacobite times. ...Mary C. Rowsell.
Waverley. ...Sir Walter Scott.

JAMAICA, WEST INDIES.

Description, History, Manners, and Customs.

Cruise of the Midge. ...Michael Scott.
Marley; or, Life of a planter in Jamaica. ...Anon.
Maroon; or, Planter life in Jamaica. ...Mayne Reid.
Meyrick's promise. ...Edith C. Phillips.
See also **WEST INDIES.**

JAMES I. OF ENGLAND. *Reigned 1603-1625.*

Afloat and ashore with Sir Walter Raleigh. ...Janet Hardy.
Arabella Stuart. ...George Payne Rainsford James.
Caged lion. ...Charlotte Mary Yonge.
Christmas, 1611. Master Sandy's snapdragon. ...Elbridge S. Brooks. (In his Storied holidays.)
Coombe Abbey. ...Selina Bunbury.
Daughter of Tyrconnell. ...Mary A. Sadlier (Mrs. James Sadlier).
Father Darcy : an historical romance. ...Anne Marsh.
For queen and king; or, Royal 'prentice. ...Henry Frith.
Fortunes of Nigel. ...Sir Walter Scott.
Guy Fawkes; or, Gunpowder treason. ...William Harrison Ainsworth.
It might have been; or, Story of the gunpowder plot. ...Emily Sarah Holt.
Langham rebels. ...Lucy Ellen Guernsey.
Masque of Ludlow. ...Anne Manning.
Miser's secret; or, Days of James I. ...Anon.
Saint Clair of the Isles. ...E. Helme.
Shepperton manor : a tale of the times of Bishop Andrewes. ...John Mason Neale. (Pseud., Aurelius Gratianus.)
Shoulder-knot. ...Benjamin F. Tefft.
Sir Ralph Willoughby. ...Sir Samuel Egerton Brydges.
Spanish match; or, Charles Stuart at Madrid. ...William Harrison Ainsworth.
Star-chamber. ...William Harrison Ainsworth.
Sweet mace : a Sussex legend of the iron times. ...George Manville Fenn.
Two penniless princesses. ...Charlotte Mary Yonge.

JAMES I. OF ENGLAND — *Continued.*
Willitoft; or, Days of James I.: a tale. ...Anon.
Vom Hofe Elisabeth und Jakobs. ...Friedrich M. von Bodenstedt.
See also **ENGLISH HISTORY.**

JAMES II. OF ENGLAND. *Reigned 1685–1688.*
Aimée: a tale of the days of James the Second. ...Agnes Giberne.
Arthur Arundel. ...Horace Smith. (Pseud., Paul Chatfield.)
Aubrey Luson; or, Field of Sedgmoor. ...Anon.
Aus dem Lehrjahren eines Strebers: Roman. ...Ebert Carlssen.
Barony. ...Anna Maria Porter.
Captain of the guard. ...James Grant.
Courtship of Morrice Buckler. ...Alfred Edward Woodley Mason.
Danvers papers. ...Charlotte Mary Yonge.
Dorothy Arden. ...J. M. Callwell.
Le duc de Monmouth: nouvelle historique. ...Anon.
Duke of Monmouth. ...Gerald Griffin.
Duke's Winton: a chronicle of Sedgmoor. ...J. R. Henslowe.
Edgar Nelthorpe; or, Fair maids of Taunton. ...Andrew Reed.
Edward Evelyn: a tale of the rebellion. ...Jane Strickland.
Fate: a tale of stirring times. ...George Payne Rainsford James.
Forester: a tale of 1688. ...M. L. Boyle.
For faith and freedom. ...Walter Besant.
Friend or foe: a tale of Sedgmoor. ...Henry C. Adams.
James der Zweite und sein Fall. ...Carl H. Dammas.
James the Second; or, Revolution of 1688. ...William Harrison
    Ainsworth.
John Deane of Nottingham. ...William Henry Giles Kingston.
Lady Betty. ...Christabel Rose Coleridge.
Last of the cavalier. ...Rose Piddington.
Life of Colonel Jack. ...Daniel Defoe. [more.
Lorna Doone: a romance of Exmoor. ...Richard Doddridge Black-
Man's foes. ...E. H. Strain.
Margery's son: a fifteenth century tale of the court of Scotland.
    ...Emily Sarah Holt.
Micah Clarke: his statement as made to his three grandchildren.
    ...Arthur Conan Doyle.
Mistress Beatrice Cope; or, Passages in the life of a Jacobite's
    daughter. ...M. E. Leclerc.
Oak staircase; or, Stories of Lord and Lady Desmond. ...Mary
    and Catherine Lee.
Outlaw. ...Anna Maria Hall.
Owain Goch: a tale of the revolution. ...Thomas Roscoe.
Pagan prince; or, Comical history of the heroick atchievements of
    the Palatine of Eboracum. ...Anon.

JAMES II. OF ENGLAND — *Continued.*
Poynings : a tale of the revolution.   ...Anon.
Reputed changeling; or, Three seventh years two centuries ago.
    ...Charlotte Mary Yonge.
Roger Willoughby ; or, Times of Benbow.   ...William Henry Giles
    Kingston.
Rosamund Fane ; or, Prisoners of Saint James.   ...Mary and
    Catherine Lee.
Scottish cavalier.   ...James Grant.
Spaewife ; or, Queen's secret : a tale of the days of Elizabeth.
    ...John Boyce.
Stronge of Nitherstronge : a tale of Sedgmoor.   ...Edith J. May.
Tales of the Jacobites.   ...Ellen E. Guthrie.
Thorndyke manor : a tale of Jacobite times.   ...Mary C. Rowsell.
Trelawny of Trelawne.   ...Anna E. Bray.
Under the blue flag : a story of Monmouth's rebellion.   ...Mary E.
    Palgrave.
Under which king ? a story of the revolution of 1688.   ...William
    Johnston.
Urith : a tale of Dartmoor.   ...Sabine Baring-Gould.
Winifred ; or, English maiden in the seventeenth century   ...Lucy
    Ellen Guernsey.
See also **ENGLISH HISTORY**.

JAMES V. OF SCOTLAND.   *Reigned 1513–1542.*
Braes of Yarrow.   ...H. Gibbon.
'X jewel : a Scottish romance.   ...Frederick Moncrieff.

JAPAN.
Description, History, Manners, and Customs.
Curse of Koshiu.   ...Lewis S. Wingfield.
Honda the Samurai.   ...William E. Griffis.
Japanese marriage.   ...Douglas Sladen.
Japs at home.   ...Douglas Sladen.
Mito Yashiki : a tale of old Japan.   ...Arthur C. Maclay.
Schoolboy days in Japan.   ...André Laurie.

JERUSALEM.
Barabbas.   ...Marie Corelli.
Broken walls of Jerusalem and the rebuilding of them.   ...Susan
    Warner. (Pseud., Elizabeth Wetherell.)
Crusades.   ...Thomas A. Archer and Charles L. Kingsford.
Doom of the holy city.   ...Lydia Hoyt Farmer.
For the temple : a tale of the fall of Jerusalem.   ...George Alfred
    Henty.
Foure prentises of London.   ...Thomas Heywood.

JERUSALEM — *Continued.*

Last days of Jerusalem. ...Antoinette K. Klitsche de La Grange.

Naomi; or, Last days of Jerusalem. ...Mrs. J. B. Peploe (Webb).

Prince of the House of David; or, Three years in the Holy City: scenes in the life of Jesus. ...Joseph Holt Ingraham.

Story of the other wise man. ...Henry Vandyke.

Zillah. ...Horace Smith. (Pseud., Paul Chatfield.)

See also **JEWS** and **PALESTINE**.

JESUITS.

Ascanio. ...Alexandre Dumas.

Ascanio. ...Alexandre Dumas. English translation.

Christian Vellacott the journalist: a story of royalism, Jesuitism, and republicanism. ...Hugh S. Scott. (Pseud., Merriman H. Seton.)

Fatal revenge; or, Family of Montoris. ...Charles Robert Maturin. (Pseud., Dennis Jasper Murphy.)

Father Laval the Jesuit missionary. ...J. M'Sherry.

Jesuit ring. ...A. A. Hayes.

John Drummond Fraser. ...Philalethes, pseud.

John Inglesant. ...Joseph Henry Shorthouse.

Last look: a tale of the Spanish inquisition. ...William Henry Giles Kingston.

Loyola. ...Eduard Duller.

Princess Viarna; or, Spanish inquisition in the reign of Emperor Charles V. ...T. Pictou.

Rose: a tale of the Spanish inquisition. ...Derwent Trenorne.

Saint Leon: a tale of the sixteenth century. ...Caleb Williams Godwin.

Spanish brothers: a tale of the sixteenth century. ...Deborah Alcock.

Spanish cavalier: a story of Seville. ...Charlotte Tucker. (Pseud., A. L. O. E.)

Westward Ho! or, Voyages and adventures of Sir Amyas Leigh. ...Charles Kingsley.

Word, only a word. ...Georg Moritz Ebers. English translation.

Ein Wort. ...Georg Moritz Ebers.

JEWELS. *A Selection.*

Crown jewels. ...Emma L. Moffett.

Dangerous jewels. ...Mary Bramston.

Danvers jewels. ...Anon.

Diamond coterie. ...Lawrence L. Lynch, pseud.

Diamond hunters of South Africa. ...Alfred W. Drayson.

Doings of Raffles Haw. ...Arthur Conan Doyle.

JEWELS — *Continued.*

Jewel mysteries I have known.   ...Max Pemberton.

Jewel of Ynys Galon.   ...O. Rhoscomyl.

Lacy diamonds.   ...J. G. A. Coulson.

Lady with the rubies.   ...Eugenie John.

Moonstone.   ...Wilkie Collins.

Naulahke : a story of West and East.   ...Rudyard Kipling and
      Charles Wolcott Balestier.

People at Pisgah.   ...Edwin W. Sanborn.

Rajah's sapphire.   ...M. P. Shiel.

Red diamonds.   ...Justin McCarthy.

Sir Charles Danvers.   ...Anon.

Stories from precious stones.   ...Mrs. Goddard Orpen.

Stories in precious stones.   ...Helen Zimmern.

Winifred's jewels.   ...Mary A. Bird.

X jewel : a Scottish romance.   ...Frederick Moncrieff.

JEWS.

Ahasvérus.   ...Edgar Quinet.

Aurelia ; or, Jews of Caperna gate.   ...Abel Quinton.

Aus dem Ghetto.   ...Leopold Kompert.

Belteshazzar : a romance of Babylon.   ...Edward R. Roe.

Ben Hur : a tale of the Christ.   ...Lew Wallace.

Benjamin the Jew of Granada.   ...Edward Maturin.

Broken walls of Jerusalem and the rebuilding of them.   ...Susan
      Warner.

Children of the Ghetto.   ...Isaac Zangwill.

Come forth.   ...Elizabeth Stuart Phelps Ward (Mrs. Herbert D.
      Ward) and Herbert D. Ward.

Daniel Deronda.   ...Marian Evans Lewes Cross. (Pseud., George
      Eliot.)

Earl Hubert's daughter.   ...Emily Sarah Holt.

Edict : a tale of 1492.   ...Grace Aguilar.

Emma de Lissau : a narrative of striking vicissitudes and peculiar
      trials.   ...Amelia Bristow.

Ephraim and Helah : a story of the exodus.   ...Edwin Hodder.

Escape : a tale of 1755.   ...Grace Aguilar.

Exiles in Babylon.   ...Charlotte Tucker. (Pseud., A. L. O. E.)

Fair Jewess.   ...Benjamin Leopold Farjeon.

Gabriel : a tale of the Jews in Prague.   ...Salomon Kohn. English
      translation.

Gawriel : historische Erzählung aus dem dreissigjahrigen Kriege.
      ...Salomon Kohn.

Ghetto tragedies.   ...Isaac Zangwill.

Gloria.   ...Benito Pérez Gladós.

JEWS — *Continued.*

Hammer : a story of Maccabean times. ...Alfred John Church and Richmond Seeley.

Helon's pilgrimage to Jerusalem : a picture of Judaism in the century which preceded the advent of our Saviour. ...Friedrich Abraham Strauss. English translation.

Helon's Wallfahrt nach Jerusalem, hundert neun Jahr vorder Geburt unsers Herrn. ...Friedrich Abraham Strauss.

Home and the synagogue of the modern Jew. ...Anon.

House of Israel. ...Susan Warner.

Idumean. ...J. M. Leavitt.

Jew. ...Carl Spindler.

Jew. ...Joseph Ignatius Kraszewski.

Jew of Verona. ...Antonio Breseiani.

Jews of Barnow. ...Karl Emil Franzos. English translation.

José und Benjamin. ...Adolf Franz Delitzsch.

Joseph and Benjamin. ...Adolf Franz Delitzch. English translation.

Joseph the Jew. ...Mrs. Scott.

Joshua : a story of biblical times. ...Georg Moritz Ebers. English translation.

Josua. ...Georg Moritz Ebers.

Judah's lion. ...Charlotte Elizabeth Tonna. (Pseud., Charlotte Elizabeth.)

Der Jude. ...Carl Spindler.

Die Juden von Barnow. ...Karl Emil Franzos.

Die Juden von Toledo. ...Eduard Jerrmann.

La Judia de Toledo. ...Antonio Mira de Mescua.

Judith Trachtenberg. ...Karl Emil Franzos.

Le Juif errant. ...Marie Joseph Eugène Sue.

Julian. ...William Ware.

King of the Schnorrers. ...Isaac Zangwill.

Kun en Spillemand. ...Hans Christian Andersen.

Leah. ...Mrs. Alexander S. Orr.

Legend of Thomas Didymus the Jewish sceptic. ...James Freeman Clarke.

Lionello. ...Antonio Breseiani.

Mercedes of Castile. ...James Fenimore Cooper.

Miriam Rosenbaum. ...Alfred Edersheim.

Narka the Nihilist. ...Kathleen O'Meara.

Nathan the jew. ...Herr Wagner.

Only a fiddler. ...Hans Christian Andersen. English translation.

Orphans of Lissau and other interesting narratives immediately connected with Jewish customs, domestic and religious. ...Amelia Bristow.

Orthodox. ...Dorothea Gerard.

JEWS — *Continued.*

Patriarchal times.   ...C. M. O'Keefe.

Pillar of fire ; or, Israel in bondage.   ...Joseph Holt Ingraham.

Poet and merchant.   ...Berthold Auerbach.   English translation.

Polish Jew.   ...Émile Erckmann and Alexandre Dumas.   English translation.

Prince of India.   ...Lew Wallace.

Prince of the House of David ; or, Three years in the Holy City: scenes in the life of Jesus.   ...Joseph Holt Ingraham.

Rameses.   ...E. Upham.

Rebel queen.   ...Walter Besant.

Rescued from Egypt.   ...Charlotte Tucker.   (Pseud., A. L. O. E.)

Reuben Sachs.   ...Amy Levy.

Romance of Vienna.   ...Frances Trollope.

Salathiel the immortal.   ...George Croly.

Scapegoat.   ...Hall Caine.

Scenes from the Ghetto.   ...Leopold Kompert.   English translation.

Shepherd of Bethlehem.   ...Charlotte Tucker.   (Pseud., A. L. O. E.)

Strong to suffer.   ...E. Wynne.

Stumbler in wide shoes.   ...Anon.

Tancred.   ...Benjamin D'Israeli.

Triumph over Midian.   ...Charlotte Tucker.   (Pseud., A. L. O. E.)

Uarda.   ...Georg Moritz Ebers.

Vale of cedars.   ...Grace Aguilar.

Wandering Jew.   ...Rudyard Kipling.   (In his Life's handicap.)

Wandering Jew.   ...Marie Joseph Eugène Sue.   English translation.

Women of Israel.   ...Grace Aguilar.

Yehl : a tale of the New York Ghetto.   ...A. Cahan.

Yoke of the Thorah.   ...Henry Harland.   (Pseud., Sidney Luska.)

Zipporah.   ...Mrs. M. E. Bewsher.

See also **BIBLE**, *Special History and Characters ;* **CHRIST**, and **JERUSALEM**.

JOAN OF ARC.   *Lived 1412–1431.*

L'amazone française : Jeanne d'Arc.   ...Maria Thérèse Péroux-d'Abany.

Jeanne d'Arc.   ...Caroline d'Hautefeuille.

Jeanne d'Arc ; ou, L'héroïne de Vaucouleurs.   ...René de Mont-Louis.

Jeanne d'Arc: ou, La France au XVᵉ siècle.   ...Jules Foussettc.

Jehanne la Pucelle.   ...Alexandre Dumas.

Joan of Arc.   ...Anon.

Joan of Arc, the maid of Orleans.   ...Thomas J. Serle.

Joan of Arc, the maid of Orleans.   ...Orlando W. Wight.

Joan of Arc ; or, Story of a noble life.   ...Anon.

Joan the maid, deliverer of France and England.   ...Elizabeth Charles.

JOAN OF ARC — *Continued.*
Die Jungfrau von Orleans. ...Stanislaus S. A. Grabowski.
Maid of Orleans. ...Jane Robinson.
Maid of Orleans and the great war of the English in France.
...William H. D. Adams.
La paysanne de Domremy. ...Eugénie Foa.
Personal recollections of Joan of Arc. ...Samuel Langhorne
Clemens. (Pseud., Mark Twain.)
La pucelle. ...Carl W. T. Frenzel.
Shield of the fleur de lis. ...Constance Goddard Du Bois.

JOHN OF ENGLAND. *Reigned 1199–1216.*
Arthur of Brittany. ...Peter Leicester.
Boy's adventure in the Barons' war. ...John George Edgar.
Cheshire pilgrims. ...Frances M. Wilbraham.
Constable's tower; or, Times of Magna Charta. ...Charlotte Mary
Yonge.
Heir of Rougemain. ...Elizabeth M. Stewart. (In Cloister
legends.)
House of Walderne. ...Augustine D. Crake.
How I won my spurs. ...John George Edgar.
John of England. ...Henry Curling.
John o' London. ...Somerville Gibney.
King John and the brigand's daughter. ...Anon.
Magna Charta stories. ...Arthur Gilman, ed.
Royston Gower. ...Thomas Miller.
Runnymede and Lincoln fair. ...John George Edgar.
Stephen Langton. ...Martin F. Tupper.
Stories of the city of London. ...Mrs. Newton Crosland. (Pseud.,
Camilla Toulmin.)
Vendigaid; or, Blessed one. ...Anon.

JOHN II. OF FRANCE. *Reigned 1350–1364.*
Before the dawn: a story of Paris and the Jacquerie. ...George
Dulac.
Jacquerie. ...George Payne Rainsford James.
La Jacquerie: Scènes féodales, suivies de la famille de La Carvajal.
...Prosper Mérimée.
Le roman du Mont St. Michel. ...E. Meunier.
Vassal: a story of old Normandy. ...Laura Valentine.

JONES, PAUL. *Lived 1747–1792.*
Le capitaine Paul. ...Alexandre Dumas.
Captain Paul. ...Alexandre Dumas. English translation.
John Paul Jones. ...Stanislaus S. A. Grabowski.
Pathfinder. ...James Fenimore Cooper.

JONES, PAUL — *Continued.*
Paul Jones. ...Allan Cunningham.
Paul Jones. ...Theodor Muegge.
Paul Jones, the son of the sea. ...Alexandre Dumas. English translation.
Pilot. ...James Fenimore Cooper.

JOSEPH II. OF GERMANY. *Reigned 1765-1790.*
Aus drei Kaiserzeiten. ...Johann Georg L. Hesekiel.
Der Berggeist im Riesengebirge. ...Franz Isidor Proschko.
Geschichte des weisen Danischmend. ...Christoph M. Wieland.
Der goldne Spiegel; oder, Die Könige von Scheschian. ...Christoph M. Wieland.
Josef Kaiser. ...Eduard Breier.
Joseph II. and his court. ...Clara M. Mundt. (Pseud., Louise Mühlbach.) English translation.
Joseph der Zweite. ...Arthur Korn.
Kaiser Josep II. ...Eduard Ille.
Kaiser Josef II. und das Geheimniss des Freihauses. ...Carl T. Fockt.
Kaiser Josep und der Sekretär. ...Adolf Muetzeburg.
Mit und ohne Vokation. ...Elizabeth von Grotthus.
Die Rosenkreuzer in Wien. ...Eduard Breier.
Wien und Rom. ...Eduard Breier.

JOURNALISM.
Christian Vellacott, the journalist: a story of royalism, Jesuitism, and republicanism. ...Hugh S. Scott. (Pseud., Merriman H. Seton.)
Mahattaners. ...E. S. van Zile.
New Grub street. ...George Gissing.
Shandon bells. ...William Black.
Three fates. ...Francis Marion Crawford.
Under orders : the story of a young reporter. ...Kirk Munroe.
William Blacklock, journalist : a love story of press life. ...T. Banks Maclachlan.

JUNG-STILLING, JOHANN HEINRICH. *Lived 1720-1817.*
Little schoolmaster Mark. ...Joseph Henry Shorthouse.

KANSAS.
Boy settlers. ...Noah Brooks.
For freedom's sake. ...Arthur Paterson.
Les émigrants. ...Elie B. Berthet.
In blue uniform. ...George Israel Putnam.

KAUFFMANN, MARIA ANGELICA. *Lived 1741-1807.*
Miss Angel. ...Anna Isabella Thackeray (Mrs. Richmond Ritchie).

KENTUCKY.

KEPLER, JOHANN. *Lived 1571–1630.*

KNIGHTS AND KING ARTHUR.

KNIGHTS AND KING ARTHUR — *Continued.*

History of Arthur of Little Britain.   ...Anon.

History of King Arthur and the quest of the Holy Grail.   ...Thomas Malory.

Knight of Gwynne.   ...Charles James Lever.

Knight of Mauléon.   ...Alexandre Dumas.   English translation.

Knight of Saint John.   ...Annie M. Porter.

Knight of the white feather.   ...Tasma, pseud.

Knights of the Horseshoe.   ...W. A. Caruthers.

Knights of the lion.   ...Anon.

Knights of the Pale.   ...C. M. O'Keefe.

Knights of the Swan.   ...Comtesse de Genlis.

Knights of the white cross.   ...George Alfred Henty.

Knights ransom.   ...Laura Valentine.

Magic ring.   ...Friedrich H. C. La Motte Fouqué.   English translation.

Monk and knight.   ...Frank W. Gunsaulus.

Otto the knight   ...Alice French.   (Pseud., Octave Thanet.)

Red knight of Germany.   ...Peter Boyle.

Stories of the days of King Arthur.   ...Charles H. Hanson.

Tales of King Arthur and his knights of the Round Table.   ...Mary V. Farrington.

Two knights of Delany castle.   ...Mary M. Sherwood.

Der Zauberring.   ...Friedrich H. C. La Motte Fouqué.

KOERNER, KARL THEODOR.   *Lived 1791–1813.*

Poet hero.   ...Minny Bothmer.

Theodor Körner.   ...Heribert Rau.

LABOR AND CAPITAL, CONFLICT OF.

Abandoned claim.   ...Flora H. Loughhead.

Absentee.   ...Maria Edgeworth.

Adventures of Harry Franco : a tale of the great panic.   ...Charles F. Briggs.

Agitator.   ...Clementina Black.

Angel of the village.   ...L. M. Ohorn.

Ballytubber; or, Scotch settler in Ireland.   ...Anon.

Barnaby Rudge.   ...Charles Dickens.

Better days; or, Millionaire of to-morrow.   ...T. and Anna M. Fitch.

Boycotted.   ...M. Morley.

Boycotted household.   ...Letitia M'Clintock.

Boy's revolt.   ...James Otis.

Bread-winners.   ...John Hay.

Complaining millions of men.   ...Edward Fuller.

Convict No. 25.   ...J. Murphy.

LABOR AND CAPITAL, CONFLICT OF — *Continued.*
Crucifixion of Philip Strong. ...Charles Sheldon.
Darcy and his friends. ...Joseph McKim.
Earl of Effingham. ...Lala McDowell.
Emigrants of Ahadarra. ...William Carleton.
Eustace Marchmont, a friend of the people. ...Evelyn Everett
‎   Green.
Felix Holt, the radical. ...Marion Evans Lewes Cross. (Pseud.,
‎   George Eliot.)
Golden bottle. ...Ignatius Donnelly.
Heart of Erin. ...Elizabeth Casey.
Heart of Mid-Lothian. ...Sir Walter Scott.
Heart of Tipperary. ...W. P. Ryan.
How they lived in Hampton. ...Edward Everett Hale.
Ierne. ...William Stewart.
Jason Edwards. ...Hamlin Garland.
Land claim. ...Francis F. Barritt.
Land leaguers. ...Anthony Trollope.
Live and let live. ...Catherine M. Sedgwick.
Looking backwards. 2000–1887. ...Edward Bellamy.
Looking within. ...J. W. Roberts.
Loyal and lawless. ...Ulrick R. Burke.
Man of mark. ...Anthony Hope.
Man who became a savage. ...W. T. Hornaday.
Manchester strike. ...Harriet Martineau.
March of the strikers. ...J. A. Bevan.
Men born equal. ...Henry Perry Robinson.
Mr. Butler's ward. ...Frances Mabel Robinson.
Mr. East's experiences in Mr. Bellamy's world. ...Conrad Wil-
‎   brandt.
Modern Dædalus. ...Thomas Greer.
Moondyne: a story from the under-world. ...John Boyle O'Reilly.
Murvale Eastman, Christian socialist. ...Albion Winegar Tour-
‎   gée.
My friend the boss. ...J. Stilman Smith.
Norah Moriarty. ...Amos Reade.
Pearl of Lisnadoon. ...Mrs. E. J. Ensell.
Plan of campaign. ...Frances Mabel Robinson.
Priest's blessing. ...Harriett Jay.
Put yourself in his place. ...Charles Reade.
Rival apprentices. ...Anon.
Rose O'Connor. ...Emily Fox.
Scenes in feudal times. ...R. H. Wilmot.
Seed she sowed: a tale of the great dock strike. ...Emma
‎   Leslie.

16

LABOR AND CAPITAL, CONFLICT OF — *Continued.*
Senator at sea: a story of mine and thine. ...G. F. Duysters.
Siege of Bodike. ...Edward Lester. .
Social crime. ...Minnie L. Armstronge and G. N. Sceets.
Strike in the B—— mill. ...Anon.
Tale of two cities. ...Charles Dickens.
Tame surrender. ...Charles King.
Terence M'Gowan, the Irish tenant. ...G. L.Tottenham.
Through the fray. ...George Alfred Henty.
Veiled hand: a novel of the sixties, seventies, and eighties
   ...Frederick Wicks.
See also **WORKING-CLASSES.**

LEGAL STORIES.
Attorney; or, Correspondence of John Quod. ...John T. Irving.
Austin Elliot. ...Henry Kingsley.
Bassett claim. ...Henry R. Elliot.
Bleak House. ...Charles Dickens.
Blind leaders of the blind. ...James R. Cocke.
California sketches. ...Leonard Kip.
Chief justice. ...Karl Emil Franzos.
Courage of her convictions. ...Caroline A. Huling and Therese
   Stewart, M. D.
Dead man's story. ...H. Herman.
Dugdale millions. ...W. G. Hudson. (Pseud., Barclay North.)
Foiled by a lawyer. ...Anon.
Fool of quality. ...Henry Brooke.
Graysons. ...Edward C. Eggleston.
In prison and out. ...Hannah Smith. (Pseud., Hesba Stretton.)
In the grip of the law. ...J. E. Muddock.
John Bodewin's testimony. ...Mary H. Foote.
Judge Lynch. ...George H. Jessop.
Justice in the by-ways. ...Francis C. Adams.
Last sentence. ...Maxwell Gray, pseud.
Leavenworth case. ...Anna Katharine Green.
Legal wreck. ...W. Gillette.
Martins of Cro' Martin. ...Charles James Lever.
Old judge. ...T. C. Haliburton.
Queen against Owen. ...Allen Upward.
Strange schemes of Randolph Mason. ...Melville Davisson Post.
Talis qualis. ...Gerald Griffin.
Vultures of Erin. ...N. J. Dunn.
Ways of the hour. ...James Fenimore Cooper.
Wedderburn's will. ...T. Cobb.
Who poisoned Hetty Duncan? ...J. E. Muddock.

LEGAL STORIES — *Continued.*

Whose fault? ...Ellis J. Davis.

Willy Reilly and his dear Coleen Bawn. ...William Carleton.

See also **DETECTIVE STORIES, CRIMINOLOGY,** and **FEMGE-RICHTE.**

LEGENDS.

American.

Acadian legends. ...E. W. Eaton.

Book of New England legends and folklore. ...Samuel A. Drake.

Chien d'or : a legend of Quebec. ...William Kirby.

Doom of Mamelons. ...W. H. H. Murray.

Forest and shore. ...Charles P. Ilsley.

Harpe's head : a legend of Kentucky. ...James Hall.

Haunted woods. ...Edward E. Ellis.

Hiawatha and other legends of the red American Indians. ...C. Matthews.

La jongleuse. ...H. Raymond Casgrain.

Laconia; or, Legends of the White Mountains. ...J. P. Scribner.

Legend of Christopher Columbus. ...Joanna Baillie.

Legend of the vision of Campbell of Inverawe.

Légendes des plantes et des oiseaux. ...Xavier Marmier.

Legends of " Le Detroit." ...Mary C. W. Hamlin.

Legends of Michigan. ...J. F. Littlejohn.

Legends of Montauk. ...J. A. Ayers.

Legends of New England. ...Julia Gill and Frances Lee.

Legends of New England. ...Nathaniel Hawthorne.

Legends of the middle ages. ...H. A. Guerber.

Legends of the Missouri and the Mississippi. ...M. Hopewell.

Legends of the Province house. ...Nathaniel Hawthorne.

Legends of the West. ...John Hall.

Mellichampe: a legend of the Santee. ...William Gilmore Simms.

Mike Fink. ...Emerson Bennett.

Myth of stone idol. ...W. P. Jones.

Richard Ireton : a legend of the early settlement of New England. ...M. Remick.

Rip van Winkle and the legend of Sleepy Hollow. ...Washington Irving.

Totem tales. ...W. S. Phillips. (Legends of the Indians of the Pacific Northwest.)

Washington and his generals. ...George Lippard.

Wa-wa-wauda : a legend of old Orange. ...Anon.

White chief. ...Mayne Reid.

LEGENDS — *Continued.*

Eastern.

Eastern legendary tales and oriental romances.   ...J. H. Caunter.

English.

Cloister legends.   ...Elizabeth M. Stuart.

Folk-lore of East Yorkshire.   ...J. Nicholson.

Hartland forest.   ...Anna Eliza Bray.

Legendary Yorkshire.   ...F. Ross.

Legend of Montrose.   ...Sir Walter Scott.

Legend of Reading Abbey.   ...Charles Knight.

Legend of Thomas Didymus, the Jewish sceptic.   ...James Free-man Clarke.

Legends for Lionel, in pen and pencil.   ...Walter Crane.

Tales and legends of national origin, or widely current in England from the earliest times.   ...William C. Hazlitt.

Finnish.

Finnish legends for English children.   ...R. Eivind.

French.

Count Monte-Leone.   ...Henri de Saint Georges.

Légendes de Fontainebleau.   ...Julie O. Lavergne.

Légendes de Trianon, Versailles et Saint Germain.   ...Julie O. Lavergne.

Legends of feudal days.   ...Henry W. Herbert. (Pseud., Frank Forester.)

Les mystères du peuple.   ...Marie Joseph Eugène Sue.

Rival races ; or, Sons of Joel.   ...Marie Joseph Eugène Sue. English translation.

German.

Albrecht.   ...Arlo Bates.

Arminius oder Hermann.   ...Daniel C. von Lohenstein.

Astaroth und Mentha.   ...Wilhelm Jensen.

Bissula: historischer Roman aus der Völkerwanderung.   ...Felix Dahn.

Buch der Sachsen.   ...Adolf Boettger.

Charakterbilder aus der Geschichte und Sage.   ...August W. Grube.

Deutsche Heldensagen für Schule und Volk.   ...Wilhelm Waegner.

Eudokia die Kaiserinn.   ...Ida Hahn-Hahn.

Felicitas · historischer Roman aus der Völkerwanderung.   ...Felix Dahn.

German classics from the fourth to the nineteenth centuries.   ...Friedrich Max Müller.

LEGENDS — *Continued.*

Heidenmauer. ...James Fenimore Cooper.
Hermann. ...Elizabeth Buerstersbinder.
Historical tales. ...Charles Morris.
Knights of the Swan. ...Stéphanie Félicité, Comtesse de Genlis.
Lays and legends of Germany. ...William J. Thoms.
Legends of Charlemagne. ...Thomas Bulfinch.
Legends of the Rhine. ...Thomas C. Grattan.
Legends of the Wagner drama. ...Jessie L. Weston.
Magic runes. ...Emma Leslie.
Passe Rose. ...Arthur S. Hardy.
Pépin et Charlemagne. ...Alexandre Dumas.
Popular romances of the middle ages. ...George W. Cox and
Eustace H. Jones.
Preussens Tausend und eine Nacht. ...Bernhard Hesslein.
Romances of the middle ages. ...Wilhelm Waegner. English
translation.
Romances relating to Charlemagne. ...George Ellis.
Die Römer in Deutschland. ...Joseph Marius Babo.
Sämundis Fuhrungen. ...Johann Arnold Kanne.
Scenes and characters of the middle ages. ...Edward L. Cutts.
Stories of old renown. ...Robert Hope Moncrieff. (Pseud., Ascott
R. Hope.)
Vor Zeiten. ...Theodor Storm.

Hungarian.

Drei Schlösser. ...Eduard Breier.
Hungarian tales. ...Catherine G. F. Gore.
A régi jó táblabiró's. ...Mór Jókai.
Tales and traditions of Hungary. ...Ferencz Aurel and Terézia
Pulszky.

Irish.

Emerald gems. ...Anon.
Fairy legends and traditions of the south of Ireland. ...Thomas
Crofton Croker.
Killarney legends. ...Thomas Crofton Croker.
Legendary fictions of the Irish Celts. ...Patrick Kennedy.
Legend of the little weaver. ...Samuel Lover.
Legends and stories of Ireland. ...Samuel Lover.
Legends of the lakes. ...Thomas Crofton Croker.

Japanese.

Old world of Japan : legends of the land of the gods. ..Frank
Rinder.

LEGENDS — *Continued.*

Scotch.

Legendary tales of the Highlands.   ...T. D. Lauder.
Legend of the vision of Campbell of Inverawe.   ...T. D. Lauder.
(In his Legendary tales of the Highlands.)
Legends of the Black-watch.   ...James Grant.

Spanish.

Patrañas.   ...Rachel H. Bush.

LEIBNITZ, GOTTFRIED WILHELM VON.   *Lived 1646–1716.*
Leibnitz und die beiden Kurfürstinnen.   ...Hermann Klencke.

LESSING, GOTTLOB EPHRAIM.   *Lived 1729–1781.*
Charlotte Ackermann: a theatrical romance.   ...Otto Müller.
English translation.
Charlotte Ackermann: ein Hamburger Theater-Roman.   ...Otto Müller.

LINCOLN, ABRAHAM.   *Lived 1809–1865.*
Clarence.   ...Francis Bret Harte.
Graysons.   ...Edward Eggleston.
In the boyhood of Lincoln.   ...Hezekiah Butterworth.
Union: a story of the great rebellion.   ...John R. Musick.

LIND, JENNY.   *Lived 1821–1887.*
Charles Auchester: a memorial. ...Elizabeth S. Sheppard. (Pseud., Beatrice Reynolds.)
(Miss Benette personates Jenny Lind.)

LISZT, FRANZ.   *Lived 1811–1886.*
Lucrezia Floriani Lavinia.   ...Amantine L. A. D. Dudevant.
(Pseud., George Sand.)
(Count Salvator Albani personates Liszt.)

LOLLARDS.   See **WYCLIFFE, JOHN.**

LONDON, ENGLAND.
American girl in London.   ...Sara Jeanette Duncan.
Andrew Golding: a tale of the great plague.   ...Annie E. Keeling.
Brave men of Eyam; or, Tale of the great plague.   ...Edward N. Hoare.
Bride of Bucklersbury: a tale of the Grocers' Company.   ...Elizabeth M. Stewart.
Caleb Field· a tale of the Puritans.   ...Margaret O. W. Oliphant.
Captain Jacques: a romance of the time of the plague.   ...Edward Fitzgibbon.   (Pseud., Somerville Gibney.)
Carved cartoon.   ...Austin Clare.

LONDON, ENGLAND — *Continued.*
Toil and trouble : a story of London life.   ...Mrs. Newton Crosland.
  (Pseud., Camilla Toulmin.)
Tower of London.   ...William Harrison Ainsworth.
Virginians.   ...William Makepeace Thackeray.
Westminster Abbey.   ...Emma Robinson.
When London burned : a story of restoration times and the great
  fire.   ...George Alfred Henty.
See also **ENGLAND** and **ENGLISH HISTORY.**

LOUIS IX. OF FRANCE. *Reigned 1226–1270.*
Boy crusaders : a story of the days of Saint Louis.   ...John George
  Edgar.
Good Saint Louis and his times.   ...Anna Eliza Bray.

LOUIS X. OF FRANCE. *Reigned 1314–1316.*
La tour de Nesle.   ...Frédéric Girard.

LOUIS XI. OF FRANCE. *Reigned 1461–1483.*
Anne of Geierstein.   ...Sir Walter Scott.
Charles le Téméraire.   ...Alexandre Dumas.
Cloister and the hearth.   ...Charles Reade.
Lodging for the night.   ...Robert Louis Stevenson.
Louis de la Trémoille ; ou, Les frères d'armes.   ...Justin J. E. Roy.
Mme. de Montdidier ; ou, La cour de Louis XI.   ...Maria A.
  Barthélemy-Hadot.
Marie de Bourgogne.   ...Mme. de Saint-Venant.
Mary of Burgundy.   ...George Payne Rainsford James.
Notre Dame de Paris.   ...Victor Hugo.   [lation.
Notre Dame ; or, Bellringer of Paris.   ...Victor Hugo. English trans-
Philippe-Monsieur.   ...Charles Buet.
Quentin Durward.   ...Sir Walter Scott.
Stormy life.   ...Georgiana C. L. G. Fullerton.

LOUIS XII. OF FRANCE. *Reigned 1498–1515.*
Faithful, but not famous : a historical tale.   ...Anon.
Jeanne de France.   ...Stéphanie Félicité, Comtesse de Genlis.
Le sire d'Aubigny.   ...Pierre de Lesconvel.
Spotless and fearless : the story of Chevalier Bayard.   ...Saint
  Albin Berville.
Thomassine Spinola.   ...Mme. d'Omatu.
Under Bayard's banner.   ...Henry Frith.

LOUIS XIII. OF FRANCE. *Reigned 1610–1643.*
Antoine de Bonneval : a tale of Paris in the days of St. Vincent de
  Paul.   ...William H. Anderson.
Arthur Blane ; or, Hundred cuirassiers.   ...James Grant.

LOUIS XIII. OF FRANCE — *Continued.*
Les beaux messieurs de Bois-Doré. ...Amantine L. A. D. Dudevant.
    (Pseud., George Sand.)
Black tulip. ...Alexandre Dumas. English translation.
Le capitaine Fracasse. ...Théophile Gautier.
Captain Fracasse. ...Théophile Gautier. English translation.
Cinq-Mars. ...Alfred V., Comte de Vigny. English translation.
Cinq-Mars; ou, Une conjuration sous Louis XIII. ...Alfred V.,
    Comte de Vigny.
Gallant lords of Bois-Doré. ...Amantine L. A. D. Dudevant.
    (Pseud., George Sand.) English translation.
Louise de La Vallière. ...Alexandre Dumas.
Mme. de Lafayette. ...Stéphanie Félicité, Comtesse de Genlis.
Le roi des raffinés sous Louis XIII. ...L. J. Emmanuel Gonzalès.
Shoulder knot. ...Benjamin F. Tefft.
Le siége de La Rochelle. ...Stéphanie Félicité, Comtesse de Genlis.
    English translation.
Three musketeers. ...Alexandre Dumas. English translation.
Les trois mousquetaires. ...Alexandre Dumas.
La tulipe noire. ...Alexandre Dumas.
Twenty years after. ...Alexandre Dumas. English translation.
Vicomte de Bragelonne. ...Alexandre Dumas.
Vingt ans après. ...Alexandre Dumas.

LOUIS XIV. OF FRANCE. *Reigned 1643–1715.*
Alamontade, der Galeerensklave. ...Johann Heinrich Zschokke.
Alamontade, the galley slave. ...Johann Heinrich Zschokke. Eng-
    lish translation.
Arnold Delahaize; or, Huguenot pastor. ...Francisca I. Ouvry.
Artamène. ...Madeleine de Scudéry.
Asylum Christi: a story of the dragonnades. ...Edward Gilliat.
Der Aufruhr in den Cevennen. ...Johann Ludwig Tieck.
Die beiden Condé: historische Novelle. ...Eugen H. von Dedenroth.
Belle-Rose. ...Louis Amédée Eugène Achard.
Blanche Gamond: a heroine of the faith. ...Anon.
Bourdaloue and Louis XIV.; or, Preacher and king. ...Lawrence
    Louis Félix Bungener. English translation.
Brothers: a tale of the Fronde. ...Henry W. Herbert. (Pseud.,
    Frank Forester.)
Camisard; or, Protestants of Languedoc. ...Anon.
Cardillac, the jeweller. ...Ernst Theodor Wilhelm Hoffmann.
Château de Louard: a story of the Edict of Nantes. ...Henry C.
    Coape.
Cross and the crown: a tale of the revocation of the Edict of
    Nantes. ...Deborah Alcock.

LOUIS XIV. OF FRANCE — *Continued.*

Les dernières maîtresses de Louis XIV.   ...Félix de Servan.

Dorothy Arden: a story of England and France two hundred years ago.   ...J. M. Callwell.

Driven into exile: a story of the Huguenots.   ...Charlotte Tucker. (Pseud., A. L. O. E.)

Duchess of Burgundy; or, Scenes in the Court of Louis XIV. ...Anon.

Duchess of La Vallière.   ...Stéphanie Félicité, Comtesse de Genlis. English translation.

Duchess of Mazarin : a tale of the times of Louis XIV.   ...Anon.

Duchesse de La Vallière.   ...Stéphanie Félicité, Comtesse de Genlis.

La Duchesse de Montpensier.   ...Pierre A., Vicomte de Ponson du Terrail.

Die eiserne Maske.   ...Paul V. Wichmann.

Every man his own trumpeter.   .. George Walter Thornbury.

Une fantaisie de Louis XIV.   ...Anne Bignan.

Gabrielle; or, House of Maurèze.   ...Alice M. C. Durand. (Pseud., Henry Gréville.)

Gabrielle; or, Pictures of a reign.   ...Luisa S. Costello.

Das Geheimniss der Frau von Nizza.   ...Emil M. Vacano.

Geneviève ; or, Children of Port Royal.   ...Deborah Alcock.

La guerre des femmes.   ...Alexandre Dumas.

Henri de Rohan ; or, Huguenot refugee.   ...Francisca I. Ouvry.

L'homme au masque de fer.   ...Elizabeth Guénard.

L'homme au masque de fer.   ...Chevalier de Mouhy.

How they kept the faith : a tale of the Huguenots of Languedoc. ...Grace Raymond.

Hubert Montreuil; or, Huguenot and the dragoon.   ...Francisca I. Ouvry.

Huguenot : a tale of the French Protestants.   ...George Payne Rainsford James.

Huguenot exiles ; or, Times of Louis XIV.   ...Eliza Ann Dupuy.

In the days of adversity.   ...J. Bloundelle Burton.

Isabeau's hero: a story of the revolt of the Cevennes.   ...Esmé Stuart.

Jacques Bonneval ; or, Days of the dragonnades.   ...Anne Manning.

Jean Cavalier ; ou, Les fanatiques de Cévennes.   ...Marie Joseph Eugène Sue.

Jeannette : a story of the Huguenots.   ...Frances M. Peard.

Lauzun.   ...Paul E. de Musset.

Légendes de Fontainebleau.   ...Julie O. Lavergne.

Lettre de cachet.   ...Catherine G. F. Gore.   (In her Romances of real life.)

LOUIS XIV. OF FRANCE — *Continued.*

Lord Montagu's page : an historical romance of the seventeenth century. ...George Payne Rainsford James.

Louis ; or, Doomed to the cloister : a tale of religious life in the time of Louis XIV. ...M. J. Hope.

Mme. D'Aubigné la petite créole. ...Eugénie Foa.

Der Mann mit der eisernen Maske. ...E. Koenigs.

Marchioness of Brinvilliers. ...Albert Smith.

Marie's story : a tale of the days of Louis XIV. ...Mary E. Bamford.

La marquise de Courcelles. ...Charles J. B. Jacquot.

Minister of Andouse. ...Heinrich Moewes. English translation.

Les missionnaires bottés de Louis XIV. ...Adolphe Michel.

Nameless nobleman. ...Jane Goodwin Austin. [tion.

Nanon ; or, Women's war. ...Alexandre Dumas. English transla-

Olympe de Clèves. ...Alexandre Dumas.

Der Pfarrer von Andouse. ...Heinrich Moewes.

Pignerol. ...Paul Lacroix.

Protestant leader : a novel. ...Marie Joseph Eugène Sue. English translation.

Refugees : a tale of two continents. ...Arthur Conan Doyle.

Reparation : a story of the reign of Louis XIV. and other tales. ...Georgiana C. L. G. Fullerton.

Royal captives. ...Ann Yearsley.

Royal hunt : a story of Huguenot emigration. ...Mrs. E. C. Wilson.

Saint Eustace ; or, Hundred and one. ...Vane I. Saint John.

Un sermon sous Louis XIV. ...Laurence Louis Félix Bungener.

Sister Louise ; or, Story of a woman's repentance. ...George John Whyte Melville.

Stray pearls. ...Charlotte Mary Yonge.

Twenty years after. ...Alexandre Dumas. English translation.

Vingt ans après. ...Alexandre Dumas.

LOUIS XV. OF FRANCE. *Reigned 1715-1774.*

Ancient régime. ...George Payne Rainsford James.

Beaumarchais. ...Albert E. Brachvogel.

By fire and sword : a story of the Huguenots. ...Thomas Archer.

Les catacombes de Paris. ...Élie B. Berthet.

Cerise : a tale of the last century. ...George John Whyte Melville.

Le chevalier d'Harmental. ...Alexandre Dumas.

La comtesse d'Egmont. ...Sophie Gay.

Conspirators ; or, Chevalier d'Harmental. ...Alexandre Dumas. English translation.

Une fille du régent. ...Alexandre Dumas.

In the shadow of God : sketches of life in France during the eighteenth century. ...Deborah Alcock.

LOUIS XV. OF FRANCE — *Continued.*
Jarousseau, le pasteur du désert.   ...Pierre Clément Eugène Pelletan.
Jean Jacques Rousseau.   ...Levin Schuecking.
Jean Jarousseau, pastor of the desert.   ...Pierre Clément Eugène
   Pelletan.   English translation.
John Law the projector.   ... William Harrison Ainsworth.
Leontine ; or, Court of Louis XV.   ...Catherine C. Maberly.
Louis XV. et le Cardinal de Fleury.   ...Anne Bignan.
Madame de Maintenon, pour servir de suite à l'histoire de la duchesse
   de La Vallière.   ...Stéphanie Félicité, Comtesse de Genlis.
Marquise Jeanne Hyacinthe de Saint Palaye.   ...Joseph Henry Short-
   house.
Les mémoires d'un médecin.   ...Alexandre Dumas.
Memoirs of a physician.   ...Alexandre Dumas.   English translation.
Montesquieu à Marseille.   ...Louis S. Mercier.
Olympe de Clèves.   ...Alexandre Dumas.
Peer's daughter.   ...Rosina W. L. Bulwer-Lytton.
Priest and Huguenot ; or, Persecution in the age of Louis XV.
   ...Laurence Louis Félix Bungener.   English translation.
Regent's daughter.   ...Alexandre Dumas.   English translation.
Ritter oder Dame : historische Novelle.   ...Johann Ferdinand Mar-
   tin Oskar Meding.
Roseville family : an historical tale of the eighteenth century.   ...Mrs.
   Alexander S. Orr.
Trois sermons sous Louis XV.   ...Laurence Louis Félix Bungener.
Watteau.   ...Carl W. T. Frenzel.
See also **FRENCH HISTORY.**

LOUIS XVI. OF FRANCE.   *Reigned 1774–1789.*
Adèle de Sénanges.   ...Adélaïde M. É. F. Souza-Botelho.
Blue ribbons : a tale of the last century.   ...Anna H. Drury.
Comtesse de Fargy.   ...Adélaïde M. É. F. Souza-Botelho.
Countess Eva.   ...Joseph Henry Shorthouse.
Eugène de Rotherlin.   ...Adélaïde M. É. F. Souza-Botelho.
Julian ; or, Close of an era.   ...Laurence Louis Félix Bungener.
   English translation.
Julien ; ou, La fin d'un siècle.   ...Laurence Louis Félix Bungener.
La mort de Louis XVI.   ...Armand R. M. Du Chatellier.
See also **FRENCH HISTORY, MARIA ANTOINETTE,** and **REVO-
   LUTIONS, French.**

LOUIS XVIII. OF FRANCE.   *Reigned 1814–1824.*
Count Monte-Leone : a legend of the Carbonari in France and
   Italy.   ...Henri de Saint Georges.   English translation.
Christopher Sauval ; ou, La société en France sous la restaura-
   tion.   ...Émile de Bonnechose.

LOUIS XVIII. OF FRANCE — *Continued.*
Le démon des Montchevreuil. ...Adolph de Lescure.
Doña Flor. ...Gustave Aimard.
Fudge family in Paris. ...Thomas Moore.
Les quatre âges. ...Charles Pougens.
See also **FRENCH HISTORY.**

LOUIS PHILIPPE. *Reigned 1830–1848.*
Château Morville; or, Life in Touraine. ...Emily Reed.
Los cien mil hijos de San Luis. ...Benito Pérez Galdós.
Les louves de Machecoul. ...Alexandre Dumas.
She-wolves of Machecoul. ...Alexandre Dumas. English translation.
See also **FRENCH HISTORY.**

LOUISIANA.
Autrefois : tales of old New Orleans and elsewhere. ...James A. Harrison.
Balcony stories. ...Grace King.
Bonaventure : a prose pastoral of Acadian Louisiana. ...George Washington Cable.
Doctor Sevier. ...George Washington Cable.
Earl of May-field. ...T. P. May.
Les émigrés français dans la Louisiane. ...Charles Sealsfield (formerly Karl Postl).
Flower of France : a story of old Louisiana. ...Marah Ellis Ryan.
Grandissimes. ...George Washington Cable.
Louisiana folk-tales. ...Alcee Fortier.
Madame Delphine. ...George Washington Cable.
Manhattaner in New Orleans. ...A. Oakey Hall.
Monsieur Motte. ...Grace King.
Old Creole days. ...George Washington Cable.
Panola. ...Sarah A. Dorsey. (Pseud., Filia.)
Philip Nolan's friends. ...Edward Everett Hale.
Planter's daughter. ...Eliza Ann Dupuy.
Strange true stories of Louisiana. ...George Washington Cable.
Tales of a time and place. ...Grace King.
Too strange not to be true. ...Georgiana C. L. G. Fullerton.
Towards the Gulf : a tale of New Orleans. ...Anon.
White rover. ...John H. Robinson.
Wonderful wheel. ...Mary Tracy Earle.

LUTHER, MARTIN. *Lived 1483–1546.*
Allerlei von Luther. ...Joachim L. Haupt.
Blind girl of Wittenberg. ...Carl August Wildenhahn. English translation.
Das Buch vom Doktor Luther. ...Hermann O. Nietschmann.

LUTHER, MARTIN — *Continued.*
Chronicles of the Schönberg-Cotta family.   ...Elizabeth Charles.
Count Arensberg : a tale of the days of Luther.   ...Joseph Sortain.
Fate of Castle Löwengard : a tale of the days of Luther.   ...Esmé Stuart.
Luther and the Cardinal.   ...Hermann O. Nietschmann.   English translation.
Luther.   Historisches Gedicht.   ...Gerhard Friedrich.   (Pseud., Gustav Waller.)
Luther in Rom.   Roman.   ...Levin Schuecking.
Luther in Rome; or, Corradina, the last of the Hohenstaufen. ...Levin Schuecking.   English translation.
Luther und die Seinen.   ...Franz Lubojatzky.
Luther und seine Zeit.   ...Theodor Koenig.
Martin Luther : historischer Roman.   ...Heinrich Eisenlohr.
Martin Luther : kirchengeschichtliches Lebensbild.   ...Carl August Wildenhahn.
Martin Luther und Graf Erbach.   ...Hermann O. Nietschmann.
True as steel.   ...George Walter Thornbury.
Ulric; or, Voices.   ...Theodore S. Fay.
See also **GERMAN HISTORY** and **REFORMATION, Germany.**

MAGIC.   See GHOST STORIES AND THE SUPERNATURAL.

MAGNA CHARTA.   See JOHN OF ENGLAND.

MAINE.
But a Philistine.   ...Virginia F. Townsend.
Gaut Gurley ; or, Trappers of Umbagog.   ...D. P. Thompson.
Pearl of Orr's Island : story of the coast of Maine.   ...Harriet Beecher Stowe.
Politician's daughter.   ...Myra S. Hamlin.
Richard Edney and the governor's family.   ...Sylvester Judd.
Sam Shirk : a tale of the woods of Maine.   ...George H. Devereux.
Tales of the Maine coast.   ...Noah Brooks.

MALAY ARCHIPELAGO.
Outcast of the Islands.   ...Joseph Conrad.

MAN WITH THE IRON MASK.   Titles are with **LOUIS XIV. OF FRANCE.**

MANNERS AND CUSTOMS.   See SOCIETY.

MANX LIFE.
Captain Davy's honeymoon : a Manx yarn.   ...Thomas Henry Hall Caine.
Deemster.   ...Thomas Henry Hall Caine.

MANX LIFE — *Continued.*

Green hills of the sea: a Manx story. ...Hugh C. Davidson.
Manxman. ...Thomas Henry Hall Caine.
Peveril of the Peak. ...Sir Walter Scott.

MARIA ANTOINETTE OF FRANCE. *Reigned 1774–1789.*

Un ami de la reine. ...Paul Gaulot.
Le collier de la reine. ...Alexandre Dumas.
Friend of the queen. ...Paul Gaulot. English trauslation.
Kaiser Joseph und Maria Antoinette. ...Clara M. Mundt. (Pseud.,
Louise Mühlbach.)
Der Leibpage der Maria Antoinette. ...Friedrich W. Bruckbräu.
Marie Antoinette. ...Eugen H. von Dedenroth.
Marie Antoinette. ...Carl L. Haeberlein.
Marie Antoinette. La reine de France décapitée. ...Ernst
Pitavell.
Marie Antoinette und ihr Sohn. ...Clara M. Mundt. (Pseud.,
Louise Mühlbach.)
Marie Antoinette and her son. ...Clara M. Mundt. (Pseud.,
Louise Mühlbach.) English translation.
Queen's necklace. ...Alexandre Dumas. English translation.
Two queens. ...Baron Simolin.
See also **FRENCH HISTORY; LOUIS XVI. OF FRANCE**, and
REVOLUTIONS, French.

MARIA DE MEDICIS. See **LOUIS XIII. OF FRANCE.**

MARIA THERESA OF GERMANY. *Reigned 1717–1780.*

Aus drei Kaiserzeiten. ...Johann Georg L. Hesekiel.
Citizen of Prague. ...Henrietta von Paalzow. English translation.
Elizabeth of Guttenstein. ...Caroline Pichler. English translation.
Elizabeth von Guttenstein. ...Caroline Pichler.
Kaunitz. ...Leopold Sacher-Masoch.
Maria Theresa and her fireman. ...Clara M. Mundt. (Pseud.,
Louise Mühlbach.) English translation.
Maria Theresia und ihr Ofenheizer. ...Clara M. Mundt. (Pseud.,
Louise Mühlbach.)
Maria Theresia und ihre Zeit. ...Eduard Duller.
Maria Theresia und ihre Zeit. ...Franz Lubojatzky.
Thomas Thyrnau. ...Henrietta von Paalzow.
See also **GERMAN HISTORY.**

MARRIAGE.

Amazing marriage. ...George Meredith.
American peeress. ...H. C. Chatfield Taylor.
Arranged marriage. ...Dorothea Gerard.
Artists' wives. ...Alphonse Daudet. English translation.

MARRIAGE — *Continued.*

Opinions of a philosopher. ...Robert Grant.
Our manifold natures. ...Sarah Grand, pseud.
Parting and a meeting. ...William Dean Howells.
Platonic affections. ...J. Smith.
Quick and the dead. ...Amélie Rives Chanler.
Rebel queen. ...Walter Besant.
Redemption of the Brahmin. ...R. Garbe.
Reflections of a married man. ...Robert Grant.
Revolution in Tanner's lane. ...William Hale White. (Pseud.,
    Mark Rutherford.)
Robert Atterbury. ...Jarboe.
Romance of Dollard. ...Mary Hartwell Catherwood.
Rubicon. ...E. F. Benson.
Sale of a soul. ...F. Frankfort Moore.
School for husbands. ...Rosina W. L. Bulwer-Lytton.
Soul of the bishop. ...Henrietta E. V. Stannard. (Pseud., John
    Strange Winter.)
Superfluous woman. ...Emma Brooke.
Strange elopement. ...William Clark Russell.
Things that matter. ...Gribble.
Two French marriages. ...Henrietta C. Jenkins.
Two marriages. ...Dinah Mulock Craik.
Una; or, Early marriages. ...Henriette Bowes.
Unclassed. ...George Gissing.
Wedded to a genius. ...Neil Christison.
Wife or slave? ...Mrs. Albert S. Bradshaw.
Woman who did. ...Grant Allen.
Woman with a future. ...Mrs. Andrew Dean.
Won and not one. ...Emily Lucas Blackall.
Yellow aster. ...Mrs. Mannington Caffyn. (Pseud., Iota.)
See also **RELATION OF THE SEXES**.

MARY OF ENGLAND. *Reigned 1553–1558.*

Cardinal Pole; or, Days of Philip and Mary. ...William Harrison
    Ainsworth.
Constable of the tower. ...William Harrison Ainsworth.
For the Master's sake: a story of the days of Queen Mary. ...Emily
    Sarah Holt.
Life of the dutches of Suffolke. ...Thomas Drue.
Lilian: a tale of three hundred years ago. ...George E. Sargent.
My Lady Nell. ...Emily Weaver.
Protestant: a tale of the reign of Queen Mary. ...Anna Eliza Bray.
Robin Tremayne of Bodmin: a tale of the Marian persecution.
    ...Emily Sarah Holt.

MARY OF ENGLAND — *Continued.*

Royal Merchant; or, Events in the days of Sir Thomas Gresham, during the reigns of Queens Mary and Elizabeth.   ...William Henry Giles Kingston.

Twice crowned: a story of the days of Queen Mary.   ...Harriet B. MacKeever.

Victor: a tale of the great persecution.   ...George G. Perry.

Wetherden Hall: an historical story of the days of Queen Mary.   ...Arthur Brown.

See also **ENGLISH HISTORY.**

MARY QUEEN OF SCOTS. *Lived 1542–1587.*

Abbot.   ...Sir Walter Scott.

Bothwell; or, Days of Mary Queen of Scots.   ...James Grant.

Crichton.   ...William Harrison Ainsworth.

David Rizzio; or, Scenes in Europe during the sixteenth century.   ...William H. Ireland.

Last days of Mary Stuart: a novel.   ...Emily Finch.

Maria Stuart in Scotland.   ...Bjørnstjerne Bjørnson.

Maria Stuart: historischer Roman.   ...Eugen Hermann von Dedenroth.

Mary of Lorraine: an historical romance.   ...James Grant.

Mary Queen of Scots.   ...John Galt.

Mary Stuart, Queen of Scots.   ...George M. Reynolds.

Monastery.   ...Sir Walter Scott.

Neil Willox: a story of Edinburgh in the days of Queen Marie.   ...J. Maclaren.

Norman Leslie.   ...Theodore S. Fay.

Queen's Maries: a romance of Holyrood.   ...George John Whyte Melville.

Recess; or, Tale of other times.   ...Sophia Lee.

Southenan.   ...John Galt.

Unknown to history: a story of the captivity of Mary of Scotland.   ...Charlotte Mary Yonge.

See also **ELIZABETH OF ENGLAND, ENGLISH HISTORY,** and **SCOTLAND.**

MARYLAND.

Clinton Bradshaw.   ...F. W. Thomas.

Days of Mackemie; or, Vine planted.   ...L. P. Bowen.

Doings in Maryland; or, Matilda Douglass.   ...Anon.

Entailed hat.   ...George A. Townsend.   (Pseud., Galt.)

Die Grundung von Maryland.   ...Johann Heinrich D. Zschokke.

Osborne of Arrochar.   ...Amanda M. Douglas.

Rob of the bowl.   ...J. P. Kennedy.   (Pseud., Solomon Secondthoughts.)

MARYLAND — *Continued.*
Pilate and Herod. ...H. Stanley.
See also **UNITED STATES HISTORY**.

MASSACHUSETTS.
Adventures of Ann: stories of colonial times. ...Mary E. Wilkins.
Agnes Surriage. ...Edwin L. Bynner.
Alban: a tale of the New World. ...Jedediah V. Huntington.
Bay-path: a tale of New England colonial life. ...Josiah Gilbert Holland.
Book of New England legends and folklore in prose and poetry. ...Samuel Adams Drake.
Brampton sketches. ...Mary B. Claflin.
Captain Kyd; or, Wizard of the sea. ...Joseph Holt Ingraham.
Captain Nelson: romance of colonial days. ...Samuel Adams Drake.
Children of old Park's tavern: a story of the South shore. ...Frances A. Humphrey.
Daughter of the Puritans. ...Anna B. Bensel. (In New England Magazine, 1886, p. 452.)
Elizabeth: a romance of colonial days. ...F. C. Sparhawk. (In New England Magazine, 1884–85.)
Fair Puritan: an historical romance of the days of witchcraft. ...Henry W. Herbert. (Pseud., Frank Forester.)
Gaut Gurley; or, Trappers of Umbagog. ...D. P. Thompson.
Heroes of America. ...Robert Hope Moncrieff. (Pseud., Ascott R. Hope.)
Hope Leslie; or, Early times in Massachusetts. ...Catherine M. Sedgwick.
Justice Warren's daughter. ...Olive M. Birrell.
Legends of New England. ...Julia Gill and Frances Lee.
Legends of New England. ...Nathaniel Hawthorne.
Legends of the Province house. ...Nathaniel Hawthorne.
Lionel Lincoln; or, Leaguer of Boston. ...James Fenimore Cooper.
Morton's hope; or, Memoirs of a provincial. ...John Lothrop Motley.
Naomi; or, Boston two hundred years ago. ...Eliza Buckminster Lee.
Nix's mate. ...R. Dawes.
Old Boston. ...A. De Grasse Stevens.
Oldtown folks. ...Harriet Beecher Stowe.
Pemaquid: a story of old times in New England. ..E. Prentiss.
Penelope's suitors. ...Edwin L. Bynner.
Pilgrims and Puritans. ...Nina Moore Tiffany.
Pilgrims: a story of Massachusetts. ...John R. Musick.
Plymouth and the Pilgrims. ...J. Banvard.

MASSACHUSETTS — *Continued.*

Priscilla; or, Trials for the truth: historical tale of the Puritans and Baptists.   ...J. Banvard.

Puritan and her daughter.   ...James K. Paulding.

Puritan lover.   ...Laura C. S. Fessenden.

Quabbin: the story of a small town, with outlooks upon Puritan life.   ...Francis H. Underwood.

Rebels; or, Boston before the revolution.   ...Lydia Maria Child.

Reuben and Rachel.   ...Susanna Rowson.

Richard Ireton: a legend of the early settlement of New England.   ...M. Remick.

Scarlet letter.   ...Nathaniel Hawthorne.

Shawmut.   ...C. K. True.

Spectre of the forest.   ...J. MacHenry.

Tales of the Puritans.   ...Delia Bacon.

Three generations.   ...Sarah A. Emery.

"Trustum" and his grandchildren.   ...Harriet B. Worron.

Twice told tales.   ...Nathaniel Hawthorne.

White chief among the red men.   ...J. T. Adams.

Witch of Salem.   ...John R. Musick.

Woman of Shawmut: a romance of colonial times.   ...Edmund J. Carpenter.

See also **UNITED STATES HISTORY**.

MAXIMILIAN OF MEXICO (FERDINAND MAXIMILIAN JOSEPH). *Lived 1832–1867.*

Les coupeurs de routes.   ...Gustave Aimard.

Crown jewels.   ...Emma L. Moffett.

Mexican; or, Love and land.   ...J. M. Dagnall.

See also **MEXICO**.

MAZARIN, JULES. *Lived 1602–1661.*

Duchess of Mazarin.   ...Anon.

La guerre des femmes.   ...Alexandre Dumas.

War of women.   ...Alexandre Dumas.   English translation.

MEDICAL NOVELS. *A Selection.*

Andrew Golding: a tale of the great plague.   ..Annie E. Keeling.

Arthur Mervyn.   ...Charles Brockdon Brown.

Bella; or, Cradle of liberty.   ...Marthy E. Berry.

Betrothed lovers.   ...Alessandro Manzoni.   English translation.

Black prophet: a tale of the Irish famine.   ...William Carleton.

Brave men of Eyam.   ...Edward N. Hoare.

Caleb Field: a tale of the Puritans.   ...Margaret O. W. Oliphant.

Captain Jacques.   ...Edward Fitzgibbon.   (Pseud., Somerville Gibney.)

MEDICAL NOVELS — *Continued.*
Carved cartoon. ...Austin Clare.
Cherry and violet: a tale of the great plague. ...Anne Manning.
Conspiracy of silence. ...Mrs. Colmore Dunn.
Country doctor. ...Honoré de Balzac. English translation.
Country doctor. ...Sarah Orne Jewett.
Demigod. ...Edward P. Jackson.
Le docteur Pascal. ...Émile Zola.
Doctor Ben. ...Anon.
Doctor Breen's practice. ...William Dean Howells.
Doctor Janet of Harley Street. ...Dr. Arabella Kenealy.
Doctor Lamar. ...Elizabeth Train.
Doctor Mirabel's theory. ...Rose George Dering.
Doctor of the old school. ...John Maclaren Watson. (Pseud., Ian
     Maclaren.)
Doctor Pascal. ...Émile Zola. English translation.
Doctor Plassid's patients. ...Una L. Bailey.
Doctor Sevier. ...George Washington Cable.
Elsie Venner: a romance of destiny. ...Oliver Wendell Holmes.
Gladys. ...Mary Greenleaf Darling. (Yellow fever plague.)
"God's providence" house. ...Isabel Banks.
Hard cash. ...Charles Reade.
Helen Brent, M. D.: a social study. ...Anon.
History of the plague in London, 1665. ...Daniel Defoe.
I promessi sposi. ...Alessandro Manzoni.
Island of Doctor Moreau. ...Henry G. Wells.
Jennett Cragg, the Quakeress. ...Maria Wright.
John Applegate, surgeon. ...Mary Harriett Norris.
Life and adventures of an Arkansas doctor. ...M. Lafayette Byrn.
Lily of Paris. ...John P. Simpson.
Lourdes. ...Émile Zola.
Lourdes. ...Émile Zola. English translation.
Mad Sir Uchtred of the hills. ...Samuel Rutherford Crockett.
Man from Nowhere. ...Flora Haines Longhead.
Maniac of Brussels. ...F. W. N. Bailey.
Maris Stella. ...Marie Clothide Balfour.
Mary Bunyan, the dreamer's blind daughter. ...Sallie Rochester
     Ford.
Le médecin de Campagne. ...Honoré de Balzac.
Medicine lady. ...Elizabeth T. Meade.
Les mémoires d'un médecin. ...Alexandre Dumas.
Memoirs of a physician. ...Alexandre Dumas. English transla-
     tion.
Mike: a tale of the great famine. ...Edward N. Hoare.
Mona Maclenn, medical student. ...Graham Travers, pseud.

MEDICAL NOVELS — *Continued.*

Night-lights: shadows from a doctor's reading-lamp.   ...Arthur B. Frost.

Old St. Paul's: a tale of the plague.   ...William Harrison Ains-worth.

Oliver Wyndham: a tale of the great plague.   ...Mrs. J. B. Peploe (Webb).

Passages from the diary of a late physician.   ...Samuel Warren.

Portfolio of a Southern medical student.   ...George M. Wharton.

Ralph Redman's atonement.   ...Herbert V. Mills.

Romance of a chalet.   ...Rose Murray Prior Praed (Mrs. Campbell Praed).

Rosemary: a tale of the fire in London.   ...Georgiana C. L. G. Fullerton.

Round the red lamp: being facts and fancies of medical life. ...Arthur Conan Doyle.

Saint Dunstan's clock.   ...E. Ward.

Scenes in the practice of a New York surgeon.   ...Edward H. Dixon.

Some passages in the practice of Doctor Martha Scarborough. ...Helen Campbell. (Pseud., Wheaton Campbell.)

Study in scarlet.   ...Arthur Conan Doyle.

Suicide club.   ...Robert Louis Stevenson.

Surgeon's daughter.   ...Sir Walter Scott.

Tales of a physician.   ...W. H. Harrison.

Underside of things.   ...Lilian Bell.

Veiled doctor.   ...V. A. J. Davis.

MENDELSSOHN, FELIX BARTHOLDY. *Lived 1809-1847.*

Charles Auchester: a memorial.   ...Elizabeth Sheppard. (Pseud., Beatrice Reynolds.) (Seraphael personates Mendelssohn.)

Tone masters.   ...Charles Barnard. Vol. 1.

METHODISM.

Bernicia.   ..Amelia Edith Barr.

Damnation of Theron Ware.   ...Harold Frederic.

De Vane: a story of plebeians and patricians.   ...Henry W. Hil-liard.

Diary of Mrs. Kitty Trevelyan.   ...Elizabeth Charles.

Expedition of Humphrey Clinker.   ...Tobias George Smollett.

Hallam succession.   ...Amelia Edith Barr.

Methodist.   ...Miriam Fletcher.

Nestleton Magna: a story of Yorkshire methodism.   ...James J. Wray.

Spiritual Don Quixote.   ...Richard Graves.

Two circuits.   ...J. L. Crane.

METHODISM — *Continued.*
Vicar of Wakefield. ...Oliver Goldsmith.
Wreckers and methodists. ...H. D. Lowry.

MEXICAN WAR WITH THE UNITED STATES OF 1845.
Archbishop. ...O. S. Beliisle.
Aurifodina. ...G. W. Peck.
Bernard Lisle: an historical romance embracing the period of the
    Texan revolution and the Mexican war. ...Jeremiah Clemens.
Biglow papers. ...James Russell Lowell. (First series.)
Cabin book; or, Scenes and sketches of the late American and
    Mexican war. ...Charles Sealsfield (formerly Karl Postl).
    English translation.
Das Cajütenbuch ; oder, Nationale Charakteristiken. ...Charles
    Sealsfield (formerly Karl Postl).
House of Yorke. ...Mary A. Tincker.
Humbled pride : a story of the Mexican war. ...John R. Musick.
Inez : a tale of the Alamo. ...Augusta J. Wilson (Mrs. Evans).
Life and death of Sam in Virginia. ...Anon.
Lone star of Texas. ...J. W. Dallam.
Mercedes : a tale of the Mexican war. ...Frederick Charles Las-
    celles Wraxall.
Rangers and regulators : a tale of the republic of Texas. ...Alfred
    W. Arrington.
Remember the Alamo. ...Amelia Edith Barr.
Stanhope Burleigh. ...Charles Edward Lester
Talbot and Vernon. ...McConnell.
Vidette : a tale of the Mexican war. ...N. M. Curtis.
White scalper. ...Gustave Aimard. English translation.
Wild life ; or, Adventures of the frontier : a tale of the early days
    of the Texan republic. ...Thomas M. Reid.

MEXICO.
Description, History, Manners, and Customs.

Abdalla the Moor and the Spanish knight. ...Robert Montgomery
    Bird.
Adventures of a young naturalist. ...Lucien Biart. English trans-
    lation.
Aventures d'un jeune naturaliste. ...Lucien Biart.
Aztec treasure house. ...Thomas A. Janvier.
Benito Vazquez. ...Lucien Biart.
By right of conquest. ...George Alfred Henty.
Calavar ; or, Knight of the conquest. ...Robert Montgomery Bird.
Chinampa ; or, Island home. ...Anon.
La conjuracion de Méjico, ó los hijos de Hernan Cortés. ...P. de
    la Escosura.

MEXICO — *Continued.*

Cortés ; or, Fall of Mexico. ...Robert Montgomery Bird.

Costel l'Indien ; ou, Le dragon de la Reine. ...E. L. G. de F. de Bellemare.

Les coupeurs de routes. ...Gustave Aimard.

Crown jewels. ...Emma L. Moffett.

Dolores. ...Gertrúdis Gomez Avellaneda.

Fair god ; or, Last of the 'tzins': a tale of the conquest of Mexico. ...Lew Wallace.

Francis Berrian, the Mexican patriot. ...T. Flint.

Guatimozin, ultimo emperador de Méjico. ...Gertrúdis Gomez de Avellaneda.

Heart of the world. ...H. Rider Haggard.

Hernan Cortez en Tabasco. ...Fermin del Rey.

Higher law. ...Edward Maitland. (Pseud., Herbert Ainslie.)

Historischer Roman. ...Costez Fernando.

Isidra. ...Willis Steell.

Izram. ...Charlotte Elizabeth Tonna. (Pseud., Charlotte Elizabeth.)

Malmiztic the Toltec and the cavaliers of the cross. ...W. W Fosdick.

Mexican ; or, Love and land. ...J. M. Dagnall.

Mexican prince. ...Anon.

Montezuma, the last of the Aztecs. ...E. Maturin.

Montezuma the serf. ...Joseph Henry Ingraham.

Montezuma's daughter. ...H. Rider Haggard.

El nigromántico Mejicano. ...I. M. Pusalgas.

Les nuits Mexicaines. ...Gustave Aimard.

Picture of Las Cruces. ...Frances C. Fisher Tiernan.

El Pleito de Hernan Cortés con Pánfilo de Narvaez. ...J. de Cañizares.

Puepla, oder, Die Franzosen in Mexico. ...H. O. F. Goedsche.

Queen of the lakes. ...Mayne Reid.

Rebel chief: a tale of Guerilla life. ...Gustave Aimard. English translation.

Red track: a story of social life in Mexico. ...Gustave Aimard. English translation.

Scènes de la vie militaire au Mexique. ...E. L. Gabriel de Ferry de Bellemare.

Shelby's expedition to Mexico. ...John N. Edwards.

Silver City: a story of adventure in Mexico. ...Frederick A. Ober.

Some unconventional people. ...Mrs. J. Gladwyn Jebb.

Stories of old new Spain. ...Thomas A. Janvier.

Stories of the conquest of Mexico and Peru. ...W. Dalton.

Süden und Norden. ...Charles Sealsfield (formerly Karl Postl).

Valentin Guillois. ...Gustave Aimard.

MEXICO — *Continued.*

Die Virey und die Aristokraten. ...Charles Sealsfield (formerly Karl Postl).

White chief. ...Mayne Reid.

White conquerors: a tale of Toltec and Aztec. ...Kirk Munroe.

Wild rose: a tale of the Mexican frontier. ...Francis Francis.

Yellow snake: a tale of treasure. ...William Henry Bishop.

See also **MEXICAN WAR OF 1845**.

MICHIGAN.

Les forestiers du Michigan. ...Gustave Aimard.

Les forestiers du Michigan. ...Berlioz d'Auriac.

Legends of Michigan. ...J. F. Littlejohn.

Oak openings. ...James Fenimore Cooper.

See also **UNITED STATES HISTORY**.

MILTON, JOHN. *Lived 1608–1674.*

Deborah's diary. ...Anne Manning.

John Milton and his times: historical novel. ...Max Ring. English translation.

John Milton und seine Zeit. ...Max Ring.

Maiden and married life of Mary Powell, afterward Mistress Milton. ...Anne Manning.

MINES AND MINING.

Black diamonds. ...Mór Jókai. English translation.

Blind brother: a story of the Pennsylvania coal-mines. ...Homer Greene.

Blind lead: the story of a mine. ...Josephine W. Bates.

Deep down: tale of the Cornish mines. ...Robert Michael Ballantyne.

Derrick Sterling: a story of the mines. ...Kirk Munroe.

Fortune Gulch. ...Sophie Bronson Titterington.

Gold, gold in Cariboo. ...C. Phillipps-Wolley.

In the valley of Tophet. ...Henry W. Nevinson.

Israel Mort, overman: a story of the mine. ...John Saunders.

Jerry. ...Sarah B. Elliott.

John Brent's field. ...L. Bates.

Master of the mine. ...Robert W. Buchanan.

Menhardoc. ...George Manville Fenn.

Miner's daughter: a Catholic tale. ...Cecilia M. Caddell.

Miner's right: a tale of the Australian gold-fields. ...Thomas A. Brown. (Pseud., Rolf Boldrewood.)

Miss Grace of all souls. ...William Edward Tirebuck.

Modern buccaneer. ...Thomas A. Brown. (Pseud., Ralph Boldrewood.)

Montezuma's gold mines. ...Frederick A. Ober.

MINES AND MINING — *Continued.*
Moondyne : a story from the under-world.   ...John Boyle O'Reilly.
Moonlight.   ...Daniel Beard.
Nevermore.   ...Thomas A. Brown.   (Pseud., Ralph Boldrewood.)
Philip ; or, Mollie's secret.   ...Robert T. S. Lowell.
Prairie folks.   ...Francis Bret Harte.
Stephen Dane.   ...Amanda M. Douglas.
Stone-pastures.   ...Eleanor Stuart.
Story of a mine.   ...Francis Bret Harte.
That lass o' Lowrie's.   ...Frances Hodgson Burnett.
Vicar's people.   ...George Manville Fenn.
Woman intervenes.   ...Robert Barr.

MINISTERS. *A Selection.*
Autobiography of Mark Rutherford.   ...William Hale White.
    (Pseud., Mark Rutherford.)
Country curate.   ...George Robert Gleig.
Crucifixion of Philip Strong.   ...Charles Sheldon.
Le curé de village.   ...Honoré de Balzac.
Dawsons of Glenara.   ...Henry Johnston.
Distracted young preacher.   ...Thomas Harding.
Doctor Johns.   ...Donald G. Mitchell.
First rector of Burgstead.   ...Edward L. Cutts.
For liberty's sake : the story of Robert Ferguson.   ...John B. Marsh,
Green graves of Balgowrie.   ...Jane H. Findlater.
His majesty myself.   ...William Mumford Baker.
Jarousseau, le pasteur du désert.   ...Pierre Clément Eugène Pelletan.
Jean Jarousseau, pastor of the desert.   ...Pierre Clément Eugène
    Pelletan.   English translation.
John Brent's field.   ...L. Bates.
Lincolnshire tragedy.   ...Anne Manning.
Little minister.   ...James Mathew Barrie.
Mark Rutherford, dissenting minister.   ...William Hale White.
    (Pseud., Mark Rutherford.)
Mark Rutherford's deliverance.   ...William Hale White.   (Pseud.,
    Mark Rutherford.)
Minister of Andouse.   ...Heinrich Moewes.   English translation.
Minister of the world.   ...C. Atwater Mason.
Minister's charge.   ...William Dean Howells.
Minister's wooing.   ...Harriet Beecher Stowe.
My double and how he undid me.   ...Edward Everett Hale.
New minister.   ...Kenneth Paul.
New rector.   ...Stanley J. Weyman.
Outcasts.   ...Roy Tellet, pseud.
Parish province.   ...E. M. Lynch.

MINISTERS — *Continued.*
Parson Jones. ...Florence Marryat Lean (Mrs. Francis Lean).
Parson's proxy. ...Kate W. Hamilton.
Passages in the life of the fair gospeller Mistress Anne Askew. ...Anne Manning.
Pastor and prelate. ...Roy Tellet, pseud.
Peep at No. 5. ...Elizabeth Stuart Phelps Ward (Mrs. Herbert D. Ward).
Der Pfarrer von Andouse. ...Heinrich Moewes.
Reverend gentleman. ...J. Maclaren Cobban.
Romance of a French parsonage. ...M. B. Edwards.
Scenes from clerical life. ...Marian Evans Lewes Cross. (Pseud., George Eliot.)
Shady-side; or, Life in a country parsonage. ...Martha S. Hubbell.
Soul of the bishop. ...Henrietta E. V. Stannard. (Pseud., John Strange Winter).
Stickit minister and common men. ...Samuel Rutherford Crockett.
Sunnyside. ...Elizabeth Stuart Phelps Ward (Mrs. Herbert D. Ward).
Supply at St. Agatha's. ...Elizabeth Stuart Phelps Ward (Mrs. Herbert D. Ward).
Traitor or patriot? ...Mary C. Rowsell.
Village rector. ...Honoré de Balzac. English translation.
See also **DOGMAS** and **SCEPTICISM**.

MINNESOTA.
Mystery of Metropolisville. ...Edward Eggleston.
Nakoma : a story of frontier life. ...George Huntington.

MIRABEAU, COMTE DE (GABRIEL HONORÉ RIQUETTI). *Lived 1749–1791.*
Count Mirabeau: an historical novel. ...Theodor Mundt.
Le dernier amour de Mirabeau. ...Charlotte Foucaux.
Graf Mirabeau. ...Theodor Mundt.
Mirabeau und Sophie. ...Oskar L. B. Wolff.

MISSIONARY STORIES.
Bending willows. ...Jane G. Fuller. (Home, Northwest.)
Child of the Ganges. ...Robert N. Barrett.
Cruise of the Mystery. ...Louise Seymour Houghton. (McCall mission, France.)
Father Laval; or, Jesuit Missionary. ...J. M'Sherry. (To the Indians.)
Fifine. ...Louise Seymour Houghton. (Paris workingmen.)
Kin-da-shon's wife. ...Mrs. Eugene S. Willard. (Alaska.)
Manuelita. ...Marian Calvert Wilson. (Jesuit mission in California.)

MISSIONARY STORIES — *Continued.*

Mark Thoresbury. ...G. E. Sargent. (Indian missions.)

Prassana and Kamini. ...Mrs. Mullens and others. (India.)

Remember the Alamo. ...Amelia Edith Barr. (Early Catholic missions in Florida.)

Zeinab the Punjabi. ...E. M. Wherry.

MISSISSIPPI.

Border beagles. ...William Gilmore Simms.

Flush times in Alabama and Mississippi. ...Joseph G. Baldwin.

George Mason, the young backwoodsman. ...Anon.

Mississippi scenes; or, Sketches of Southwestern life. ...Joseph B. Cobb.

See also **UNITED STATES HISTORY.**

MISSOURI.

Missouri outlaws. ...Gustave Aimard. English translation.

Les outlaws du Missouri. ...Gustave Aimard.

See also **UNITED STATES HISTORY.**

MOHAMMEDANS.

La bannière bleue: aventures d'un Musulman, d'un Chrétien, et d'un Païen a l'époque de scroisades et de la conquête mongole. ...Leon Cahun.

Betrothed. ...Sir Walter Scott.

Blue banner. ...Leon Cahun. English translation.

Boy crusaders. ...John George Edgar.

Brothers in arms: a story of the crusades. ...F. Bayford Harrison.

Cheshire pilgrims. ...Frances M. Wilbraham.

Count Robert of Paris. ...Sir Walter Scott.

Crusaders and captives. ...George E. Merrill.

Daventry. ...Julia Pardoe.

Elfrica. ...Charlotte G. Boger.

Die Fahrten Thiodolfs des Isländers. ...Friedrich H. C. La Motte Fouqué.

Fighting the Saracens; or, Boy knight. ...George Alfred Henty.

Florine, princess of Burgundy. ...William B. MacCabe.

Foure prentices of London. ...Thomas Heywood.

From bondage to freedom. ...Anon.

Heroines of the crusades. ...Charles A. Bloss.

I Lombardi alla prima crociata. ...Tommaso Grossi.

In the brave days of old: the story of the crusades. ...Henry Frith.

Ivanhoe. ...Sir Walter Scott.

Lady Sybil's choice: a tale of the crusades. ...Emily Sarah Holt.

Longbeard, Lord of London. ...Charles MacKay.

Mathilda, prince of England. ...Sophie Risteau Cottin. English translation.

MOHAMMEDANS — *Continued.*

Mathilde ; ou, Mémoires tirés de l'histoire des croisades. ...Sophie Risteau Cottin.

Maud and Mariam ; or, Fair crusader. ...Harriet B. MacKeever.

Orphans of Alsace. ...Anon.

Parcival. ...Albert E. Brachvogel.

Philip Augustus ; or, Brothers in arms. ...George Payne Rainsford James.

Prince and the page : the last crusade. ...Charlotte Mary Yonge.

Prince of India. ...Lew Wallace.

Ransom : a tale of the thirteenth century. ...Laura Jewry.

Raymond de Saint Gilles at the crusades. ...Henrietta de Witt.

Richard Cœur de Lion. ...James White.

Richard the lion-hearted. ...Robert Tomes.

Saxon's daughter : a tale of the crusades. ...Major Michel.

Shadow of the ragged stone. ...C. F. Grindrod.

Stories of the crusades. ...William E. Dutton.

Stories of the crusades. ...John Mason Neale. (Pseud., Aurelius Gratianus.)

Sword and scimetar. ...Alfred Trumble.

Tales of the early ages. ...Horace Smith. (Pseud., Paul Chatfield.)

Talisman. ...Sir Walter Scott.

Under Bayard's banner. ...Henry Frith.

Westminster cloisters : the story of a life's ambition. ...M. Bidder.

Winning his spurs. ...Elijah Kellogg.

With script and staff. ...Elia W. Peattie.

Young prophetess : a tale of the children's crusade. ...R. Leighton Gerhart.

MOLIÈRE, JEAN BAPTISTE. *Lived 1622–1673.*

Molière. ...Alexander Ungern-Sternberg.

School for husbands. ...Rosina W. L. Bulwer-Lytton.

MONEY. *A Selection.*

About money. ...Dinah Maria Mulock Craik.

Adventures of Harry Franco : a tale of the great panic. ...C. F. Briggs.

L'argent. ...Émile Zola.

L'argent des autres. ...Émile Gaboriau.

Beyond the dreams of avarice. ...Walter Besant.

Colonel Dunwoddie, millionaire. ...William Mumford Baker.

Le comte de Monte Cristo. ...Alexandre Dumas.

Count of Monte Cristo. ...Alexandre Dumas. English translation.

Cynthia Wakeham's money. ...Anna Katharine Green.

Dame Fortune smiled. ...Willis Barnes.

Doctor Warwick's daughters. ...Rebecca Harding Davis.

MONEY — *Continued.*

Financial atonement.   ...B. B. West.

Gerard.   ...Margaret E. Braddon.

Golden bottle.   ...Ignatius Donnelly.   (Pseud., Edmund Boisgilbert
    M. D.)

Happiness of being rich.   ...Hendrik Conscience.   English trans-
    lation.

Heiress of Bruges.   ...Thomas C. Grattan.

Heiress of Kilorgan.   ...Mary A. Sadlier (Mrs. James Sadlier).

Henry Powers (banker).   ...Robert B. Kimball.

Het geluk van ryk te Byn.   ...Hendrik Conscience.

Katharine Lauderdale.   ...Francis Marion Crawford.

King Mammon and the heir apparent.   ...George A. Richardson.

Memoirs of a millionaire.   ...Lucia T. Ames.

Money.   ...Émile Zola.   English translation.

Nimble dollar.   ...Clara M. Thompson.

Perils of Pearl street, including a taste of the dangers of Wall
    street.   ...Asa Greene.

Powder and gold.   ...Levin Schuecking.   English translation.

Puebla, oder die Franzosen in Mexico.   ...H. O. F. Goedsche.

Pulver und Gold.   ...Levin Schuecking.

Queen money.   ...Ellen Warner Olney Kirk.   (Pseud., Harry
    Hayes.)

Quicksands of Pactolus.   ...Horace A. Vachell.

Rich Miss Riddell.   ...Dorothea Gerard.

Der Schatz des Inka.   ...Franz Hoffmann.

Sham gold.   ...Stephanie Wohl.

Silver baron.   ...Carleton Waite.   ("Silver craze of 1896.")

Six thousand tons of gold.   ...H. R. Chamberlain.

Stockings full of money.   ...Mary Kyle Dallas.

Story of a cañon.   ...Beveridge Hill.

Stumbler in wide shoes.   ...Anon.

Treasure of the Incas.   ...Franz Hoffmann.   English translation.

Undercurrents of Wall street.   ...Richard B. Kimball.

Wenzel's inheritance.   ...Annie Lucas.

MONMOUTH REBELLION, 1685.   See **REBELLIONS.**

MORE, SIR THOMAS.   *Lived 1478-1535.*

Alice Sherwin: a tale of the days of Sir Thomas More.   ...C. J. M.

Darnley; or, Field of the cloth of gold.   ...George Payne Rainsford
    James.

Household of Sir Thomas More.   ...Anne Manning.

Lettice Eden: a tale of the last days of Henry the Eighth.   ...Emily
    Sarah Holt.

MORMONISM.
Apples of Sodom : a story of Mormon life. ...Anon.
Bar-sinister. ...Jeannette R. Walworth.
Button's Inn. ...Albion Winegar Tourgee.
California Crusoe. ...Anon.
Deborah the advanced woman. ...Mary Ives Todd.
Fate of Madame La Tour. ...Mrs. A. G. Paddock.
Female life among the Mormons. ...Ward.
Humbled pride : a story of the Mexican war. ...John R. Musick.
In the toils. ...Cornelia Paddock.
John Brent. ...Theodore Winthrop.
Les Mormons. ...Paul Duplessis.
Salt Lake fruit. ...Anon.
Study in scarlet. ...Arthur Conan Doyle.

MOZART, JOHANN CHRYSOSTOM WOLFGANG GOTTLIEB. *Lived 1756–1791.*
Daponte und Mozart. Roman. ...Julius M. Grosse.
In the days of Mozart. ...Lily Watson.
Mozart. Der junge Arzt. ...Joseph Gobiet.
Mozart : a biographical romance. ...Heribert Rau.
Tone-masters. ...Charles Barnard. Vol. 1.

MUSIC. *A Selection.*
Alcestis. ...Mrs. Cornish.
Appassionata : a musician's story. ...Elsa D'Esterre Keeling.
As it was written. ...H. Harland.
Bach ; or, Fortunes of an idealist. ...Albert E. Brachvogel. English translation.
Beethoven : historischer Roman. ...Heribert Rau.
Den Bergtekne. ...Kristofer Nagel Janson.
Blind musician. ...V. Korolenko. English translation.
Charles Auchester : a memorial. ...Elizabeth S. Sheppard. (Pseud., Beatrice Reynolds.)
Compensation. ...A. M. H. Brewster.
Consuelo. ...Amantine L. A. D. Dudevant. (Pseud., George Sand.)
Consuelo. ...Amantine L. A. D. Dudevant. (Pseud., George Sand.) English translation.
Cord from a violin. ...Winifred Agnes Haldane.
Counterparts. ...Elizabeth S. Sheppard. (Pseud., Beatrice Reynolds.)
Countess of Rudolstadt. ...Amantine-L. A. D. Dudevant. (Pseud., George Sand.)
Crown of thorns. ...Allen Upward.
Daponte und Mozart. Roman. ...Julius W. Grosse.
Daughter of music. ...Mrs. Colmore Dunn.

MUSIC — *Continued.*

MUSIC — *Continued.*

Seppi. ...Franz Hoffmann.

Soprano. ...Charles Barnard.

Spellbound fiddler : a Norse romance. ...Kristofer Nagel Janson. English translation.

Stories of the Wagner opera. ...H. A. Guerber.

Teacher of the violin and other tales. ...John Henry Shorthouse.

That fiddler fellow. ...Horace Hutchinson.

Thekla : a story of Viennese musical life. ...William Armstrong.

Tone-masters. ...Charles Barnard.

Trilby. ...George Du Maurier.

Trilby. ...George Du Maurier. English translation.

Unrequited love. ...Ludwig Nohl.

Le violin de faïence. ...Jules F. F. H. Fleury. (Pseud., Champfleury.)

Wagner story-book : firelight tales of the great music dramas. ...William H. Frost.

Yanko the musician. ...Henryk Sienkiewicz. English translation.

Zegelda Romanief. ...Charles Barnard.

MYSTERIES.

Antoinette ; or, Marl-pit mystery. ...Georges Ohnet.

Big bow mystery. ...Isaac Zangwill.

Cave secret. ...Burton Saxe.

Le crime d'Orcival. ...Émile Gaboriau.

Danvers jewels and Sir Charles Danvers. ...Anon.

David Poindexter's disappearance and other tales. ...Julian Hawthorne.

Disappearance of Mr. Derwent. ...Thomas Cobb.

Doctor Phœnix's skeleton. ...Fewistesh, pseud.

L'École des Robinsons. ...Jules Verne.

Girl in checks. ...J. W. Daniel.

Godfrey Morgan : a California mystery. ...Jules Verne. English translation.

Great Chin episode. ...Paul Cushing.

Great Porter square : a mystery. ...Benjamin Leopold Farjeon.

L'homme au masque de fer. ...Chevalier de Mouhy.

Isabel de Bavière. ...Alexandre Dumas.

Isabel of Bavaria, Queen of France. ...Alexandre Dumas. English translation.

Jewel mysteries I have known. ...Max Pemberton.

Leech club. ...George W. Owen.

Lost Stradivarius. ...J. Meade Falkner.

Mill mystery. ...Anna Katharine Green (Mrs. Rolfe).

Mountain mystery. ...Lawrence L. Lynch, pseud.

MYSTERIES — *Continued.*
Mysteries and miseries of San Francisco. ...Anon.
Mysteries of the backwoods. ...Thomas B. Thorpe.
Mystery of Allenwold. ...Elizabeth Van Loon.
Mystery of Killard. ...Richard Dowling.
Mystery of Orcival. ...Émile Gaboriau. English translation.
Mystery of Paul Chadwick. ...J. W. Postgate.
Mystery of Sasassa valley. ...Arthur Conan Doyle.
Mystery of the Campagna. ...Albert Degen.
Mystery of the locks. ...Edgar W. Howe.
Mystery of the Patrician club. ..Albert D. Vandam.
Mystery of the Unseen; or, Supernatural stories of English life.
...Gilbert Campbell.
Mystery of Witch-face Mountain. ...Mary N. Murfree. (Pseud.,
Charles Egbert Craddock.)
Strange disappearance. ...Anna Katharine Green.
Tales of adventure, mystery, and imagination. ...Edgar Allan Poe.
Tales of mystery. ...Edgar Allan Poe.
Tales of mystery. ...George Saintsbury, ed.
Tragic mystery. ...Julian Hawthorne.
Witch Winnie's mystery. ...Elizabeth W. Champney.
Witness to the deed. ...George Manville Fenn.
See also **DETECTIVE STORIES.**

NAPOLEON I. OF FRANCE. *Reigned 1799–1814.*
Adventures of a casket. ...Justin J. E. Roy. English translation.
Adventures of Brigadier Gerard. ...Arthur Conan Doyle.
Aims and obstacles. ...George Payne Rainsford James.
Alice de Beaurepaire. ...Victorien Sardou.
Les aventures d'une Cassette. ...Justin J. E. Roy.
Bailén. ...Benito Pérez Galdós.
Ben Brace: the last of Nelson's Agamemnons. ...Frederick
Chamier.
Bivouac; or, Stories of the Peninsular war. ...William H.
Maxwell.
Les blancs et les bleus. ...Alexandre Dumas.
Blockade: an episode of the fall of the first French empire
...Émile Erckmann and Alexandre Chatrian. English translation.
Le blocus. ...Émile Erckmann et Alexandre Chatrian.
Boulogne und Austerlitz. ...Ferdinand Stolle.
Boy of the first empire. ...Elbridge S. Brooks.
Cádiz. ...Benito Pérez Galdós.
Le capitaine Richard. ...Alexandre Dumas.
Catherine. ...Frances M. Peard.
Charles O'Malley, the Irish dragoon. ...Charles James Lever.

NAPOLEON I. OF FRANCE — *Continued.*

Citizen Bonaparte. ...Émile Erckmann and Alexandre Chatrian. English translation.

Le citoyen Bonaparte. ...Émile Erckmann et Alexandre Chatrian.

Die Clubisten in Mainz. ...Heinrich J. Koenig.

Compagnons de Jéhu. ...Alexandre Dumas.

Companions of Jehu. ...Alexandre Dumas. English translation.

Le comte de Monte Cristo. ...Alexandre Dumas.

Conscript: an historical novel of the days of the first Napoleon. ...Alexandre Dumas. English translation.

Conscript: an historical novel of the days of the first Napoleon. ...Émile Erckmann and Alexandre Chatrian. English translation.

Count of Monte Cristo. ...Alexandre Dumas. English translation.

Courtship by command.. ...M. M. Blake.

Czar. ...Deborah Alcock.

Ein deutsches Reiterleben. ...Julius von Wickede.

Diana's crescent. ...Anne Manning.

Der Domherr. ...Jodocus D. H. Temme.

Don Alonso; ou, l'Espagne : histoire contemporaine. 1788–1823. ...Narcisse A., Comte de Salvandy.

Édouard; ou, siège de Saragosse. ...Léopold Méry.

1805; oder die Franzosen zum erstenmal in Wien. ...Eduard Breier.

1812: ein historischer Roman. ...Ludwig Rellstab.

1813. ...Ferdinand Stolle.

Eighty years ago. ...Harriett Morton.

Empress Josephine. ...Clara M. Mundt. (Pseud., Louise Mühlbach.) English translation.

Enriqueta Faber. ...Andrés Clemente Vazquez.

Episodes of French history during the consulate and the first empire. ...Julia Pardoe.

El equipaje del rey José. ...Benito Pérez Galdós.

Der Feldmarschall Blücher und der Pfarrer Kretzschmar. ...Philipp F. W. Oertel.

Les fiancés de 1812. ...Joseph Doutre.

Fiddler of Lugau. ...Margaret Roberts.

France et Marie. ...Hyacinthe J. A. Thabaud.

Franz Alzeyer. ...Paul Heyse.

Die Franzosen in Berlin. 1806–8. ...Friedericke H. Unger.

Eine Freundin Napoleons. ...Auguste W. Lorenz.

General Rapp und die Belagerung von Danzig. ...Maria von Roskowska.

Grand cross of the legion. ...Henry C. Adams.

Great shadow and beyond the city. ...Arthur Conan Doyle.

Her heart was true : a story of the Peninsular war. ...E. M. Cuttim.

Hidden treasures. ...Frederick Hardmann.

NAPOLEON I. OF FRANCE — *Continued.*

Vaterländische Romane.   ...Johann Georg L. Hesekiel.

Verschollen.   ...Johann Ferdinand Martin Oskar Meding.   (Pseud., Gregor Samarow.)

Die Völkerschlacht bei Leipzig.   ...Joseph von Hinsberg.

Vor Saalfeld bis Aspern.   ...Heinrich J. Koenig.

War and peace.   ...Lyeff Nikolaievich Tolstoi.   English translation from the Russian.

Waterloo.   ...Émile Erckmann and Alexandre Chatrian.   English translation.

Waterloo: suite du conscrit de 1813.   ...Émile Erckmann et Alexandre Chatrian.

White lies.   ...Charles Reade.

Whites and the blues.   ...Alexandre Dumas.   English translation.

Young Breton volunteer.   ...Frances M. Wilbraham.

Young buglers : a tale of the Peninsular war.   ...George Alfred Henty.

Young Muscovite.   ...Mrs. M. E. Bewsher.

Zaragoza.   ...Benito Pérez Galdós.

See also **BATTLES, Waterloo.**

NAPOLEON II. OF FRANCE.   *Reigned 1848–1851.*

Louis Napoleon.   ...Eugen H. von Dedenroth.

Louis Napoleon.   ...Julius Gundling.

Napoleon II.   ...Carl J. Braun.

NAPOLEON III. OF FRANCE.   *Reigned 1852–1870.*

L'assommoir.   ...Émile Zola.

Cäsar und Napoleon III.   Roman.   ...Julius Gundling.

Gervaise : the natural and social life of a family under the second empire.   ...Émile Zola.   English translation.

Histoire d'un crime.   ...Victor Hugo.

History of a crime.   ...Victor Hugo.   English translation.

Louis Napoleon : Roman und Geschichte.   ...Eugen H. von Dedenroth.

Malgré tout.   ...Amantine L. A. D. Dudevant.   (Pseud., George Sand.)

Member for Paris : a tale of the second empire.   ...Grenville Murray.

Monsieur de Camors.   ...Octave Feuillet.

Napoleon III.: Fortsetzung von Louis Napoleon.   ...Julius Gundling.

Napoleon III. und sein Hof in Anekdoten und Charakterzügen.   ...Julius Gundling.

Parisians.   ...Edward George Earle Lytton Bulwer-Lytton.

Les Rougon-Macquart : histoire naturelle et sociale d'une famille sous le second empire.   ...Émile Zola.

NAPOLEON III. OF FRANCE — *Continued.*
Rumor. ...Elizabeth S. Sheppard. (Beatrice Reynolds.)
Salon in the last days of the empire. ...Kathleen O'Meara.
See also **FRANCO-GERMAN WAR, 1870–1871,** and **SIEGES, Paris, 1870–1871.**

NEGROES.
Balaam and his master, and other sketches and stories. ...Joel Chandler Harris.
Beatrice of Bayou Teche. ...Alice Ilgenfritz Jones.
Black diamonds gathered in the darkey homes of the South. ...Edward A. Pollard.
Bricks without straw. ...Albion Winegar Tourgee.
Doty Dontcare : a story of the garden of the Antilles. ...Mary F Foster. (Negro insurrection of 1878.)
Fool's errand. ...Albion Winegar Tourgee.
Hour and the man. ...Harriet Martineau.
In all shades. ...Grant Allen.
Mrs. Merriam's scholars. ...Edward Everett Hale.
Negro slave. ...Washington Nash.
Question of color. ...F. C. Philips.
Samantha on the race problem. ...Marietta Holley. (Pseud., Josiah Allen's Wife.)
Theresa at San Domingo. ...Mme. A. Fresneau.
Trooper Peter Halket of Mashonaland. ...Olive Schreiner. (Pseud., Ralph Iron.)
Uncle Tom's cabin ; or, Life among the lowly. ...Harriet Beecher Stowe.
Without blemish. ...Jeanette R. Walworth. •
See also **SLAVERY** and **SOUTH, U. S. A.**

NELSON, HORATIO. *Lived 1758–1805.*
Ben Brace, the last of Nelson's Agamemnons. ...Frederick Chamier.
Diana's crescent. ..Anne Manning.

NERO, CLAUDIUS CÆSAR DRUSUS GERMANICUS. *Lived 37–68.*
Beric the Briton. ...George Alfred Henty.
Burning of Rome ; or, Story of the days of Nero. ...Alfred John Church.
Darkness and dawn. ...Frederick William Farrar.
Greek maid at the court of Nero. ...Franz Hoffmann. English translation.
Julia of Baioe. ...J. W. Brown.
Nero. ...Ernst Eckstein.
Quo Vadis. ...Henryk Sienkiewicz.
Quo Vadis : narrative of the time of Nero. ...Henryk Sienkiewicz. English translation. •

NERO, CLAUDIUS CÆSAR DRUSUS GERMANICUS — *Continued.*
Unlaid ghost : a story in metempsychosis.   ...Joseph Vila Prichard.
See also **ROME.**

NETHERLANDS.

Description, History, Manners, and Customs.

Agnes and Carel.   ...John B. de Liefde.
Alarum for London ; or, Siege of Antwerp.   ...Anon.
Amnesty ; or, Duke of Alba in Flanders.   ...Charles F. Ellerman.
Amulet.   ...John B. de Liefde.
Baes Gansendonck.   ...Hendrik Conscience.
Baroness.   ...Frances M. Peard.
Beggars ; or, Founders of the Dutch republic.   ...John B. de Liefde.
Black tulip.   ...Alexandre Dumas.   English translation.
Blacksmith of Antwerp.   ...Georgiana G. L. G. Fullerton.   (In her
    Seven stories.)
Blacksmith's daughter.   ...Anon.
Blue pavilions.   ...Arthur T. Q. Couch.   (Pseud., Q.)
De Boerekryg.   ...Hendrik Conscience.
De Burgemeester van Luik.   ...Hendrik Conscience.
Burgomaster's family ; or, Weal and woe in a little world.   ...E. C.
    W. van Walrée.   (Pseud., Christine Muller.)
Burgomaster's wife.   ...Georg Moritz Ebers.   English translation.
By England's aid ; or, Freeing of the Netherlands.   ...George Alfred
    Henty.
Campaign in Holland.   ...Anon.
Captain-General : being the story of the attempt of the Dutch to
    colonize New Holland.   ...William J. Gordon.
Cloister and the hearth.   ...Charles Reade.
Constance Aylmer : a story of the seventeenth century.   ...Helen
    F. Parker.
Count Hugo.   ...Hendrik Conscience.   English translation.
Count Renneberg's treason : a tale of the siege of Steenwyk.
    ...Harriette E. Burch.
Days of Prince Maurice.   ...Mary O. Nutting.
Fisherman's daughter.   ...Hendrik Conscience.   English translation.
Der fliegende Holländer.   ...Albert E. Brachvogel.
For faith and Fatherland.   ...Mary Bramston.
Die Frau Bürgemeisterin.   ...Georg Moritz Ebers.
Galama ; or, Beggars.   ...John B. de Liefde.
God's fool.   ...J. Schwartz.
Greater glory.   ...J. van der Poorsen-Schwartz.   (Pseud., Maartens
    Maartens.)
Hans Brinker ; or, The silver skates.   ...Mary Mapes Dodge.
Harrington ; or, Exiled royalist.   ...Frederick S. Bird.

NETHERLANDS — *Continued.*

Vera; or, War of the peasants.   ...Hendrik Conscience.   English translation.

Village innkeeper.   ...Hendrik Conscience.   English translation.

Walter's escape; or, Capture of Breda.   ...John B. de Liefde.

White hoods: an historical romance.   ...Anne Eliza Bray.

White hoods: a tale of the free city of Ghent: a legend of the fourteenth century.   ...Max Eking.

Wind and wave fulfilling his word.   ...Harriette E. Burch.

Within sea-walls.   ...E. H. Walshe and George E. Sargent.

Within the walls: a tale of the siege of Haarlem.   ...Mary Doig.

Word, only a word.   ...Georg Moritz Ebers.   English translation.

Ein Wort.   ...Georg Moritz Ebers.

See also **REFORMATION in Netherlands**; **REPUBLICS, Dutch.**

NEW ENGLAND.   *A Selection.*

Description, History, Manners, and Customs.

About an old New England church.   ...Gerald Stanley Lee.

Achsah: a New England life-study.   ...William M. F. Round.

Annals of Brookdale: a New England village.   ...Frances Greenough.

Arthur Bonnicastle.   ...Josiah Gilbert Holland.

Autobiography of a New England farmhouse.   ...Nathan H. Chamberlain.

Bay-path: a tale of New England colonial life.   ...Josiah Gilbert Holland.

Blithedale romance.   ...Nathaniel Hawthorne.

Book of New England legends and folklore.   ...S. A. Drake.

Brampton sketches.   ...Mary B. Claflin.

Brother Jonathan; or, New Englanders.   ...John Neal.

Doctor Grimshaw's secret.   ...Nathaniel Hawthorne.

Down Easters.   ...John Neal.

Ethan Allen; or, King's men.   ...Melville, pseud.

Eunice: a story of domestic life in New England.   ...Mary M. Robertson.

Fawn of the pale faces.   ...John P. Brace.

Fools of Nature.   ...Alice Brown.

Gilead guards.   ...Mrs. O. W. Scott.

House of seven gables.   ...Nathaniel Hawthorne.

In old New England: the romance of a colonial fireside.   ...Hezekiah Butterworth.

Jesuit ring: a romance of Mount Desert.   ..Augusta A. Hayes.

Kinley hollow.   ...G. H. Hollister.

Laconia; or, Legends of the White Mountains.   ...J. P. Scribner

Last Penacook.   ...A. B. Berry.

NEW ENGLAND — *Continued.*

Legends of New England. ...Julia Gill and Frances Lee.
Legends of New England. ...Nathaniel Hawthorne.
Little Moccasin ; or, Along the Madawaska. ...John Neal.
May Flower and miscellaneous writings. ...Harriet Beecher Stowe.
May Martin and other tales of the Green Mountains. ...D. P. Thompson.
Meadow grass : tales of New England life. ...Alice Brown.
Miss Gilbert's career. ...Josiah Gilbert Holland.
Mosses from an old manse. ...Nathaniel Hawthorne.
Myself: a romance of New England Life. ...E. Emery.
New England nun and other stories. ...Mary E. Wilkins.
Norwood. ...Henry Ward Beecher.
Old homestead : a story of New England farm life. ...Ann S. W. Stephens.
Old New England days. ...Sophie E. Damon.
Oldtown folks. ...Harriet Beecher Stowe.
One summer. ...Blanche Willis Howard.
Patience Strong's outings. ...Adeline D. T. Whitney.
Pearl of Orr's Island. ...Harriet Beecher Stowe.
Pemaquid: a story of old times in New England. ...E. Preston.
Pembroke. ...Mary E. Wilkins.
Peter and Polly. ...Annie D. Green.
Poganuc people. ...Harriet Beecher Stowe.
Quabbin: New England life in the early part of this century. ...Francis II. Underwood.
Red rover. ...James Fenimore Cooper.
Regicides : a tale of early colonial times. ...Frederick Hull Cogswell.
Richard Ireton : a legend of the early settlement of New England. ...M. Remick.
Rise of Silas Lapham. ...William Dean Howells.
Romance of the charter oak. ...William Seton.
Snow image and other twice-told tales. ...Nathaniel Hawthorne.
Standish of Standish. ...Jane Goodwin Austin.
Steadfast. ...Rose Terry Cooke.
Stories from New England life. ...Martha Russell.
Story of a child. ...Mary Deland.
Summer in Leslie Goldthwaite's life. ...Adeline D. T. Whitney.
Tales of New England. ...Sarah O. Jewett.
Tales of the White Hills. ...Nathaniel Hawthorne.
Timothy's Quest. ...Kate Douglas Wiggin (Mrs. George Riggs).
Tom Sylvester. ...T. R. Sullivan.
Under a colonial rooftree. ...Aria S. Huntington.

NEW ENGLAND — *Continued.*
Village watch-tower. ...Kate Douglas Wiggin (Mrs. George Riggs).
Voyage of discovery: a novel of American society. ...Aïdé Hamilton.
Winthrop family: a story of New England life fifty years ago. ...Clara A. Willard.
See also individual New England States.

NEW FOREST, ENGLAND.
Children of the New Forest. ...Frederick Marryat.
Craddock Norvell: tale of the New Forest. ...Richard Doddridge Blackmore.
Hide and seek: a story of the New Forest in 1647. ...E. E. Cooper (Mrs. Frank Cooper).

NEW HAMPSHIRE.
Amber star and a fair half-dozen. ...Mary Louise Dickinson.
Lisbeth Wilson: a daughter of New Hampshire hills. ...Eliza N. Blair.
Winterborough. ...Eliza O. White.

NEW JERSEY.
Haunted hearts. ...Maria S. Cummins.
His little royal highness. ...Ruth Ogden.
Miss Eaton's romance: a story of the New Jersey shore. ...Richard Allen.

NEW ORLEANS, LOUISIANA.
Autrefois: tales of old New Orleans and elsewhere. ...James A. Harrison.
Doctor Sevier. ...George Washington Cable.
Lady Jane. ...Celia V. Jamison.
Manhattaner in New Orleans. ...A. Oakey Hall.
Towards the Gulf: a tale of New Orleans. ...Anon.

NEW YORK.
Begum's daughter. ...E. L. Bynner.
Book of Saint Nicholas. ...J. K. Paulding.
Bride of the northern wilds. ...N. M. Curtis.
Brise-de-Mai; ou, Les trappeurs de l'Hudson. ...V. Lamy.
Claudius, the cowboy of Ramapo Valley. ...P. Demarest Johnson.
Constance Aylmer: a story of the seventeenth century. ...Helen F. Parker.
Deutsche Pioniere. ...Friedrich Spielhagen.
Doctor Grattan. ...William A. Hammond.
Dolly Dillenbeck. ...James L Ford.
Dominie Freylinghausen. ...Florence Wilford.

NEW YORK — *Continued.*

Dutchman's fireside. ...J. K. Paulding.
Faith doctor : a story of New York. ...Edward Eggleston.
Farnell's folly. ...John T. Trowbridge.
First of the Knickerbockers. ...P. H. Myers.
Gallagher and other stories. ...Richard Harding Davis.
Green pastures and Piccadilly. ...William Black.
Hoboken : a romance of New York. ...Theodore S. Fay.
Hotspur : a tale of the old Dutch manor. ...M. T. Walworth.
In Leisler's time. ...Elbridge Streeter Brooks.
Johnson manor. ...James Kent.
King of the Hurons. ...P. H. Myers.
Knickerbocker's history of New York. ...Washington Irving.
Last days of Knickerbocker life in New York. ...A. C. Dayton.
Layman's story ; or, Experiences of John Laicus in a country parish. ...Lyman Abbott.
Legends of Montauk. ...J. A. Ayres.
Lucy Arlyn. ...John T. Trowbridge.
Old New York ; or, Democracy in 1689. ...Elizabeth Oakes Smith.
Roderick Hume : the story of a New York teacher. ...C. W. Bardeen.
Satanstoe. ...James Fenimore Cooper.
Seth's brother's wife : a study of life in greater New York. ...Harold Frederic.
Stormcliff. ...M. T. Walworth.
Story of Helen Troy. ...Constance Cary Harrison.
Queen money. ...Ellen Warner Olney Kirk. (Pseud., Henry Hayes.)
Their wedding journey. ...William Dean Howells.
Tulip place : a story of New York. ...Virginia W. Johnson.
Way of the hour. ...James Fenimore Cooper.
White satin and homespun. ...Katharine Trask.
See also **NEW YORK CITY, NEW YORK.**

NEW YORK CITY, NEW YORK.

Annals of the empire city. ...Joseph H. Ingraham.
Artie : a story of the streets and town. ...George Ade.
Bow of orange ribbon. ...Amelia Edith Barr.
Boy's revolt. ...James Otis Kaler. (Pseud., James Otis.)
Debutante in New York society. ...Rachel Buchanan.
Fortune hunter : a novel of New York society. ...Anna C. Ritchie.
Golden justice. ...William H. Bishop.
House of a merchant prince. ...William H. Bishop.
In Leisler's time. ...Elbridge Streeter Brooks.
Katharine Lauderdale. ...Francis Marion Crawford.
Metropolitans. ...Jeanie Drake.

NEW YORK CITY, NEW YORK — *Continued.*

Midge.   ...Henry C. Bunner.
New York family.   ...Edgar Fawcett.
Norman Leslie.   ...Theodore S. Fay.
Otto's inspiration.   ...Mary H. Ford.
Pearl and emerald.   ...Robert Edward Francillon.
Queen money.   ...Ellen Warner Olney Kirk. (Pseud., Harry Hayes.)
Seth's brother's wife : a study of life in greater New York.   ...Harold Frederic.
Sweet bells out of tune.   ...Constance Cary Harrison (Mrs. Burton Harrison).
Sword of Damocles.   ...Anna Katharine Green.
Tulip place : a story of New York.   ...Virginia W. Johnson. (Pseud., Cousin Virginia.)

NEW ZEALAND.

Description, History, Manners, and Customs.

Divers.   ...Hume Nisbet.
Half a hero.   ...Anthony H. Hawkins. (Pseud., Anthony Hope.)
Majesty of man.   ...Alien, pseud.
Maori and settler : a story of the New Zealand war.   ...George Alfred Henty.
Supple Jack : a romance of Maoriland.   ...R. Ward.
Web of the spider.   ...H. B. M. Watson.

NONCONFORMITY.

Autobiography of Mark Rutherford.   ...William Hale White. (Pseud., Mark Rutherford.)
Catherine Furze.   ...William Hale White. (Pseud., Mark Rutherford.)
Ida Vane.   ...Andrew Reed.
Mariam's schooling.   ...William Hale White. (Pseud., Mark Rutherford.)
Mark Rutherford's deliverance.   ...William Hale White. (Pseud., Mark Rutherford.)
Revolution in Tanner's lane.   ...William Hale White. (Pseud., Mark Rutherford.)
See also **ENGLISH HISTORY** and **PERSECUTION.**

NORMAN CONQUEST.

Andreds-Weald ; or, House of Michelham : a tale of the Norman conquest.   ...Augustine D. Crake.
Behind the veil : a tale of the days of William the Conqueror.   ...Emily Sarah Holt.
Bishop's daughter.   ...Anon.

NORMAN CONQUEST — *Continued.*

Camp of refuge. ...Charles MacFarlane.

Diane. ...Katharine S. Macquoid.

Fitz-Alwyn. ...Elizabeth M. Stewart.

Harold the boy earl. ...J. Frederick Hodgetts.

Harold the last of the Saxon kings. ...Edward George Earle Lytton Bulwer-Lytton.

Hereward the wake. ...Charles Kingsley.

Legend of Heading Abbey. ...Charles Knight.

Rufus ; or, Red king. ...James Grant.

Siege of Norwich Castle. ...M. M. Blake.

Stanfield Hall. ...J. Frederick Smith.

Thira ; or, Cairn Braich. ...Anon.

Third chronicle of Æscendum. ...Augustine D. Crake.

William the Conqueror. ...Charles J. Napier.

Wolf the Saxon : a story of the Norman conquest. ...George Alfred Henty.

Wulf the Saxon. ...Ralph Peacock.

NORTH CAROLINA.

Dike shanty. ...Maria Louise Pool.

Fisher's river. ...H. E. Taliaferro.

Girl in checks. ...J. W. Daniel.

Land of the sky ; or, Adventures in mountain by-ways. ...Frances C. Fisher. (Pseud., Christian Reid.)

Lunsford Lane. ...William G. Hawkins.

Mary Barker. ...Charles Vernon.

Summer idyl. ...Frances C. Fisher. (Pseud., Christian Reid.)

NORWAY.

Description, History, Manners, and Customs.

Adam Schrader. ...Jonas Lauritz Edemil Lie. English translation.

Adventures of Olaf Trygg Veson, king of Norway. ...Pamelia M. Reed (Mrs. Joseph J. Reed).

Afraja. ...Theodor Muegge.

Afraja ; or, Life and love in Norway. ...Theodor Muegge. English translation.

Against heavy odds : a tale of Norse heroism. ...Hjalmar H. Boyesen. English translation.

Barque Future ; or, Life in the far North. ...Jonas Lauritz Edemil Lie. English translation.

Den Bergtekne. ...Kristofer Nagel Janson.

Boyhood in Norway : stories of boy-life in the land of the midnight sun. ...Hjalmar H. Boyesen. English translation.

Bridal march and other stories. ...Bjørnstjerne Bjørnson. English translation.

NORWAY — *Continued.*
Visionary ; or, Pictures from Nordland.   ...Jonas Lauritz Edemil Lie. English translation.
See also **VIKINGS.**

NUREMBERG, BAVARIA.
Cooper of Nuremberg.   ...Ernst T. W. Hoffmann.
Franz Sternbald's Wanderungen.   ...Johann Ludwig Tieck.
Die Gred.   Roman aus dem alten Nuremberg.   ...Georg Moritz Ebers.
In the fire of the forge : a romance of old Nuremberg.   ...English translation.
Lia : a tale of Nuremberg.   ...Esmé Stuart.
Margery : a tale of old Nuremberg.   ...Georg Moritz Ebers. English translation.
Norica, das sind Nürnberggische Novellen aus alter Zeit.   ...August Hagen.
Norica; or, Tales of Nuremberg from the olden times.   ...August Hagen. English translation.
Nürnberger Tand.   Eine Geschichte aus dem fünfzehnten Jahrhundert.   ...Ludovika Hesekiel.

OBERAMMERGAU, BAVARIA.
Am Kreuz : ein Passionsroman aus Oberammergau.   ...Wilhelmine von Hillern.
On the cross.   (Passion play.)   ...Wilhelmine von Hillern. English translation.

OCCULTISM.   See **THEOSOPHY.**

OHIO.
Banks of the Ohio.   ...James K. Paulding.
Bart Ridgeley : a story of Northern Ohio.   ...Albert G. Riddle.
Breadwinners : a social study.   ...John Hay.
Clovernook children.   ...Alice Cary.
Clovernook ; or, Recollections of our neighborhood in the West.   ...Alice Cary.
Clovernook papers.   ...Alice Cary.
Counterfeiters of the Cuyahoga.   ...William T. Coggeshall.
East and West : a story of new-born Ohio.   ...Edward Everett Hale.
Good investment : a tale of the upper Ohio.   ...William J. Flagg.
League of the Ohio.   ...Emerson Bennett.
Leni Leoti.   ...Emerson Bennett.
Mike Fink.   ...Emerson Bennett.
Portrait : a romance of Cuyahoga Valley.   ...Anon.
Rocky Fork.   ...Mary Hartwell Catherwood.

OLD PRETENDER'S REBELLION. See **REBELLIONS, Old Pretender's.**

OPPORTUNITY.

Aims and obstacles. ...George Payne Rainsford James.

Destiny. ...Susan Edmonston Ferrier.

Le fils du Titien. ...Alfred de Musset.

Himself his worst enemy. ...A. P. Brotherhead.

Mere cypher. ...Mary Angela Dickens.

My time, and what I 've done with it. ...F. C. Burnand.

Second opportunity of Mr. Staplehurst. ...William Pett Ridge.

This one thing I do. ...Anna Maria Porter.

What will he do with it ? ...Edward George Earle Lytton Bulwer-Lytton.

OREGON.

Bridge of the Gods. ...F. H. Balch.

Twice bought : a tale of the Oregon gold fields. ...Robert Michael Ballantyne.

Le Whip-poor-will ; ou, Les pionniers de l'Orégon. ...Amédée Bouis.

PALESTINE.

Description, History, Manners, and Customs.

Ahasvérus. ...Edgar Quinet.

Boy crusaders. ...John George Edgar.

Count Robert of Paris. ...Sir Walter Scott.

Doom of the holy city. ...Lydia Hoyt Farmer.

For the temple : a tale of the fall of Jerusalem. ...George Alfred Henty.

Hammer : a story of Maccabean times. ...Alfred John Church and Richmond Seeley.

Hebrew heroes. ...Charlotte Tucker. (Pseud., A. L. O. E.)

Helon's pilgrimage to Jerusalem : a picture of Judaism in the century which preceded the advent of our Saviour. ...Friedrich Abraham Strauss. English translation.

Helon's Wallfahrt nach Jerusalem, hundert neun Jahr vor der Geburt unsers Herrn. ...Friedrich Abraham Strauss.

Idumean. ...J. M. Leavitt.

Ivanhoe. ...Sir Walter Scott.

Joshua : a story of biblical times. ...Georg Moritz Ebers. English translation.

Josua. ...Georg Moritz Ebers.

Julian. ...William Ware.

King of Tyre : a tale of the times of Ezra and Nehemiah. ...James M. Ludlow.

Lady Sybil's choice : a tale of the crusades. ...Emily Sarah Holt.

Leah. ...Mrs. Alexander S. Orr.

PARIS, FRANCE — *Continued.*

On the edge of the storm.   ...Margaret Roberts.

Out of Bohemia.   ...Gertrude Christian Fosdick.

Owen Tudor.   ...E. Robinson.

Paris.   ...Émile Zola.   English translation.

Parisians.   ...Edward George Earle Lytton Bulwer-Lytton.

La patrie en danger.   ...Émile Erckmann et Alexandre Chatrian.

Princess Sonia.   ...Julia Magruder.

Quinze ans de bagne.   ...Louis Ulbach.

Red republic.   ...Robert W. Chambers.

Red spell.   ...Francis Gribble.

Le roman d'un brave homme.   ...Edmond About.

Les Rougon-Macquart : histoire naturelle et sociale d'une famille sous le second empire.   ...Émile Zola.

Saint Katharine's by the tower.   ...Walter Besant.

States General; or, Beginnings of the great French revolution. ...Émile Erckmann and Alexandre Chatrian.   English translation.

Story of an honest man.   ...Edmond About.   English translation.

Tale of two cities.   ...Charles Dickens.

Thirty years of Paris and my literary life.   ...Alphonse Daudet. English translation.

Tom Sylvester.   ...Thomas R. Sullivan.

Trente ans de Paris.   ...Alphonse Daudet.

Trilby.   ...George Du Maurier.

Trilby.   ...George Du Maurier.   English edition.

Tuileries.   ...Catherine C. F. Gore.

La Vendée : an historical romance.   ...Anthony Trollope.

Le ventre de Paris.   ...Émile Zola.

Wild fire.   ...George Walter Thornbury.

Within iron walls: a tale of the siege of Paris.   ...Annie Lucas.

World's verdict.   ...Marsh Hopkins, Jr.

Year one of the republic.   ...Émile Erckmann and Alexandre Chatrian.   English translation.

See also **FRANCE, FRENCH HISTORY, REPUBLICS,** French, **REVOLUTIONS,** French, and **SIEGES,** Paris.

PARIS STUDENT LIFE.

Atelier du Lys; or, Art students in the reign of terror.   ...Margaret Roberts.

Daughter of to-day.   ...Sara Jeannette Duncan (Mrs. Everard Cotes).

Guenn : a wave on the Breton coast.   ...Blanche Willis Howard.

Out of Bohemia.   ...Gertrude Christian Fosdick.

Princess Sonia.   ...Julia Magruder.

PARIS STUDENT LIFE — *Continued.*
Trilby. ...George Du Maurier.
Trilby. ...George Du Maurier. English edition.
World's verdict. ...Marsh Hopkins, Jr.

PASSION PLAY AT OBERAMMERGAU.
Am Kreuz: ein Passionsroman aus Oberammergau. ...Wilhelmine von Hillern.
On the cross. ...Wilhelmine von Hillern. English translation.

PATRIOTISM. *A Selection.*
Charles Morton. ...Mary S. B. D. Shindler.
Emigrants. ...George Imlay.
For faith and fatherland. ...Mary Bramston.
Francis Berrian, the Mexican patriot. ...T. Flint.
Lucia Dare. ...Sarah A. Dorsey. (Pseud., Filia.)
Man without a country. ...Edward Everett Hale.
Mountain patriots. ...Mrs. Alexander S. Orr.
Patriot and tory one hundred years ago. ...Julia M. Wright.
Patriot prince. ...Harriette E. Burch.
Patriotism at home. ...I. H. Anderson.
Philip Nolan's friends. ...Edward Everett Hale.
Under the yoke. ...Ivan Vazoff. English translation from the Russian.
Union: a story of the great rebellion. ...John R. Musick.

PEASANT LIFE. *A Selection.*

Bulgarian.
Under the yoke. ...Ivan Vazoff. English translation from the Russian.

French.
Contes de la Montagne. ...Émile Erckmann et Alexandre Chatrian.
Les États Généraux. Émile Erckmann et Alexandre Chatrian.
Peasant and prince. ...Harriet Martineau.
Red cockade. ...Stanley J. Weyman.
States General; or, Beginning of the great French revolution. ...Émile Erckmann and Alexandre Chatrian.

German.
Auf der Höhe. ...Berthold Auerbach.
Florian Geyer. ...Wilhelm R. Hellern.
Der Heiland von der Rhon. ...Paul Lippert.
In the olden time. ...Margaret Roberts.
Lienhard und Gertrud. ...Johann H. Pestalozzi.
On the heights. ...Berthold Auerbach. English translation.
Der prophet. ...Theodor Muegge.
Trübe Tage. ...Wilhelm Koch.

PEASANT LIFE — *Continued.*

Irish.

Across an Ulster bog. ...M. A. Hamilton.

At the rising of the moon. ...Frank Mathew.

Bog-land studies. ...Jane Barlow.

Gems of the bog : a tale of the Irish peasantry. ...Jane D. Chaplin.

Glenveigh ; or, Victims of vengeance : a tale of Irish peasant life in the present. ...Patrick S. Cassidy.

Irish idylls. ...Jane Barlow.

Manor of Glenmore ; or, Irish peasant. ...Dennis B. Kelly.

Molly and Kitty ; or, Peasant life in Ireland. ...Olga Eschenbach.

Nun's curse. ...Charlotte Eliza Lawson Cowan Riddell (Mrs. J. H. Riddell).

Phil Purcel and other tales of Ireland. ...William Carleton.

Tales of the Irish peasantry. ...Anna Maria Hall (Mrs. S. C. Hall).

Traits and stories of the Irish peasantry. ...William Carleton.

Ulic O'Donnell ; or, Irish peasant's progress. ...Denis Holland.

Where the Atlantic meets the land. ...Caldwell Lipsett.

Russian.

Annals of a sportsman. ...Ivan Sergevitch Turgenef. English translation from the Russian.

Highway of sorrow. ...Hannah Smith. (Pseud., Hesba Stretton.)

Scotch.

Daughter of Fife. ...Amelia Edith Barr.

Harry Muir : a story of Scottish life. ...Margaret O. W. Oliphant.

Laurie Todd ; or, Settlers in the woods. ...John Galt.

Muckle Jock and other stories of peasant life in the North. ...Malcolm McLennan.

Peasant life : sketches of the villages and field laborers in Glenaldie. ...Malcolm McLennan.

Scottish peasant's fireside tales. ...Alexander Bethune.

Tales of the Scottish peasantry. ...Henry Duncan.

Sicilian.

Cavalleria Rusticana. ...Giovanni Verga.

PEDAGOGY.

Cap and gown comedy : a schoolmaster's stories. ...Robert Hope Moncrieff. (Pseud., Ascott R. Hope.)

Chautauquans. ...John Habberton.

Diplomatic disenchantments. ...Edith Bigelow.

District school as it was. ...Warren Burton.

Frontier schoolmaster. ...C. Thomas.

His masters : a story of school forty years ago. ...S. S. Pugh.

PEDAGOGY — *Continued.*
Hoosier school-boy.   ...Edward Eggleston.
Hoosier schoolmaster.   ...Edward Eggleston.
Lal.   ...William Alexander Hammond.
Little schoolmaster Mark.   ...Joseph Henry Shorthouse.
Locke Amsden; or, Schoolmaster.   ...Daniel P. Thompson.
Log schoolhouse on the Columbia.   ...Hezekiah Butterworth.
Patriot schoolmaster.   ...Hezekiah Butterworth.
Phil Vernon and his schoolmasters.   ...Byron A. Brooks.
Professor.   ...Charlotte Brontè.   (Pseud., Currer Bell.)
Robert Urquhart.   ...Gabriel Setoun.
Roderick Hume.   ...C. W. Bardeen.
Romance of a schoolmaster.   ...Edmonde Amicis.
School experiences of a fag at a private and public school.   ...George
    Melly.
Schoolmaster and his son.   ...Carl H. Caspari.   English translation.
Der Schulmeister und sein Sohn : eine Erzählung aus dem dreissig-
    jähr Kriege.   ...Carl H. Caspari.
Thing that hath been.   ...A. H. Gilkes.
Well out of it.   ...John Habberton.
Wrong man.   ...Dorothea Gerard.
Yankee school-teacher in Virginia.   ...Lydia W. Baldwin.

PENINSULAR WAR.   1808-1814.
Adventures of an aide-de-camp ; or, Campaign in Calabria.   ...James
    Grant.
Alice Lorraine.   ...Richard Doddridge Blackmore.
Bailén.   ...Benito Pérez Galdós.
Bivouac ; or, Stories of the Peninsular war.   ...William Hamilton
    Maxwell.
Cadiz.   ...Benito Pérez Galdós.
Charles O'Malley, the Irish dragoon.   ...Charles James Lever.
Don Alonso ; ou, l'Espagne hist. contemporaine, 1788–1823.   ...Nar-
    cisse A., Comte de Salvandy.
El equipage del rey José.   ...Benito Pérez Galdós.
Felix Alvarez ; or, Manners in Spain.   ...Alex. R. C. Dallas.
Gerona.   ...Benito Pérez Galdós.
Her heart was true : a story of the Peninsular war.   ...E. M. Cuttim.
Juan Martin el empecinado.   ...Benito Pérez Galdós.
King's own borderers.   ...James Grant.
Napoléon en Chamartin.   ...Benito Pérez Galdós.
El 19 de Marzo y el 2 de Mayo.   ...Benito Pérez Galdós.
Phantom regiment ; or, Stories of Ours.   ...James Grant.
Romance of war; or, Highlanders in Spain.   ...James Grant.
Saragossa ; or, Houses of Castello and De Arno.   ...E. A. Archer.

PERSECUTIONS — *Continued.*

Camp on the Severn.   ...Augustine D. Crake.

Cave in the hills.   ...Anon.

Constance Sherwood: an autobiography of the sixteenth century.
...Georgiana C. L. G. Fullerton.

Daughters of Pola.   ...Anon.

Diotima: eine culturhistorische Novelle aus der Zeit der diocletianischen Verfolgung.   ...Wilhelm Tangermann. `

Egyptian Wanderers.   ...John Mason Neale.   (Pseud., Aurelius Gratianus.)

Exiles of Lucerne.   ...John R. Macduff.

Farm of Aptonga: a story for children, of the time of Saint Cyprian.
...John Mason Neale.   (Pseud., Aurelius Gratianus.)

For the Master's sake : a story of the days of Queen Mary.   ...Emily Sarah Holt.

Godfrey Brenz: a tale of persecution.   ...Sarah J. Jones.

History of Nicolas Muss.   ...Charles DuBois Melly.

House of Yorke.   ...Mary A. Tincker.

In Editha's days.   ...Mary E. Bamford.

Isoult Barry of Wynscote: a tale of Tudor times.   ...Emily Sarah Holt.

Jacques Bonneval; or, Days of the dragonnades.   ...Anne Manning.

Jarousseau, le pasteur du désert.   ...Pierre Clément Eugène Pelletan.

Jean Jarousseau, pastor of the desert.   ...Pierre Clément Eugène Pelletan.   English translation.

Justice Warren's daughter.   ...Olive M. Birrell.

Die letzten Humanisten.   ...Adolf Stern.

Martyrs of Carthage.   ...Mrs. J. B. Peploe (Webb).

Martyrs of Spain.   ...Elizabeth Charles.

Mary Bunyan, the dreamer's blind daughter.   ...Sallie Rochester Ford.

Men of the Moss-Hags.   ...Samuel Rutherford Crockett.

Money God.   ...Abel Quinton.

Priest and the Huguenots; or, Persecution in the age of Louis XV.
...Laurence Louis Félix Bungener.   English translation.

Protestant.   ...Anna Eliza Bray.

Richard Hume.   ...George E. Sargent.

Robin Tremayne of Bodmin: a tale of the Marian persecution.
...Emily Sarah Holt.

Three hundred years ago.   ...William Henry Giles Kingston.

Tor Hill.   ...Horace Smith.   (Pseud., Paul Chatfield.)

Trois sermons sous Louis XV.   ...Laurence Louis Félix Bungener.

Twice crowned.   ...Harriette B. MacKeever.

Valerius: a Roman story.   ...John Gibson Lockhart.

PERSECUTIONS — *Continued.*

Victor: a tale of the great persecutions. ...George G. Perry.

Villegagnon: a tale of the Huguenot persecution. ...William Henry Giles Kingston.

Within sea walls; or, How the Dutch kept the faith. ...E. H. Walshe and George E. Sargent.

Word, only a word. ...Georg Moritz Ebers. English translation.

Ein Wort. ...Georg Moritz Ebers.

See also **PURITANS, QUAKERS, and WALDENSES.**

PERSIA.
Description, History, Manners, and Customs.

Eine Aegyptische Königstochter. ...Georg Moritz Ebers.

Arabian nights entertainment. ...Anon.

Delaphaine. ...Mansfield Tracy Walworth.

Egyptian princess. ...Georg Moritz Ebers. English translation.

Emir Malek, prince of the assassins. ...Anon.

Hussein the hostage; or, Boy's adventure in Persia. ...G. Norway.

Julamerk. ...Mrs. J. B. Peploe (Webb).

Shaving of Shagpat: an Arabian entertainment and Farina. ...George Meredith.

Thousand and one days: a collection of Persian tales. ...Justin McCarthy.

Zoroaster. ...Francis Marion Crawford.

PERU.
Description, History, Manners, and Customs.

Coitlan: a tale of the Inca world. ...Anson Uriel Hancock.

Hoffnungen in Pérou. ...Ernst Baron von Bibra.

Inca queen; or, Lost in Peru. ...J. Evelyn.

Les Incas; ou, La destruction de l'empire du Pérou. ...J. F. Marmontel.

Incas; or, Destruction of the empire of Peru. ...J. F. Marmontel. English translation.

Letters written by a Peruvian princess. ...F. H. de Graffigny. English translation.

Lettres d'une Péruvienne. ...F. H. de Graffigny.

Manco the Peruvian chief. ...William Henry Giles Kingston.

Secret of the Andes: a romance. ...F. Hassaurek.

Stories of the conquest of Mexico and Peru. ...W. Dalton.

Under the Southern cross. ...Anne Manning.

PHILADELPHIA, PENNSYLVANIA.

Arthur Mervyn. ...Charles Brockden Brown.

Miss MacRéa: Roman historique. ...Michel René Hillard d'Auberteuil.

PHILADELPHIA, PENNSYLVANIA — *Continued.*

Old bell of independence; or, Philadelphia in 1776.   ...Henry C. Watson.

Page from the colonial history of Philadelphia.   ...Blackbeard.

Quaker soldier; or, British in Philadelphia: a romance of the revolution.   ...J. Richter Jones.

Sons and daughters.   ...Ellen W. Kirk.   (Pseud., Henry Hayes.)

PHILIP VI. OF FRANCE (VALOIS). *Reigned 1328-1350.*

Le bâtard de Manléon.   ...Alexandre Dumas.

Die Belagerung von Calais.   ...Carl Weichselbaumer.

Blanche d'Évreaux.   ...Amélie J. Candeille.

Eustache de Saint Pierre; or, Surrender of Calays.   ...Henry W. Herbert.   (Pseud., Frank Forester.)

Half brothers; or, Head and the hand.   ...Alexandre Dumas. English translation.

Robert d'Artois.   ...John Humphrey Saint Aubyn.

Le siège de Calais.   ...Claudine A. G. Tencin.

PHILIPPE D'ORLEANS. See **LOUIS XV. OF FRANCE.**

PHILOSOPHICAL NOVELS. *A Selection.*

Alkahest; or, Home of Cläes.   ...Honoré de Balzac.   English translation.

Charenton; or, Follies of the age.   ...J. H. Lelarge Lourdoueix. English translation.

Emigrants; or, History of an expatriated family.   ...George Imlay.

Études philosophiques.   ...Honoré de Balzac.

Few days in Athens.   ...Frances Wright d'Arusmont.

Les folies du siècle.   ...J. H. Lalarge Lourdoueix.

Hypatia.   ...Charles Kingsley.

Leibnitz und die beiden Kurfürstinnen.   ...Hermann Klencke.

Louis Lambert.   ...Honoré de Balzac.

Louis Lambert.   ...Honoré de Balzac.   English translation.

Macpherson, the great confederate philosopher.   ...Alfred C. Hills.

Magic skin.   ...Honoré de Balzac.   English translation.

Opinions of a philosopher.   ...Robert Grant.

La peau de chagrin.   ...Honoré de Balzac.

Poet and prince.   ...Berthold Auerbach.

Séraphita.   ...Honoré de Balzac.

Seraphita.   ...Honoré de Balzac.   English translation.

Spinoza.   ...Berthold Auerbach.

PIKE'S PEAK, COLORADO.

Literary courtship under the auspices of Pike's Peak.   ...Anna Fuller.

Pasque flowers from Pike's Peak.   ...Susan T. Dunbar.

PILGRIMAGE. *A Selection.*
Cheshire pilgrims. ...Frances M. Wilbraham.
Diary of a pilgrimage. ...Jerome K. Jerome.
Eva and pilgrims of the Rhine. ...Edward George Earle Lytton Bulwer-Lytton.
Faith White's letter-book. ...Mrs. M. H. Whiting.
Florine, princess of Burgundy. ...William B. MacCabe.
Ivanda; or, Pilgrim's quest. ...Claude Bray.
Lady Sybil's choice: a tale of the crusades. ...Emily Sarah Holt.
Lancashire witches. ...William Harrison Ainsworth.
Little pilgrim. ...Margaret O. W. Oliphant.
Moravian Indian boy. ...Anon.
Norseman's pilgrimage. ...Hjalmar H. Boyesen. English translation.
Pilgrims. ...W. Carlton Dawe.
Strangers and pilgrims. ...Mary E. Braddon.
Their pilgrimage. ...Charles Dudley Warner.
Two little pilgrims' progress. ...Frances Hodgson Burnett.
See also **CRUSADES** and **PLYMOUTH.**

PIRACY. See **BUCCANEERS.**

PLAGUES, FAMOUS. *A Selection.*
Andrew Golding: a tale of the great plague. ...Annie E. Keeling. (England, 1665.)
Betrothed. ...Alessandro Manzoni. English translation. (Milan.)
Brave men of Eyam; or, Tale of the great plague. ...Edward N. Hoare. (England, 1665.)
Caleb Field: a tale of the Puritans. ...Margaret O. W. Oliphant. (England, 1665.)
Captain Jacques: a romance of the time of the plague. ...Edward Fitzgibbon. (Pseud., Somerville Gibney.) (England, 1665.)
Carved cartoon. ...Austin Clare. (England, 1665.)
Cherry and violet: a tale of the great plague. ...Anne Manning. (England, 1665.)
" God's providence house." ...Isabel Banks. (England, 1665.)
History of the great plague in London in 1665. ...Daniel Defoe.
I promessi sposi. ...Alessandro Manzoni. (Milan.)
Jeannett Cragg, the Quakeress. ...Maria Wright. (England, 1665.)
Old Saint Paul's: a tale of the plague and the fire. ...William Harrison Ainsworth. (England, 1665.)
Oliver Wyndham; or, Tale of the great plague. ...Mrs. J. B. Peploe (Webb).
Ralph Redman's atonement. ...Herbert V. Mills. (England, 1598.)
Rosemary: a tale of the fire in London. ...Georgiana C. L. G. Fullerton. (England, 1665.)

PLAGUES, FAMOUS — *Continued.*
Saint Dunstan's clock: a story of 1666.   ...E. Ward.
Two years ago.   ...Charles Kingsley.

PLYMOUTH, MASSACHUSETTS.
Betty Alden, the first-born daughter of the Pilgrims.   ...Jane Goodwin Austin.
Daughter of the Puritans.   ...Anna B. Bensel. (In New England Magazine, May, 1886.)
David Alden's daughter, and other stories of colonial times.   ...Jane Goodwin Austin.
Doctor Le Baron and his daughters.   ...Jane Goodwin Austin.
Faith White's letter book.   ...M. H. Whiting.
Golden hair: a tale of the Pilgrim Fathers.   ...Frederick C. L. Wraxall.
Hobomok.   ...Lydia Maria Child.
Justice Warren's daughter.   ...Olive M. Birrell.
Little Pilgrims at Plymouth.   ...Frances A. Humphrey.
May-pole of Merrymount.   ...Nathaniel Hawthorne.
Merrymount: a romance of the Massachusetts colony.   ...John Lothrop Motley.
Nameless nobleman.   ...Jane Goodwin Austin.
New world planted; or, Adventures of the forefathers of New England.   ...J. Croswell.
Peep at the Pilgrims in 1636.   ...Harriet V. Cheney.
Pictures of the olden time as shown in the fortunes of a family of pilgrims.   .. E. H. Sears.
Pilgrims of New England: a tale of the early American settlers.   ...Mrs. J. B. Peploe (Webb).
Plymouth and the Pilgrims.   ...J. Banvard.
Priscilla; or, Trials for the faith: historical tale of the Puritans and the Baptists.   ...J. Banvard.
Puritan and the Quaker.   ...Rebecca I. Beach.
Seeking a country; or, Home of the Pilgrims.   ...Edward N. Hoare.
Standish of Standish.   ...Jane Goodwin Austin.
White chief among the red men; or, Knight of the golden Melice.   ...J. T. Adams.
See also **WAR OF KING PHILIP OF POKANOKET, 1607.**

POCAHONTAS.   *Lived 1595–1616.*
Captain Smith and the Princess Pocahontas: an Indian tale.   ...J. Davis.   [Cooke.
My Lady Pocahontas: a true relation of Virginia.   ...John Esten
Pocahontas: a legend with historical and traditional notes.   ...Mary W. Moseby.
Youth of the old Dominion.   ...S. Hopkins.
See also **VIRGINIA.**

POLAND.

Description, History, Manners, and Customs.

Bez Dogmatu. ...Henryk Sienkiewicz.
Blue roses. ...Charlotte L. H. Dempster.
Casimir Maremma. ...Sir Arthur Helps.
Deluge. ...Henryk Sienkiewicz. English translation.
Iermola. ...Joseph Ignatius Kraszewski.
In the old, old château : a story of Russian Poland. ...R. H. Savage.
Iza : a story of life in Russian Poland. ...R. H. Savage.
Janko Muzykant. ...Henryk Sienkiewicz.
Jermola : obrazki wiejskie. ...Joseph Ignatius Kraszewski.
Jew. ...Joseph Ignatius Kraszewski. English translation.
Lezsko the bastard. ...Alfred Austin.
Lost cause. ...W. W. Aldred.
Maid of Warsaw. ...Ernst Jones.
Modern vassal. ...John Wilmer.
Der Neue Hiob. ...Leopold von Sacher-Masoch.
New Job. ...Leopold von Sacher-Masoch. English translation.
Ognien i mieczem. ...Henryk Sienkiewicz.
Orthodox. ...Dorothea Gerard.
Pan Michael. ...Henryk Sienkiewicz. English translation.
Polish tales. ...Catherine G. F. Gore.
En Polsk Familie. ...Johann Carsten von Hauch.
Potop. ...Henryk Sienkiewicz.
Thaddeus of Warsaw. ...Jane Porter.
With fire and sword. ...Henryk Sienkiewicz. English translation.
Without dogma. ...Henryk Sienkiewicz. English translation.
Wizard king. ...David Ker.
Yanko, the musician. ...Henryk Sienkiewicz. English translation.

POLITICS. *A Selection.*

American.

American politician. ...Francis Marion Crawford.
Among the law-makers. ...Edmond Alton.
Archbishop. ...O. S. Bellisle. (" Know-nothings.")
Avery Glibun. ...Robert H. Newell.
Cabinet minister. ...Catherine G. F. Gore.
Christian Vellacott, the journalist : a story of royalism, Jesuitism, and republicanism. ...Hugh S. Scott. (Pseud., Merriman H. Seton.)
Conspirators. ...Eliza Ann Dupuy.
Demagogue. ...David R. Locke. (Pseud., Petroleum Nasby.)
Democracy. ...Clarence King.
Famous victory. ...Anon.

POLITICS — *Continued.*

Five hundred majority; or, Days of Tammany. ...Willys Niles.

Humbled pride: a story of the Mexican war. ...John R. Musick.

"I'm fur 'im"; or, Solid for Mulhooly. ...Rufus E. Shapley.

Life and death of Sam in Virginia. ...Gardiner. (" Know-nothings.")

Looking backward, 2000–1887. ...Edward Bellamy.

Love or a name. ...Julian Hawthorne.

Mr. East's experiences in Mr. Bellamy's world. ...Conrad Wilbrandt.

Monikins. ...James Fenimore Cooper.

Politician's daughter. ...Myra S. Hamlin.

Professor Conant. ...Lucius Seth Huntington.

Regicides. ...Frederick Hull Cogswell.

Rivals. ...Jeremiah Clemens.

Silver Baron. ...Carlton Waite. (Silver craze of 1896.)

Stanhope Burleigh. ...Charles Edward Lester. (" Know-nothings.")

Story of Rodman Heath; or, Mugwumps. ...By one of them.

Through one administration. ...Frances Hodgson Burnett.

English.

Alton Locke. ...Charles Kingsley.

Coningsby. ...Benjamin Disraeli.

Corruption. ...Percy White.

De Vere. ...Robert P. Ward.

Endymion. ...Benjamin Disraeli.

Illustrations of political economy. ...Harriet Martineau.

Lothair. ...Benjamin Disraeli.

Marcella. ...Mary Augusta Ward (Mrs. Humphry Ward).

Professor Conant. ...Lucius Seth Huntington.

Right honorable: a romance of society and politics. ...Justin McCarthy and Rose Murray Prior Praed (Mrs. Campbell Praed).

Sybil; or, Two nations. ...Benjamin Disraeli.

Themaine. ...Robert P. Ward.

Yeast: a problem. ...Charles Kingsley.

French.

Citoyenne Jacqueline: a woman's lot in the great French revolution. ...Henrietta Keddie. (Pseud., Sarah Tytler.)

Contes de la Montagne. ...Émile Erckmann et Alexandre Chatrian.

French wines and politics. ...Harriet Martineau.

Histoire d'un homme du peuple. ...Émile Erckmann et Alexandre Chatrian.

Member from Paris. ...Grenville Murray.

Scènes de la vie politique. ...Honoré de Balzac.

Scenes from the political life. ...Honoré de Balzac. English translation.

POLITICS — *Continued.*

German.

Cipher dispatch. ...K. von Bayer. (Pseud., Robert Byr.) English translation.

For scepter and crown. ...Johann Ferdinand Martin Oskar Meding. (Pseud., Gregor Samarow.) English translation.

Eine geheime Depesche. ...K. von Bayer. (Pseud., Robert Byr.)

Um Scepter und Kronen. ...Johann Ferdinand Martin Oskar Meding. (Pseud., Gregor Samarow.)

Vanished emperor. ...Percy Andreae.

Waldfried. ...Berthold Auerbach.

Hungarian.

Dr. Dumany's wife. ...Mór Jókai. English translation.

Eyes like the sea. ...Mór Jókai. English translation.

Irish.

Adventures of Mick Callighim, M. P.: a story of home rule. ...W. R. Anckettill.

Ballytubber; or, Scotch settler in Ireland. ...Anon.

Boycotted: a story. ...M. Morley.

Boycotted household. ...Letitia McClintock.

Convict No. 25; or, Clearances of West Meath. ...J. Murphy.

Croppy: a tale of 1798. ...Michael Benim.

Darcy and his friends: an Irish tale. ...Joseph McKim.

Doreen: the story of a singer. ...Ada Ellen Bayly. (Pseud., Edna Lyall.)

Earl of Effingham. ...Lalla McDowell.

Fair Saxon. ...Justin McCarthy.

Falcon family; or, Young Ireland. ...Marmion W. Savage.

Grace Cassidy; or, Repealers. ...Marguerite Gardiner.

Heart of Erin: an Irish story of to-day. ...Elizabeth Casey. (Pseud., E. Owens Blackburne.)

Heart of Tipperary. ...W. P. Ryan.

Ierne. ...William Stewart Trench.

Irish Widow's son; or, Pikemen of '98. ...Con. O'Leary.

Knight of Gwynne. ...Charles James Lever.

Land leaguers. ...Anthony Trollope.

Lloyd Pennant: a tale of the West. ...Ralph Neville.

Loyal and lawless. ...Ulick R. Burke.

Mr. Butler's ward. ...Frances Mabel Robinson.

Modern Dædalus. ...Thomas Greer.

My lords of Strogue: a chronicle of Ireland from the convention till the union. ...Lewis S. Wingfield.

Nevilles of Garretstown: a tale of 1760. ...Anna Marsh.

POLITICS — *Continued.*

Norah Moriarty; or, Revelations of modern Irish life. ...Amos Reade.

Olive Lacy : a tale of the Irish rebellion of 1798. ...Anna Argyle.

Pearl of Lisnadoon; or, Glimpse of our Irish neighbours. ...Mrs. E. J. Ensell.

Phineas Finn, the Irish member. ...Anthony Trollope.

Plan of campaign : a story of the fortune of war. ...Frances Mabel Robinson.

Priest's blessing; or, Poor Patrick's progress from this world to a better. ...Harriet Jay.

Ridgeway : an historical romance of the Fenian invasion of Canada. ...Scian Dubh.

Rose de Blaquiere. ...Anna Maria Porter.

Rose O'Connor : a story of the day. ...Emily Fox. (Pseud., Toler King.)

Siege of Bodike : a prophecy of Ireland's future. ...Edward Lester.

Terence M'Gowan, the Irish tenant. ...G. L. Tottenham.

Valentine M'Clutchy, the Irish agent; or, Chronicles of Castle Cumber. ...William Carleton.

Newfoundlandic.

Under the great seal. ...Joseph Hatton.

Polish.

Bez Dogmatu. ...Henryk Sienkiewicz.

Maid of Warsaw. ...Ernest Jones.

Thaddeus of Warsaw. ...Jane Porter.

Without dogma. ...Henryk Sienkiewicz. English translation.

Russian.

Condemned as a Nihilist : a story of escape from Siberia. ...George Alfred Henty.

Female Nihilist. ...Ernst Lavigne. English translation.

In the dwellings of silence. ...Walker Kennedy.

Ivan Vejeeghen. ...Thaddei Bulgarin.

Narka the Nihilist. ...Kathleen O'Meara.

Le roman d'un Nihiliste. ...Ernst Lavigne.

POMPEII.

Last days of Pompeii. ...Edward George Earle Lytton Bulwer-Lytton.

Slave girl of Pompeii. ...Emily Sarah Holt.

PORTUGAL.

Description, History, Manners, and Customs.

Agnes Surriage. ...Edwin L. Bynner.

Don Sebastian. ...Anna Maria Porter.

PORTUGAL — *Continued.*

Foreign tales and traditions. ...George G. Cunningham.

Ignez de Castro. ...Stéphanie Félicité, Comtesse de Genlis. English translation.

Inès de Castro. ...Stéphanie Félicité, Comtesse de Genlis.

Ines; or, Bride of Portugal. ...Isabella Harwood. (Pseud., Ross Neil.)

Lucta de gigantes. ...Camillo Castello Branco.

Ni rey ni roque. ...Patricio de la Escosura.

O, ou, o terremoto do 1755. ..Marquez de Pombal.

Prime minister. ...William Henry Giles Kingston.

Stories of Torres Vedras. ...John G. Milligen.

Talba. ...Anna Eliza Bray.

PRISONS AND PRISONERS.

Algerine captive. ...Royall Tyler.

Border spy. ...Colonel Hazeltine.

Boscobel. ...William Harrison Ainsworth.

Caged lion. ...Charlotte Mary Yonge.

Le comte de Monte-Cristo. ...Alexandre Dumas.

Confessions of a convict. ...Julian Hawthorne.

Convict. ...George Payne Rainsford James.

Convict No. 25. ...John Murphy.

Count of Monte Cristo. ...Alexandre Dumas. English translation.

Crusades and captives : a tale of the children's crusade. ...George E. Merrill.

Fardorougha, the miser. ...William Carleton.

French prisoners. ...Edward Bertz.

Highway of sorrow. ...Hannah Smith. (Pseud , Hesba Stretton.)

His broken sword. ...Winnesheik Louise Taylor.

His natural life. ...Marcus Clarke.

Historical tales; or, Romance of reality. ...Charles Morris.

L'homme au masque de fer. ...Chevalier de Mouhy.

In prison and out. ...Hannah Smith. (Pseud., Hesba Stretton.)

It is never too late to mend. ...Charles Reade.

Little Dorrit. ...Charles Dickens.

Memoirs of Jane Cameron, female convict. ...Frederick William Robinson.

Les miserables. ...Victor Hugo. English translation.

Les misérables. ...Victor Hugo.

Moondyne: a story from the under-world. ...John Boyle O'Reilly.

Oliver Ellis; or, Fusiliers. ...James Grant.

Pickwick papers. ...Charles Dickens.

Prison princess. ...Arthur Griffiths.

Prisoners of the Border. ...Hamilton Meyers.

PRISONS AND PRISONERS — *Continued.*
Prisoners of the mill.   ...Colonel Hazeltine.
Refugees.   ...Arthur Conan Doyle.
Rogue's march.   ...Ernest Hornung.
Time's revenges.   .. David Christie Murray.
Uline's escape; or, Hid with the nuns.   ...Mrs. Alexander S. Orr.
Unknown to history.   ...Charlotte Mary Yonge.

PROTESTANTISM.
Abbot.   ...Sir Walter Scott.
Bolsover castle: a tale from Protestant history of the sixteenth cen·
     tury.   ...Anon.
Boycotted household.   ...Letitia McClintock.
By fire and sword : a story of the Huguenots.   ...Thomas Archer.
Camisard; or, Protestants of Languedoc.   ...Anon.
Chillon ; or, Protestants of the sixteenth century.   ...Jane L.
     Willyams.
Christlich oder Päpstlich ?   ...Eduard Jost.
Claude the colporteur.   ...Anne Manning.
Constance Sherwood: an autobiography of the sixteenth century.
     ...Georgiana C. L. G. Fullerton.   ·
Faithful but not famous.   ...Anon.
For the Master's sake : a story of the days of Queen Mary.   ...Emily
     Sarah Holt.
Headsman ; or, Abbaye des Vignerous.   ..James Fenimore Cooper.
Isoult Barry of Wynscote : a tale of Tudor times.   ...Emily Sarah
     Holt.
Jacques Bonneval ; or, Days of the dragonnades.   ...Anne Manning.
Jarousseau le pasteur du désert.   ...Pierre Clément Eugène Pelletan.
Jean Jarousseau, the pastor of the desert.   ...Pierre Clément
     Eugène Pelletan.   English translation.
Kenilworth.   ...Sir Walter Scott.
Monastery.   ...Sir Walter Scott.
Mountain patriots : a tale of the reformation in Savoy.   ...Mrs.
     Alexander S. Orr.
Priest and the Huguenot; or, Persecution in the age of Louis XV.
     ...Laurence Louis Félix Bungener.   English translation.
Protestant.   ...Anna Eliza Bray.
Robin Tremayne of Bodmin : a tale of the Marian persecution.
     ...Emily Sarah Holt.
Spanish barber.   ...Anne Manning.
Three hundred years ago.   ...William Henry Giles Kingston.
Trois sermons sous Louis XV.   ...Laurence Louis Félix Bungener.
Twice crowned : a story of the days of Queen Mary.   ...Harriet B.
     MacKeever.

PROTESTANTISM .— *Continued.*

Uline's escape; or, Hid with the nuns. ...Mrs. Alexander S. Orr.

Der Untergang der Protestanten in Oberösterreich. ...Franz Lubojatzky.

Westminster Abbey. ...Edith Robinson,

Won and not one. ...Emily Lucas Blackall.

See also **HUGUENOTS** and **PERSECUTIONS**.

PRUSSIA.

Description, History, Manners, and Customs.

Der alte Dessauer. ...Franz Lubojatzky.

Der alte Fritz und seine Zeit. ....Clara M. Mundt. (Pseud., Louise Mühlbach.)

L'argent des autres. ...Émile Gaboriau.

Auf Befehl des Königs. ...Clarissa Lohde.

Aus den Tagen zweier Könige. ...Friedrich Adami.

Der Bachtanz zu Langenselbold. ...Philipp F. W. Oertel.

Die Bären von Augustusburg : eine Erzählung aus der sächsischen Geschichte des 18 Jahrhunderts. ...Carl Gustav Nieritz.

Bears of Augustusburg. ...Carl Gustav Nieritz. English translation.

Belfry of St. Jude. ...Esmé Stuart.

Berlin and Sans-Souci. ...Clara M. Mundt. (Pseud., Louise Mühlbach.) English translation.

Bischof und König. ...Mariam Tenger.

Bombardier H. and Corporal Dose. ...Friedrich W. Hacklaender.

Die Brautschau Friedrichs des Grossen. ...Julius Bacher.

Cabanis. Vaterländischer Roman. ...Georg Wilhelm H. Haering.

Comedy of terrors. ...James De Mille.

Cyrus. ...Christoph M Wieland.

Der Domherr. ...Jodocus D. H. Temme.

Dorothea Cappel. ...Emilie Friederike S. Lohmann.

Elizabeth von Guttenstein. ...Caroline Pichler.

Elsie's dowry : a tale of the Franco-Prussian war. ...Emma Leslie.

Enemy's friendship. ...S. M. S. Clarke.

Fifteen years. ...Thérèse Albertine Luise von Jacob Robinson. (Pseud., Talvi.) English translation.

Frederick the Great and his court. Clara M. Mundt. (Pseud., Louise Mühlbach.) English translation.

Friedrich der Grosse. . ...Eugen H. von Dedenroth.

Friedrich der Grosse und sein Hof. ...Clara M. Mundt. (Pseud., Louise Mühlbach.)

Fünfzehn Jahre. ...Thérèse Albertine Luise von Jacob Robinson. (Pseud., Talvi.)

Gate and the glory beyond it. ...Onyx, pseud.

PRUSSIA — *Continued.*

Gräfin Lichtenan.   ...Robert Springer.

Der grosse Kurfürst in Preussen.   ...Ernst A. A. G. Wichert.

Histoire du plébiscite.   ...Émile Erckmann et Alexandre Chatrian.

Historische Novellen über Friedrich II. von Preussen und seine Zeit.   ...Joseph E. C. Bischoff.

Holderlin.   ...Heribert Rau.

Johann Gotzkowsky, der Kaufmann von Berlin.   ...Clara M. Mundt. (Pseud., Louise Mühlbach.)

Kentucky's love; or, Roughing it about Paris.   ...Edward King.

Leonie; or, Light out of darkness.   ...Annie Lucas.

Louisa of Prussia and her time.   ...Clara M. Mundt. (Pseud., Louise Mühlbach.)   English translation.

Margaret Muller : story of the late war in France.   ...Eugenie Bersier.

Max Kromer: a story of the siege of Strasburg.   ...Hannah Smith. (Pseud., Hesba Stretton.)

Merchant of Berlin.   ...Clara M. Mundt. (Pseud., Louise Mühlbach.)   English translation.

Öfverste Stobée, ella holsteinska partiet under frihetstiden.   ...Hermann Bjursten.

Old Fritz and the new era.   ...Clara M. Mundt. (Pseud., Louise Mühlbach.)   English translation.

Other people's money.   ...Émile Gaboriau.   English translation.

Parisians.   ...Edward George Earle Lytton Bulwer-Lytton.

Powder and Gold.   ...Levin Schuecking.   English translation.

Die Prinzessin von Wolfenbüttel.   ...Johann Heinrich D. Zschokke.

Pulver und gold.   ...Levin Schuecking.

Le roman d'un brave homme.   ...Edmond About.

Six years ago.   ...James Grant.

Die Soldaten Friedrichs des Grossen.   ...Julius von Wickede.

Sophie Charlotte, die philosophische Königin.   ...Julius Bacher.

Story of an honest man.   ...Edmond About.   English translation.

Story of the plébiscite.   ...Émile Erckmann and Alexandre Chatrian.   English translation.

Twins of Saint-Marcel.   ...Mrs. Alexander S. Orr.

Valentin : a French boy's story of the Sedan.   ...Henry Kingsley.

Vaterländische Romane.   ...Johann Georg L. Hesekiel.

Der Vetter im Consistorium.   ...Philipp F. W. Oertel.

Von Fels zum Meer.   ...Hans von Zollern.

Vor hundert Jahren.   ...Carl W. Ritter von Martine.

Within iron walls: a tale of the siege of Paris.   ...Annie Lucas.

Workman and soldier: a tale of Paris life during the siege and rule of the commune.   ...James F. Cobb.

Young Franc-Tireurs and their adventures in the Franco-Prussian war.   ...George Alfred Henty.

PSYCHOLOGICAL NOVELS. *A Selection.*

Beggars all. ...Lily Dougall.
Blood Royal. ...Grant Allen.
Can such things be ? ...Ambrose Bierce.
Conscience. ...Hector Malot.
Doctor Mirabel's theory. ...Ross George Dering.
Dreams of the dead. ...E. Staunton, pseud.
In the first person. ...Maria Louise Pool.
John Applegate, surgeon. ...Mary Harriett Norris.
Moral dilemma. ...Annie Thompson.
Nephelé. ...Francis William Bourdillon.
Quality of mercy. ...William Dean Howells.
Les roches blanches. ...Édouard Rod.
Statement of Stella Maberly. ...Thomas Anstey Guthrie. (Pseud., F. Anstey.)
Tragic comedians. ...George Meredith.
Wages of sin. ...Mrs. William Harrison. (Pseud., Lucas Malet.)
Wedded to a genius. ...Neil Christison.
White rocks. ...Édouard Rod. English translation.
Wish. ...Hermann Sudermann. English translation.
Der Wunsch. ...Hermann Sudermann.

PURITANS.

Alice Lisle: a tale of Puritan times. ...Richard King.
Betty Alden, the first-born daughter of the Pilgrims. ...Jane Goodwin Austin.
Caleb Field: a tale of the Puritans. ...Margaret O. W. Oliphant.
Daughter of the Puritans. ...Anna B. Bensel. (In New England Magazine, 1886, page 452.)
Fair Puritan: an historical romance of the days of witchcraft. ...Henry W. Herbert. (Pseud., Frank Forester.)
Micah Clarke, his statement as made to his three grandchildren. ...Arthur Conan Doyle.
Pilgrims and Puritans. ...Nina Moore Tiffany. ·
Priscilla; or, Trials for truth: historical tale of the Puritans and Baptists. ...J. Banvard.
Puritan and his daughter. ...James K. Paulding.
Puritan and the Quaker. ...Rebecca G. Beach.
Puritan lover. ...Laura C. S. Fessenden.
Puritan's grave. ...William P. Scargill.
Shawmut. ...C. K. True.
Standish the Puritan. ...Eldred Grayson.
Tales of the Puritans. ...Delia Bacon.
See also **MASSACHUSETTS, NEW ENGLAND,** and **PLYMOUTH.**

QUAKERS.

Clayton's rangers; or, Quaker partisans. ...E. H. Williamson.
Colonial wooing. ...C. Conrad Abbott.
Fighting Quakers. ...Augustine Duganne.
Friend Olivia. ...Amelia Édith Barr.
Fugitives; or, Quaker scout of Wyoming: a tale of the massacre of 1778. ...Edward S. Ellis.
Hope's heart bells. ...Sara L. Oberholtzer.
Jennett Cragg the Quakeress. ...Maria Wright.
Justice Warren's daughter. ...Olive M. Birrell.
Lost illusion. ...Keith Leslie, pseud.
Nathan the Quaker. ...Robert Montgomery Bird.
Puritan and the Quaker. ...Rebecca G. Beach.
Quaker girl of Nantucket. ...Mary and Catherine Lee.
Quaker home. ...George Fox Tucker.
Quaker idyls. ...Mrs. S. M. H. Gardner.
Quaker partisans. ...E. H. Williamson.
Quaker soldier; or, British in Philadelphia: a romance of the revolution. ...J. Richter Jones.
Rachel Stanwood. ...Lucy Gibbons Morse.
Story of Kennett. ...Bayard Taylor.
True hero; or, Story of William Penn. ...William Henry Giles Kingston.
Twin heroes. ...F. A. Reed.

QUEBEC, CANADA.

" Le chien d'or ": a legend of Quebec. ...William Kirby.
Seats of the mighty. ...Gilbert Parker.

RAMESES OF EGYPT (SESOSTRIS).

Ephraim and Helah: a story of the Exodus. ...Edwin Hodder.
Pillar of fire; or, Israel in bondage. ...Joseph Holt Ingraham.
Rameses. ...Edward Upham.
Rescued from Egypt. ...Charlotte Tucker. (Pseud., A. L. O. E.)
Uarda: Roman aus dem alten Aegypten. ...Georg Moritz Ebers.
Uarda: a romance of ancient Egypt. ...Georg Moritz Ebers.

REBELLIONS. *A Selection.*

Bacon's (Virginia). 1676.

Century too soon: a story of Bacon's rebellion. ...John R. Musick..
Hansford. ...Saint George Tucker.
White aprons. ...Maud Wilder Goodwin.

Brittany. Seventeenth Century.

Noble sacrifice. ...Paul H. C. Feval.

REBELLIONS — *Continued.*

India, Sepoy. 1857.

Begumbagh : a tale of the Indian mutiny. ...George Manville Fenn.

First love and last love : a tale of the Indian mutiny. ...James Grant.

Lost in the jungle : a story of the Indian mutiny. ...Augusta Marryat.

On the face of the waters : a tale of the mutiny. ...Flora Annie Steel.

On to the rescue. ...William Gordon Stables.

Seeta. ...Meadows Taylor.

Stretton. ...Henry Kingsley.

Jamaica. 1865.

Meyrick's promise. ...E. C. Phillips.

Monmouth's. 1685.

Aubrey Luson ; or, Field of Sedgemoor. ...Anon.

Barony. ...Anna Maria Porter.

Danvers papers. ...Charlotte Mary Yonge.

Dorothy Arden. ...J. M. Callwell.

Le duc de Monmouth : Novelle historique. ...Anon.

Duke of Monmouth. ...Gerald Griffin.

Duke's Winton : a chronicle of Sedgemoor. ...J. R. Henslowe.

Edgar Nelthorpe ; or, Fair maids of Taunton. ...Andrew Reed.

Fate : a tale of stirring times. ...George Payne Rainsford James.

For faith and fatherland. ...Walter Besant.

Friend or foe ? a tale of Sedgemoor. ...Henry C. Adams.

In Taunton town : a story of the rebellion of James, Duke of Monmouth, in 1685. ...Evelyn Everett Green.

Micah Clarke : his statement made to his three grandchildren. ...Arthur Conan Doyle.

Mistress Dorothy Marvin. ...J. C. Snaith.

Oak staircase ; or, Stories of Lord and Lady Desmond. ...Mary and Catherine Lee.

Stronge of Netherstronge : a tale of Sedgemoor. ...Edith J. May.

Trelawny of Trelawne. ...Anna Eliza Bray.

Under the blue flag : a story of the Monmouth rebellion. ...Mary E. Palgrave.

Urith : a tale of Dartmoor. ...Sabine Baring-Gould.

Winifred ; or, English maiden in the seventeenth century. ...Lucy Ellen Guernsey.

Old Pretender's. 1715–1716.

Dorothy Forster. ...Walter Besant.

Father Clement : a Roman Catholic story. ...Grace Kennedy.

REBELLIONS — *Continued.*
For the king.   ...Charles Gibbon.
Henry Smeaton   ...George Payne Rainsford James.
Lucy Arden ; or, Hollywood Hall.   ...James Grant.
Preston fight; or, Insurrection of 1715.   ...William Harrison Ainsworth.
Rebel's wooing; or, 1715: a tale of the rebellion.   ...Lawrence Goodchild.
Rob Roy.   ...Sir Walter Scott.
Stories of the Scotch rebellion of 1716 and 1745.   ...A. D. Fillan.
Tales of the white cockade.   ...Barbara Hutton.
Winifred, Countess of Nithsdale: a tale of the Jacobite war. ...Barbarina Brand.

Young Pretender's. 1745-1746.

Adventures of Dennis.   ...Mary Bramston.
Allan Breck.   ...George R. Gleig.
Ascanius; or, Young adventurer : containing a partial history of the rebellion in Scotland in 1745-1746.   ...Anon.
Bonnie Prince Charlie : a tale of Fontenoy.   ...George Alfred Henty.
Chevalier: a romance.   ...Katharine Thompson.
Dangerous guest : a story of 1745.   ...Frances Brown.
Earl's path : a narrative founded on the historical events of 1745. ...Sidney Corner.
For James or George : a school-boy's tale of 1745.   ...Henry C. Adams.
Gerald Fitzgerald "the Chevalier."   ...Charles James Lever. (Pseud., Cornelius O'Dowd.)
Good old times : a story of the Manchester rebels of '45.   ...William Harrison Ainsworth.
Highland chronicle.   ...Samuel Bayard Dod.
In the king's name; or, Cruise of the " Kestrel."   ...George Manville Fenn.
Irish cavalier; or, 1745.   ...Elizabeth M. Stewart.
Kidnapped : being memoirs of the adventures of David Balfour in 1751.   ...Robert Louis Stevenson.
Die letzten Stuarts.   ...Franz F. Wangenheim.
Mysteries of Deepdene Manor : a romance of the days of the Pretender.   ...Frank Manduit.
Old manor house.   ...Charlotte Smith.
Out in the forty-five; or, Duncan Keith's vow.   ...Emily Sarah Holt.
Ponsonby : a tale of troublous times.   ...Anon.
Prince Charlie the young chevalier.   ...Meredith Johnes.
Redgauntlet.   ...Sir Walter Scott.

REFORMATION — *Continued.*

Germany.

Allerlei von Luther.   ...Joachim L. Haupt.
Blind girl of Wittenberg.   ...Carl August Wildenhahn.   English
    translation.
Das Buch vom Doktor Luther.   ...Hermann O. Nietschmann.
Chronicles of the Schonberg Cotta family.   ...Elizabeth Charles.
Count Arensberg : a tale of the days of Luther.   ...Joseph Sortain.
Count Erbach : a story of the reformation.   ...Hermann O. Nietsch-
    mann.   English translation.
Count Ulrich von Lindburg : a tale of the reformation.   ...William
    Henry Giles Kingston.
Duke Christian of Luneburgh ; or, Traditions from the Hartz.
    ...Jane Porter.
Duke Christopher : a story of the reformation.   ...Fanny H. Chris-
    topher.
Fate of castle Löwengard : a story of the days of Luther.   ...Esmé
    Stuart.
Forester's daughter : a tale of the reformation.   ...Anon.
Hans Sachs.   ...Adolf Friedrich Furschau.
Joachim Slütter.   ...Julius von Wickede.
Katharine von Schwarzburg.   ...Carl G. Berneck.
Die Kronenwächter.   ...Ludwig A. von Arnim.
Lukas Cranach : historischer Roman.   ...Hermann Klencke.
Luther and the cardinal.   ...Hermann O. Nietschmann.   English
    translation.
Luther in Rom.   Roman.   ...Levin Schueking.
Luther in Rome ; or, Corradina the last of the Hohenstaufen.
    ...Levin Schuecking.   English translation.
Luther und die Seinen.   ...Franz Lubojatzky.
Luther und seine Zeit.   ...Theodor Koenig.
Luther's Brautfahrt.   ...Joseph E. C. Bischoff.
Margarethe : a tale of the sixteenth century.   ...Emma Leslie.
Martin Luther : historischer Roman.   ...Heinrich Eisenlohr.
Martin Luther : kirchengeschichtliches Lebensbild.   ...Carl August
    Wildenhahn.
Paul Gerhardt : an historical tale of the Lutherans and Reformed in
    Brandenburg under the great elector.   ...Carl August Wilden-
    hahn.   English translation.
Paul Gerhardt : kirchengeschichtliches Lebensbild aus der Zeit des
    grossen Kürfursten.   ...Carl August Wildenhahn.
Tales from Alsace.   ...Anne Manning.
Thomas Müntzer.   Roman.   ...Theodor Mundt.
True as steel.   ...George Walter Thornbury.

REFORMATION — *Continued.*
Ulric; or, Voices. ...Theodore S. Fay.
See also **LUTHER, MARTIN.**

Italy.

From dark to dawn in Italy. ...Anon.
Struggle in Ferrara. ...William Gilbert.

Netherlands.

Aurelia; oder, Die Martyrer von Gorkum. ...M. Lehmann.
Brighter days; or, Story of Catherine Jans. ...Helen C. Chapman.
Jacqueline: a story of the reformation in Holland. ...Janet Hardy.
Olden's mission: a tale of the famine in Leyden. ...Anon.
Patriot prince: William the Silent. ...Harriette E. Burch.
Rosalia Vanderiver. ...Anon.
Rudolph of Rosenfeldt; or, Leaven of the reformation. ...John W. Spear.
Soldier's ward; or, Saved for martyrdom. ...E. Gerdes.
Those dark days; or, Diaries of two Netherland girls. ...Helen C. Chapman.
Walter Harmsen: a tale of the reformation times in Holland. ...E. Gerdes.
Willem de Eerste of de grondlegging der Nederlandsche vryheid. ...Johannes Nomsz.
William, Prince of Orange; or, King and his hostage. ...Titus M. Merriman.

Scotland.

Arthur Erskine's story: a tale of the days of Knox. ...Deborah Alcock.
Dark year of Dundee: a tale of the Scottish reformation. ...Deborah Alcock.
David the scholar; or, From dark to dawn: a Scottish story. ...Anon.
Magdalen Hepburn: a story of the Scottish reformation. ...Margaret O. W. Oliphant.
Marmion: historical romance. ...Henry W. Grosette.
· No cross, no crown. ...Caroline E. Davis.

Switzerland.

Calvin: culturhistorischer Roman. ...Theodor Koenig.
Chillon; or, Protestants of the sixteenth century. ...Jane L. Willyams.
City and the castle: a story of the reformation in Switzerland. ...Annie Lucas.
Geneva's shield: a story of the Swiss reformation. ...William M. Blackburn.

REFORMATION — *Continued.*

Good old times : a tale of Auvergne.   ...Anne Manning.

Hermit of Livry : a tale of the days of Calvin.   ...Emma Leslie.

Louis Belat : a tale of the reformation in Savoy.   ...Mrs. Alexander S. Orr.

Mountain patriots : a tale of the reformation in Savoy.   ...Mrs. Alexander S. Orr.

Nikolaus Manuel : Roman aus der Zeit der schweizerischen Glaubenskampfe.   ...Ludwig Eckardt.

Story of a noble life ; or, Zurich and its reformer, Ulrich Zwingli.   ...Janet Hardy.

True to the end.   ...H. S. Burrage.

Ulrich Zwingli.   ...Theodor Koenig.

. **REIGN OF TERROR, FRENCH HISTORY.   1793-1794.**

L'ami du peuple.   ...J. M. Gassier Saint Amand.

L'amour sous la terreur.   ...Adolphe de Lescure.

Atelier du Lys ; or, Art student in the reign of terror.   ...Margaret Roberts.

At the sign of the guillotine.   ...Herold Spender.

Charlotte Corday.   ...Henri Alphonse Esquiros.

Charlotte Corday : an historical novel.   ...Henri Alphonse Esquiros. English translation.

Charlotte Corday.   ...Carl W. T. Frenzel.

Charlotte Corday.   ...Rose E. Temple.

Le chevalier de Maison-Rouge.   ...Alexandre Dumas.

Chevalier de Maison-Rouge : a tale of the reign of terror.   ...Alexandre Dumas.   English translation.

Dead marquise.   ...Leonard Kip.

Un épisode sous la Terreur.   ...Honoré de Balzac.

Episode under the terror.   ...Honoré de Balzac.   English translation.

La famille de l'émigré.   ...J. P. Fourquet d'Hachette.

Le faubourg Saint Antoine.   ...Tony Révillon.

Friend of the people : a tale of the reign of terror.   ...Mary C. Rowsell.

Ingénue.   Alexandre Dumas.

Ingenue ; or, Death of Marat.   ...Alexandre Dumas.   English translation.

In the reign of terror : the adventures of a Westminster boy.   ...George Alfred Henty.

Das letzte Bankett der Girondisten.   ...Charles Nodier.

Life scenes in the reign of terror.   ...Washington Frothingham.

Maid of Normandy.   ...Edmund J. Eyre.

Noblesse oblige.   ...Margaret Roberts.

REIGN OF TERROR, FRENCH HISTORY. 1793-1794 — *Continued.*

Red cap and blue jacket. ...George Dunn.

Red Republic. ...Robert W. Chambers.

Reign of terror: diary of a volunteer of the year II. of the French revolution. ...Anon.

Reign of terror. ...Catherine G. F. Gore. (In her Romances of real life.)

Robespierre. ...Carl F. A. Wartenburg.

1794: a tale of the terror. ...Charles John de Ricault. (Pseud., Charles d'Héricault.) English translation.

Thermidor. ...Charles John de Ricault. (Pseud., Charles d'Héricault.)

Thorndyke manor: a tale of Jacobite times. ...Mary C. Rowsell.

Wild fire. ...George Walter Thornbury.

Young Marmaduke: a story of the reign of terror. ...William Henry Davenport Adams.

See also **REVOLUTION, French.**

RELATION OF THE SEXES.

Bachelor maid. ...Constance Cary Harrison (Mrs. Burton Harrison).

Bundle of life. ...John Oliver Hobbes, pseud.

Clara Hopgood. ...William Hale White. (Pseud., Mark Rutherford.)

Country doctor. ...Sarah Orne Jewett.

Daughters of Danaus. ...Mona Card.

Emancipated. ...Geo. Gissing.

George Mandeville's husband. ...Anon.

Heavenly twins. ...Sarah Grand, pseud.

Ideala: a study from life. ...Sarah Grand, pseud.

I forbid the banns. ...F. Frankfort Moore.

Is this your son, my Lord? ...Helen H. Gardener.

Jenny's case. ...Ellen F. Pinsent.

Keynotes. ...George Egerton, pseud.

Mighty atom. .. Marie Corelli.

My opinions and Betsey Bobbet's: designed as a beacon light to guide women to life, liberty, and happiness. ...Marietta Holley. (Pseud., Josiah Allen's wife.)

Odd women. ...George Gissing.

Our manifold nature. ...Sarah Grand, pseud.

Pray you, sir, whose daughter? ...Helen H. Gardener.

Rebel queen. ...Walter Besant.

Sinner's comedy. ...John Oliver Hobbes, pseud.

Some emotions and a moral. ...John Oliver Hobbes, pseud.

Story of a modern woman. ...Ella Hepworth Dixon.

Study in temptation. ...John Oliver Hobbes, pseud.

RELATION OF THE SEXES — *Continued.*

Superfluous woman.   ...Anon.

Unclassed.   ...George Gissing.

Vashti, old and new.   ...Marvel Kayve.

Vawder's understudy : a study in platonic affection.   ...James Knapp
Reeves.

Whose was the blame ?   ...Mrs. J. Gregor.

Wife or slave ?   ...Mrs. Albert S. Bradshaw.

Woman regained.   ...George Barlow.

Wreckage.   ...H. Crackanthorpe.

Yellow aster.   ...Mrs. Mannington Caffyn.   (Pseud., Iota.)

See also **MARRIAGE.**

REPUBLICS.   *A Selection.*

Dutch.   1565-1604.

Beggars ; or, Founders of the Dutch republic.   ...John B. de Liefde.

By England's aid ; or, Freeing of the Netherlands.   ...George
Alfred Henty.

By pike and dyke : a tale of the rise of the Dutch republic.
...George Alfred Henty.

Galama ; or, Beggars.   ...John B. de Liefde.

Gideon Florensz.   ...Anna L. G. Bosboom.

Leycester in Nederland.   ...Anna L. G. Bosboom.

Liberators of Holland.   ...Elizabeth Charles.   (In her Martyrs of
Spain.)

De vrouwen van het Leycestersche tijdvak. ...Anna L. G. Bosboom.

Within sea walls ; or, How the Dutch kept the faith.   ...E. H.
Walshe and G. E. Sargent.

French.   First Republic.   1792-1804.

Der achtzehnte Brumaire.   ...Schmidt-Weissenfels (Eduard Schmidt).

Adèle : a tale of France.   ...E. Randall.

Adrian and Thecla ; or, Friendship in adversity.   ...Anon.

L'an I. de la république, 1793.   ...Émile Erckmann et Alexandre
Chatrian.

Anna Saint Ives.   ...Thomas Holcroft.

Antoinette : a tale of the ancien régime.   ...Anna Eliza Bray.

Aubert Dubayet ; or, Sister republics.   ...Charles E. A. Gayarre.

Les aventures de Saturnin Fichet.   ...Melchior Frédéric Soulié.

Behind the hedge.   ...Pauline de Witt.   (In her Dames of high
estate.)

Bellah : a tale of La Vendée.   ...Octave Feuillet.   English translation.

Blameless knights ; or, Lützen and La Vendée.   ...Alice H. F.
Byng, Viscountess Enfield.

REPUBLICS — *Continued.*

Un ménage de garçon.   ...Honoré de Balzac.
Nannette and her lovers.   ...Talbot Gwynne.
'93.   ...Victor Hugo.   English translation.
Ninety-three.   ...John W. Lyndon.
On the edge of the storm.   ...Margaret Roberts.
Orphan of La Vendée.   ...Anna Eliza Bray.
Peasant and the prince.   ...Harriet Martineau.
Le premier amour de Lord Saint Albans.   ...Charles John de
    Ricault.   (Pseud., Charles d'Héricault.)
Quatrevingt-treize.   ...Victor Hugo.
Un roman sous la révolution.   ...Einoel, pseud.
Saint Katharine's by the tower.   ...Walter Besant.
Sidney Smith : historischer Roman.   ...Robert Springer.
Sous le directoire.   ...Charlotte Foucaux.
Souvenirs d'un Bas-Breton.   ...Émile Souvestre.
Tale of two cities.   ...Charles Dickens.
Two brothers.   ...Honoré de Balzac.   English translation.
Two mothers.   ...J. M. Joy.
Les va-nu-pieds.   ...Léon A. Cladel.
La Vendée : an historical romance.   ...Anthony Trollope.
Veva ; or, War of the peasants.   ...Hendrik Conscience.   English
    translation.
Votaries of reason.   ...Henry C. Adams.   (In his Tales upon
    texts.)
Year one of the republic.   ...Émile Erckmann and Alexandre
    Chatrian.   English translation.
Zeitgedichte, von 1789–1803.   ...J. W. L. Gleim.
See also **REIGN OF TERROR** and **REVOLUTIONS**.

France.   Second Republic.   1848–1852.

Edmond Dantes.   ...Alexandre Dumas.
Edmond Dantes.   ...Alexandre Dumas.   English translation.
Five-chimney farm.   ...Mary A. M. Hoppus (Mrs. Marks).
Histoire d'un homme du peuple.   ...Émile Erckmann et Alexandre
    Chatrian.
Man of the people.   ...Émile Erckmann and Alexandre Chatrian.
    English translation.

France.   Third Republic.   1870.

L'année terrible.   ...Victor Hugo.
Aus dem Paris der dritten Republik.   ...Paul Lindenberg.
Aus den Tagen der Commune.   ...Stanislaus S. A. Grabowski.
Le chien perdu et la femme fusillée.   ...Arsène Houssaye.
Kentucky's love ; or, Roughing it about Paris.   ...Edward King.

REPUBLICS — *Continued.*
Léon Michel Gambetta. ...Eugen Bernard.
Napoléon IV.; chronique de l'avenir. ...Matthieu de Boulogne.
Odile: a tale of the Commune. ...Mrs. Frank Pentrill.
Parisians. ...Edward George Earle Lytton Bulwer-Lytton.
Workman and soldier: a tale of Paris life during the siege and
rule of the Commune. ...James F. Cobb.

Helvetian.

Rose of Disentis. ...Johann Heinrich D. Zschokke. English
translation.
Die Rose von Disentis. ...Johann Heinrich D. Zschokke.

Ideal Republics.

Commonwealth of Oceana. ...James Harrington.
Looking backwards, 2000–1887. ...Edward Bellamy.
Looking within. ...J. W. Roberts.
Man of mark. ...Anthony Hope Hawkins. (Pseud., Anthony
Hope.)
Mr. East's experiences in Mr. Bellamy's world. ...Conrad Wil-
brandt.

RESTORATION IN ENGLAND. 1660–1685.
Agnes Beaumont: a true story of the year 1670. ...Marian
Caldecott.
Aphra Behn. ...Clara M. Mundt. (Pseud., Louise Mühlbach.)
Aphra Behn. ...Clara M. Mundt. (Pseud., Louise Mühlbach.)
English translation.
At the sign of the Blue Boar: a story of the reign of Charles II.
...Emma Leslie.
Beyond the seas: being the surprising adventures and ingenious
opinions of Ralph, Lord St. Keyne. ...Oswald Crawfurd.
Boscobel: a narrative of the adventures of Charles II. after the
battle of Worcester. ...George W. Dodd.
Captain Jacques: a romance of the time of the plague. ...Edward
Fitzgibbon. (Pseud., Somerville Gibney.)
Captain of the guard. ...James Grant.
Cavaliers of England; or, Times of the revolutions of 1642 and
1648. ...Henry W. Herbert. (Pseud., Frank Forester.)
Champion court: the days of the ejectment. ...Emma J. Worboise.
Claude Duval: a romance of the days of Charles II. ...Henry D.
Miles.
Court at Tunbridge in 1664. ...Catherine G. F. Gore.
Courtier in the days of Charles II. ...Catherine G. F. Gore.
Courtier of the days of Charles II. ...Mrs. Gordon.

RESTORATION IN ENGLAND. 1660–1685 — *Continued.*
Dame Rebecca Berry; or, Court scenes in the reign of Charles the
    Second. ...Elizabeth I. Spencer.
Drifted and sifted : a domestic chronicle of the seventeenth century.
    ...Miss McLaren.
La duchesse de Châtillon. ...François T. M. d'Arnaud.
Dutch in the Medway. ...Charles Macfarlane.
For liberty's sake; the story of Robert Ferguson. ...John B. Marsh.
Hero in the strife. ...Louise C. Silke.
Ida Vane : a tale of the restoration. ...Andrew Reed, Jr.
In the East country with Sir Thomas Brown. ...Emma Marshall.
In the golden days. ...Ada E. Bayly. (Pseud., Edna Lyall.)
In the service of Rachel, Lady Russell. ...Emma Marshall.
Janet and her father. ...Mary E. Bamford.
Jennett Cragg, the Quakeress. ...Maria Wright.
Karl II. ; oder, der lustige Monarch. ...Johann Lenz.
Lord Montagu's page : an historical romance of the seventeenth
    century. ...George Payne Rainsford James.
Maid of honor. ...Catherine G. F. Gore. (In her Romances of real
    life.)
Mary Bunyan, the dreamer's blind daughter. ...Sallie Rochester
  - Ford.
Mary Hollis : a romance of the days of Charles II. and William,
    Prince of Orange. ...Hendrik J. Schimmel. English translation.
Memoirs of a lady-in-waiting. ...J. D. Fenton.
Merry monarch ; or, England under Charles II. ...William Henry
    Davenport Adams.
On both sides of the sea. ...Elizabeth Charles.
Pattie Durant : a tale of 1662. ...Ellen Clacy. (Pseud., Cycla.)
Peter the apprentice : a tale of the restoration. ...Emma Leslie.
Peveril of the peak. ...Sir Walter Scott.
Le prophète irlandois. ...Charles de Saint-Évremond.
Puritan's grave. ...William P. Scargill.
Puritan's wife. ...Max Pemberton.
Robber. ...George Payne Rainsford James.
Das Roggenhaus Komplott. ...Georg Hiltl.
Rupert Aubrey of Aubrey Chase : an historical tale of 1681.
    ...Thomas John Potter.
Russell : a tale of the reign of Charles II. ...George Payne Rains-
    ford James.
Rye house plot. ...George W. M. Reynolds.
Saint Dunstan's clock : a story of 1666. ... E. Ward.
Saint Valentine's day, 1664. ...Elbridge S. Brooks. (In his Storied
    holidays.)
Shepherd of Grove Hall : a tale of 1662. ...Anon.

RESTORATION IN ENGLAND. 1660–1685 — *Continued.*

Silas Verney, being the story of his adventures in the days of King Charles the Second. ...Edgar Pickering.

Sir Ralph Esher; or, Memoirs of a gentleman of the court of Charles the Second. ...Leigh Hunt.

Stanfield Hall. ...J. Frederick Smith.

Talbot Harland. ...William Harrison Ainsworth.

Through unknown ways; or, Journal-books of Dorothea Trundel. ...Lucy Ellen Guernsey.

Traitor or patriot? a tale of the Rye house plot. ...Mary C. Rowsell.

True hero; or, Story of William Penn. ...William Henry Giles Kingston.

Truth; or, Persis Clareton. ...Charles B. Tayler.

Two swords; being a story of old Bristol. ...Emma Marshall.

Wearyholm; or, Seedtime and harvest: a tale of the restoration of Charles II. ...Emily Sarah Holt.

Whitefriars; or, Days of Charles II. ...Jane Robinson.

Winchester in the time of Thomas Ken. ...Emma Marshall.

REVOLUTION.

America. 1775–1789.

Alamance; or, Great and final experiment. ...Calvin Wiley.

Amelia; or, Faithless Briton. ...Anon.

American hunter: a tale from incidents which happened during the war with America. ...Anon.

American revolution. ...Chauncey C. Hotchkiss.

American spy; or, Freedom's early sacrifice: a tale of the revolution. ...Jeptha R. Simms.

Arnold; or, British spy: a tale of treason and treachery. ...Joseph Holt Ingraham.

Ashleigh: a tale of the revolution. ...Eliza Ann Dupuy.

Bald eagle; or, Last of the Ramarpaughs. ...Elizabeth Oakes Smith.

Bastonnais: a tale of the American invasion of Canada. ...John Lesperance.

Betrothed of Wyoming. ...Anon.

Betsey's bedquilt: a story of 1777–78. ...Clara F. Guernsey.

Black Hollow; or, Dragoon's bride. ...N. C. Iron.

Black-plumed riflemen. ...Newton M. Curtis.

Blanche of Brandywine; or, September the eleventh, 1777. ...George Lippard.

Blue jackets of '76. ...Willis J. Abbot.

Bonnybel Vane, embracing the history of Henry Saint-John, gentleman. ...John Esten Cooke.

Boston conspiracy; or, Royal police: a tale of 1773–78. ...John H. Robinson.

REVOLUTION — *Continued.*

Boston tea party; and other stories of the American revolution. ...Henry C. Watson.

Bow of orange ribbon : a romance of New York. ...Amelia Edith Barr.

Bride of Fort Edward, founded on an incident of the revolution. ...Delia S. Bacon.

British partizan : a tale of the olden time. ...Mary E. Davis.

Burton ; or, Sieges. ...Joseph Holt Ingraham.

Buttonwoods ; or, Refugees of the revolution. ...Anon.

Cabin book; or, Scenes and sketches of the late American and Mexican war. ...Charles Sealsfield (formerly Karl Postl). English translation.

Das Cajütenbuch, oder, nationale Charakteristiken. ...Charles Sealsfield (formerly Karl Postl).

Camp Charlotte : a tale of 1774. ...Anon.

Camp-fires of the revolution. ...Henry C. Watson.

Le capitaine Paul. ...Alexandre Dumas.

Captain Molly, the story of a brave woman. ...Ellen T. H. Putnam. (Pseud., Thrace Talmon.)

Captain Paul. ...Alexandre Dumas. English translation.

Chain-bearer. ...James Fenimore Cooper.

Charles Morton ; or, Young patriot. ...Mary S. B. D. Shindler.

Chauncey Judd ; or, Stolen boy : a tale of the revolution. ...Israel P. Warren.

Choir invisible. ...James Lane Allen.

Claudius, the cowboy of Ramapo valley. ...P. Demarest Johnson.

Clayton's rangers ; or, Quaker partisans. ...E. H. Williamson.

Conspirator. ...Eliza Ann Dupuy.

Daughters of the revolution. ...Charles C. Coffin.

Daughters of the revolution and their times. ...Kate Chopin.

Doom of the Tory's guard. ...Newton M. Curtis.

Double masquerade : a romance of the revolution. ...Charles R. Talbot.

Dutch dominie of the Catskills ; or, Times of the " Bloody Brandt " : a tale of the revolution. ...David Murdock.

Eagle of Washington. ...Burkitt J. Newman.

Edwin Brothertoft. ...Theodore Winthrop.

Ellen Grafton, the lily of Lexington ; or, Bride of liberty. ...Benjamin Barker.

Emma Corbett ; or, Miseries of civil war. ...Samuel J. Pratt.

Ethan Allen ; or, King's men. ...Melville, pseud.

Ethel Hamilton; or, Lights and shadows of the war of independence. ...Anna T. Sadlier.

Eutaw. ...William Gilmore Simms.

REVOLUTION — *Continued.*

Legends of the West.   ...James Hall.

Linwoods.   ...Catherine Maria Sedgwick.

Lionel Lincoln; or, Leaguer of Boston.   ...James Fenimore Cooper.

Lone star of Texas.   ...J. W. Dallam.

Maid of Esopus; or, Trials and triumphs of the revolution.   ...I. N. Iron.

Margaret Moncrieffe, the first love of Aaron Burr: a romance of the revolution.   ...Charles Burdett.

Marrying by lot: a tale of the primitive Moravians.   ...Charlotte B. Mortimer.

Mehetabel: a story of the revolution.   ...Mrs. H. C. Gardner.

Mellichampe: a legend of the Santel.   ...William Gilmore Simms.

Meredith; or, Mystery of the Meschianza: a tale of the American revolution.   ...Anon.

Merton: a tale of the revolution.   ...Eliza Ann Dupuy.

Miss MacRéa: Roman historique.   ...Michel René Hilliard d'Auberteuil.

Monarchist: an historical novel.   ...John B. Jones.

Monody on Major Andrè.   ...Anna Seward.

My comrades: adventures in the Highlands and legends of the neutral ground.   ...Howard Hinton.

Nazarene; or, Last of the Washingtons.   ...George Lippard.

Near to nature's heart.   ...Edward Payson Roe.

Oath of Marion: a tale of the revolution.   ...Charles J. Peterson.

Old Put.   ...Justin Jones.   (Pseud., Harry Hazel.)

Old revolutionary soldier.   ...Joseph Alden.

Old stone house.   ...Joseph Alden.

Out of a besieged city.   ...Charles W. Hutson.

Overland.   ...J. W. DeForest.

Partisan.   William Gilmore Simms.

Patriot and tory one hundred years ago.   ...Julia M. Wright.

Paul and Persis; or, Revolution struggle in the Mohawk valley.   ...Mary E. Brush.

Paul Ardenheim, the monk of Wissahikon.   ...George Lippard.

Paul Jones: a romance.   ...Allen Cunningham.

Paul Jones.   ...Theodor Muegge.

Paul Redding: a tale of the Brandywine.   ...T. Buchanan Read.

Pemberton; or, One hundred years ago.   ...Henry Peterson.

Peter and Polly; or, Home life in New England a hundred years ago.   ...Annie D. Greene.

Pride of Lexington.   ...W. Seton.

Quaker soldier; or, British in Philadelphia.   ...J. Richter Jones.

Ralph the drummer boy: a story of the days of Washington.   ...Théophile L. Rousselet.

REVOLUTION — *Continued.*

Ralphton; or, Young Carolinian of 1776. ...A. H. Brisbane.

Rangers; or, Tory's daughter. ...D. P. Thompson.

Rebels and tories; or, Blood of the Mohawk. ...Lawrence Labree.

Rejected wife. ...Ann S. Stephens.

Revolutionary times: sketches of our country one hundred years ago. ...Edward Abbott.

Rivals. ...Jeremiah Clemens.

Rosalie du Pont; or, Treason in the camp. ...Emerson Bennett.

Rudolph of Rosenfeldt; or, Leaven of the Reformation. ...J. W. Spear.

Russell and Sidney. ...Eliza Leslie.

Saratoga: a tale of 1787. ...Anon.

Saratoga: a tale of the revolution. ...Anon.

Les scalpeurs blancs. ...Gustave Aimard.

Scout. ...William Gilmore Simms.

Septimius Felton. ...Nathaniel Hawthorne.

Seventy-six. ...John Neal.

Sketches of western adventure. ...John A. McClung.

Spinning-wheel stories. ...Louisa May Alcott.

Spy: a tale of the neutral ground. ...James Fenimore Cooper.

Standish the Puritan: a tale of the American revolution. ...Eldred Grayson, pseud.

Stories of the Old Dominion from the settlement to the end of the revolution. ...John Esten Cooke.

Stories of the revolution. ...Josiah Priest.

Storm mountain. ...Edward S. Ellis.

Story of a Hessian: a tale of the revolution in New Jersey. ...Lucy Ellen Guernsey.

Thankful Blossom: a tale of the Jerseys. ...Francis Bret Harte.

Three girls of the Revolution. ...Lucy Ellen Guernsey.

Tory spy; or, Britisher " done brown." ...J. Colfort Clinton.

True stories of American wars. ...Anon.

True stories of the American fathers. ...Rebecca MacConkey.

True stories of the days of Washington. ...Anon.

True to the old flag: a tale of the American war of independence. ...George Alfred Henty.

Virginians. ...William Makepeace Thackeray.

Walter Thornley; or, Peep at the past. ...Susan R. Sedgwick.

Washington and his generals; or, Legends of the revolution. ...George Lippard.

Washington and his men. ...George Lippard.

Washington and seventy-six. ...Lucy Ellen Guernsey.

Watchfires of '76. ...Samuel Adams Drake.

Water waif: a story of the revolution. ...Elizabeth S. Bladen.

REVOLUTION— *Continued.*

Wau-nan-gee ; or, Massacre at Chicago : a romance of the American revolution.   ...John Richardson.

White scalper : a story of the Texas war.   ...Gustave Aimard. English translation.

Woodcraft.   ...William Gilmore Simms.

Yankee champion ; or, Tory and his league.   ...Sylvanus Cobb, Jr.

Yorktown: an historical romance.   ...Anon.

Young rebels : a story of the battle of Lexington.   ...Robert Hope Moncrieff.   (Pseud., Ascott R. Hope.)

England.   1685–1688.

Aimée : a tale of the days of James the Second.   ...Agnes Giberne.

Arthur Arundel.   ...Horace Smith.   (Pseud., Paul Chatfield.)

Barony.   ...Anna Maria Porter.

Danvers papers.   ...Charlotte Mary Yonge.

Duke of Monmouth.   ...Gerald Griffin.

Edgar Nelthorpe ; or, Fair maids of Taunton.   ...Andrew Reed.

Fate : a tale of stirring times.   ...George Payne Rainsford James.

James II.; or, Revolution of 1688.   ...William Harrison Ainsworth.

John Deane.   ...William Henry Giles Kingston.

Lady Betty.   ...Christabel R. Coleridge.

Last of the cavaliers.   ...Rose Piddington.

Oak staircase ; or, Stories of Lord and Lady Desmond.   ...Mary and Catherine Lee.

Outlaw.   ...Anna Maria Hall.

Owain Goch : a tale of the revolution.   ...Thomas Roscoe.

Roger Willoughby ; or, Times of Benbow.   ...William Henry Giles Kingston.

Stronge of Netherstronge : a tale of Sedgemoor.   ..Edith J. May.

Winifred ; or, English maiden in the seventeenth century.   ...Lucy Ellen Guernsey.

France.   1789–1795.

Albert Lunel ; or, Château of Languedoc.   ...Henry Brougham.

L'an 1 de la république.   ...Émile Erckmann et Alexandre Chatrian.

Andrée de Taverny.   ...Alexandre Dumas.

Angeli Pisani.   ...Lord Strangford.

Ange Pitou.   ...Alexandre Dumas.

Anna Saint Ives.   ...Thomas Holcroft.

Annette : a tale.   ...William F. Deacon.

Arundel : a tale of the French revolution.   ...Francis Vincent.

Atelier du Lys ; or, Art student in the reign of terror.   ...Margaret Roberts.

At the sign of the guillotine.   ...Harold Spender.

REVOLUTION — *Continued.*

Barnave. ...Jules G. Janin.

Beaumarchais. ...Albert E. Brachvogel.

Bellah : Roman. ...Octave Feuillet.

Birth and education. ...Marie Sophie Schwartz. English translation.

De Boerenkryg. ...Hendrik Conscience.

Blockade : an episode of the fall of the first empire. ...Émile Erckmann and Alexandre Chatrian. English translation.

Le Blocus. ...Émile Erckmann et Alexandre Chatrian.

Börd och Bildning. ...Marie Sophie Schwartz.

La chanoinesse. ...André Theuriet.

Charlotte Corday : an historical novel. ...Henry Alphonse Esquiros.

Le chemin de France. ...Jules Verne.

Le chevalier de Maison-Rouge. ...Alexandre Dumas.

Le chevalier de Maison-Rouge : a tale of the reign of terror. ...Alexandre Dumas. English translation.

Child of the revolution. ...Margaret Roberts.

Citoyenne Jacqueline : a woman's lot in the great French revolution. ...Henrietta Keddie. (Pseud., Sarah Tytler.)

La comtesse de Charny. ...Alexandre Dumas.

Contes de la Montagne. ...Émile Erckmann et Alexandre Chatrian.

Count Mirabeau : an historical novel. ...Theodor Mundt. English translation.

Countess de Charny. ...Alexandre Dumas. English translation.

Country in danger, 1792; or, Episodes of the great French revolution. ...Émile Erckmann and Alexandre Chatrian. English translation.

Dead heart : a tale of the Bastile. ...Charles Gibbon.

Dead marquise. ...Leonard Kip.

Le dernier amour de Mirabeau. ..Charlotte Foucaux.

Duchenier; or, Revolt of La Vendée. ..John Mason Neale. (Pseud., Aurelius Gratianus.)

Eighty years ago. ...Harriet Morton.

Embassy : an historical romance. ...Louis A. Chamerovzow.

Les États Généraux. ...Émile Erckmann et Alexandre Chatrian.

Exile's trust : a tale of the French revolution. ...Frances Browne.

Le faubourg Saint Antoine. ...Tony Révillon.

Flight to France; or, Memoirs of a dragoon. ...Jules Verne. English translation.

Four years in a cave. ...Anon.

French wines and politics. ...Harriet Martineau.

Gabrielle Andrée. ...Sabine Baring-Gould.

Graf Mirabeau. ...Theodor Mundt.

History of a French artisan during the last revolution. ...**George Payne Rainsford James.**

REVOLUTION — *Continued.*

In exitu Israel.   ...Sabine Baring-Gould.

Ingénue.   ...Alexandre Dumas.

Ingenue; or, Death of Marat.   ...Alexandre Dumas.   English translation.

Invalide; or, Pictures of the French revolution.   ...Carl Spindler.

Julian; or, Close of an era.   ...Laurence Louis Félix Bungener. English translation.

Julien; ou, La fin d'un siècle.   ...Laurence Louis Félix Bungener.

Lady of Provence; or, Humbled and healed: a tale of the first French revolution.   ...Charlotte Tucker. (Pseud., A. L. O. E.)

Last of the Kerdrecs.   ...William Minturn.

Louis Duval: a tale of the French revolution.   ...Anon.

Madame Thérèse.   ...Émile Erckmann et Alexandre Chatrian.

Madame Therese; or, Volunteers of '92.   ...Emile Erckmann and Alexandre Chatrian.   English translation.

Maid of honor: a tale of the dark days of France.   ...Lewis S. Wingfield.

Marston; or, Soldier and statesman.   ...George Croly.

Maurice Tiernay, the soldier of fortune.   ...Charles James Lever.

Mirabeau und Sophie.   ...Oskar L. B. Wolff.

Monsieur Jack: a tale of old war-time.   ...Alfred H. Engelbach.

Ninety-three.   ...Victor Hugo.   English translation.

Ninety-three: the story of the French revolution of my French tutor.   ...Anon.

Nobleman of '89: an episode of the French revolution.   ...Abel Quinton.

Noblesse oblige.   ...Margaret Roberts.

On the edge of the storm.   ...Margaret Roberts.

Orphan of La Vendée.   ...Anna Eliza Bray.

La patrie en danger.   ...Émile Erckmann et Alexandre Chatrian.

Peasant and the prince.   ...Harriet Martineau.

Philip of Lutetia; or, Revolution of 1789.   ...Louis A. Chamerovzow.

Poor little Gaspard's drum: a tale of the French revolution. ...Alfred H. Engelbach.

Princess.   ...Mrs. Cecil Fane.

La prise de la Bastille.   ...Leon Petitdidier.

La prise de la Bastille, fait historique.   ...Pierre M. Parein.

Quatrevingt-treize.   ...Victor Hugo.

Red cap and blue jacket.   ...George Dunn.

Red cockade.   ...Stanley J. Weyman.

Reds of the Midi.   ...Felix Gras.

Reign of terror.   ...Catherine G. F. Gore.

Rosine.   ...George John Whyte Melville.

Saint Katharine's by the tower.   ...Walter Besant.

REVOLUTION — *Continued.*

Seven times in the fire. ...C. Maud Battersby.

Six years later. ...Alexandre Dumas. English translation.

States General, 1789: or, Beginning of the great French revolution. ...Émile Erckmann and Alexandre Chatrian. English translation.

Tale of the French revolution. ...Francis O. Giffard.

Tale of two cities. ...Charles Dickens.

Through rough waters. ...Frances M. Peard.

Tuileries. ...Catherine G. F. Gore.

Under the meteor flag. ...William J. C. Lancester. (Pseud., Harry Collingwood.) .

La Vendée: an historical romance. ...Anthony Trollope.

Vera; or, War of the peasants. ...Hendrik Conscience. English translation.

Votaries of reason. ...Henry C. Adams. (In his Tales upon texts.)

When Greek meets Greek. ...Joseph Hatton.

White and red. ...J. R. Henslowe.

Wild fire. ...George Walter Thornbury.

Will and a way. ...Georgiana C. L. G. Fullerton.

Year one of the republic, 1793. ...Émile Erckmann and Alexandre Chatrian. English translation.

Young Marmaduke: a story of the reign of terror. ...William H. D. Adams.

Zanoni. ...Edward George Earle Lytton Bulwer-Lytton.

See also **REIGN OF TERROR** and **REPUBLICS, French First.**

France. Revolution of 1830.

Algier und Paris im Jahre, 1830. ...Ludwig Rellstab.

Les barricades de 1830. ...Anon.

1830. Roman und Geschichte. ...Anton J. Gross Hoffinger.

1830. Roman und Geschichte. .. Julius Gundling.

Les étudiants, épisode de la révolution de 1830. ...Louise Maignaud.

Maniac of Brussels. ...F. W. N. Bayley.

Two friends. ...Marguerite Gardiner.

France. Revolution of 1848.

Edmond Dantes. ...Alexandre Dumas.

Edmond Dantes. ...Alexandre Dumas. English translation.

Five-chimney-farm. ...Mary A. M. Hoppus (Mrs. Marks).

Histoire d'un homme du peuple. ...Émile Erckmann et Alexandre Chatrian.

Man of the people: a tale of 1848. ...Émile Erckmann and Alexandre Chatrian. English translation.

REVOLUTION — *Continued.*
Germany. Revolution of 1848.

Andreas Burns und seine Familie.   ...Philipp Lange. (Pseud., Philipp Galen.)

Eine deutsche Revolution; oder, Der Karneval von 1848.   ...Eugen H. von Dedenroth.

Durch Nacht zum Licht.   ...Friedrich Spielhagen.

1848; oder, Nacht und Licht.   ...Franz Lubojatzky.

1849; oder des Königs Maienblüte.   ...Franz Lubojatzky.

Hohensteins.   ...Friedrich Spielhagen. English translation.

Margaret von Ehremberg, the artist's wife.   ...William and Mary Howitt.

Otto der Schutz.   ...Johann Gottfried Kinkel.

Problematic characters.   ...Friedrich Spielhagen. English translation.

Problematische Naturen.   ...Friedrich Spielhagen.

Die Ritter vom Geiste.   ...Carl F. Gutzhow.

Schleswig-Holstein Meer-umschlugen.   ...Carl von Kessel.

Through night to light.   ...Friedrich Spielhagen. English translation.

Die von Hohenstein.   ...Friedrich Spielhagen.

Greek. 1821.

Amygdala : a tale of the Greek revolution.   ...Mrs. Edmonds.

Anthea : true history of the war of Greek independence   ...Cecile Cassavetti.

Hungary. 1848.

Hungarian girl.   ...Mariam Tenger.

Interrupted wedding.   ...Anne Manning.

Politikai divatok.   ...Mór Jókai.

RICHARD I. OF ENGLAND.   See **CRUSADES.**

RICHARD II. OF ENGLAND.   *Reigned 1377–1399.*

Abbess of Shaftesbury; or, Days of John of Gaunt.   ...Anon.

Chronicles of Yate Court : a narrative of 1399.   ...Charles Charlton.

Claribel, the sea maid : a tale of the Fishmongers' Company.   ...Elizabeth M. Stewart.

Eleanor and I : a tale of the days of King Richard II.   ...Mary E. Bamford.

Hubert Ellis : a tale of Richard the Second's days.   ...F. Davenant.

John of Gaunt.   ...James White.

Last wolf : a tale of England in the fourteenth century.   ...Anne Mercier.

Otterbourne.   ...Edward Duras.

Queen Phillippa and the hurrer's daughter : a tale of the Haberdashers' Company.   ...Elizabeth M. Stewart.

RICHARD II. OF ENGLAND — *Continued.*
Stories from English history during the middle ages. ...Maria Hack.
True stories of the times of Richard II. ...Henry P. Dunster.
Wardship of Steepcoombe. ...Charlotte Mary Yonge.
See also **INSURRECTIONS. England. Wat Tyler's, 1381.**

RICHARD III. OF ENGLAND. *Reigned 1483–1485.*
Bosworth-field. ...John Beaumont.
Bosworth-field; or, Fate of the Plantagenet. ...Paul Leicester.
Last of the Plantagenets. ...William Haseltine.
Richard of York; or, White rose of England. ...Anon.
Richard Plantagenet. ...J. Frederick Hodgetts.
Siballa the sorceress; or, Flower-girl of London: a tale of the days
    of Richard III. ...W. H. Peck.
Woodman. ...George Payne Rainsford James.

RICHELIEU, CARDINAL, ARMAND JEAN DUPLESSIS. *Lived
    1585–1642.*
Cinq-Mars; ou, une conjuration sous Louis XIII. ...Alfred Victor,
    Comte de Vigny.
Cinq-Mars; or, Conspiracy under Louis XIII. ...Alfred Victor,
    Comte de Vigny. English translation.
Duplessis. ...E. Robinson.
Letters writ by a Turkish spy. ...Giovanni P. Marana.
De l'Orme. ...George Payne Rainsford James.
Richelieu: a tale of France. ...George Payne Rainsford James.
Richelieu; or, Broken heart: an historical tale. ...Catherine G. F.
    Gore.
Richelieu; or, Conspiracy. ...Edward George Earle Lytton Bulwer-
    Lytton.
Le Roy de Polexandre. ...Marin Gomberville.
Under the red robe. ...Stanley J. Weyman.

ROBIN HOOD. See **HOOD, ROBIN.**

ROB ROY (ROBERT MACGREGOR, OF SCOTLAND). *Lived 1671–
    1734.*
Adventures of Rob Roy. ...James Grant.
Rob Roy. ...Sir Walter Scott.

ROME.

Description, History, Manners, and Customs.

Antonina; or, Fall of Rome. ...Wilkie Collins.
Antinous: a romance of ancient Rome. ...Adolf Hausrath. (Pseud.,
    George Taylor.) English translation.
Antinous: historischer Roman aus der römischen Kaiserzeit.
    ...Adolf Hausrath. (Pseud., George Taylor.)

ROME — *Continued.*

Numa Pompilius. ...Jean Pierre Claris de Florian.

One that wins: story of a holiday in Italy. ...Anon.

Pietro Ghisleri. ...Francis Marion Crawford.

Pomponia ; or, Gospel in Cæsar's household. ...Mrs. J. B. Peploe (Webb).

Probus; or, Rome in the third century. ...William Ware.

Prusias : Roman aus dem letzten Jahrhundert der römischen Republik. ...Ernst Eckstein.

Quintus Claudius. ...Ernst Eckstein. English translation.

Rienzi, the last of the Roman tribunes. ...Edward George Earle Lytton Bulwer-Lytton.

Roman exile. ...G. Gajani.

Roman singer. ...Francis Marion Crawford.

Rome. ...Émile Zola.

Rome. ...Émile Zola. English translation.

Sant' Ilario. ...Francis Marion Crawford.

Stories of old Rome. ...Charles H. Hanson.

Struggle for Rome. ...Felix Dahn.

Theban legion: story of the time of Diocletian. ...William M. Blackburn.

Three Roman girls. ...Mary E. Bamford.

Tolla. ...Edmond François Valentin About.

Tolla : a tale of modern Rome. ...Edmond François Valentin About. English translation.

Truce of God : a tale of the eleventh century. ...George H. Miles.

Valerius : a Roman story. ...John Gibson Lockhart.

Victory of the vanquished. ...Elizabeth Charles.

Villa of Claudius. ...Edward Lewes Cutts.

Vittoria Coronna. ...Anon.

Wien und Rom. ...Eduard Breier.

See also **ITALIAN HISTORY** and **ITALY.**

RUSSIA.

Description, History, Manners, and Customs.

Annals of a sportsman. ...Ivan Sergevitch Turgenef. English translation.

Baron Munchausen's narrative of his marvellous travels and campaigns in Russia. ...Rudolf Erick Raspe.

Beauty and the beast. ...Bayard Taylor.

Black and gold. ...Patten Saunders.

Blind musician. ...Vladimir Korolenko.

Boyar of the Terrible : a romance of the court of Ivan the Cruel, first Tzar of Russia. ...Frederick Wishaw.

By order of the Czar. ...Joseph Hatton.

RUSSIA — *Continued.*
Castle foam.   ...Harry W. French.
Countess Jamina: an historical novel from Russian life.   ...Gustav Genrychowitch Taube.   English translation.
Crime and punishment.   ...Fedor M. Dostoevsky.   English translation.
Czar and Sultan : adventure of a British lad in the Russo-Turkish war of 1877–78.   ...Archibald Forbes.
Czar Paul.   ...Theodor Mundt.
·Czarina.   ...Barbara Holland.
Daughter of an empress.   ...Clara M. Mundt.   (Pseud., Louise Mühlbach.)   English translation.
Delaplaine.   ...Mansfield Tracy Walworth.
Deluge.   ...Henryk Sienkiewicz.   English translation.
Dmitri : a romance of old Russia.   ...Francis W. Bain.
Dosia : a Russian story.   ...Alice M. Durand.   (Pseud., Henry Gréville.)
Female Nihilist.   ...E. Lavigna.
Green book ; or, Freedom under the snow.   ...Mór Jókai.   English translation.
Gun-maker of Moscow : a tale of the empire under Peter the Great.   ...Sylvanus Cobb, Jr.
Helen and Olga.   ...Anne Manning.
Highway of sorrow.   ...Hannah Smith.   (Pseud., Hesba Stretton.)
In the dwelling of silence : a romance of Russia.   ...Walker Kennedy.
Ivan de Biron ; or, Russian court in the middle of last century.   ...Sir Arthur Helps.
Iwan.   ...S. F. E. Meyer.
Janko Muzykant.   ...Henryk Sienkiewicz.
Kenneth ; or, Rear-guard of the grand army.   ...Charlotte Mary Yonge.
Last day of the carnival.   ...J. Kostromiten.
Little blue lady.   ...Mitchell.
Long exile and other stories for children.   ...Lyeff Nikolaievich Tolstoi.   English translation.
Mam'zelle Eugenie.   ...Alice M. Durand.   (Pseud., Henry Gréville.)
Michael and Theodora.   ...Amelia Edith Barr.
Myths and folk-tales of the Russians, Western Slaves, and Magyars.   ...Jeremiah Curtin.
Narka, the Nihilist.   ...Kathleen O'Meara (Grace Ramsay).
Ogniem i mieczem.   ...Henryk Sienkiewicz.
On the red staircase.   ...M. Imlay Taylor.
Peter the Great's negro.   ...Aleksandr Sergueievitch Pushkin.

RUSSIA — *Continued.*

Poor folk. ...Fedor M. Dostoevsky. English translation.

Potop. ...Henryk Sienkiewicz.

Prince of Balkistan. ...Allen Upward.

Princess Mazaroff. ...Joseph Hatton.

Princess Tarakanova. ...G. P. Danilevski. English translation.

Prinzessin Tartaroff ; oder, Die Tochter einer Kaiserin. ...Clara M. Mundt. (Pseud., Louise Mühlbach.)

Prophet of the Caucasus. ...Captain Spencer.

Romance of war. ...L. Bellstab.

Rudin. ...Ivan Sergevitch Turgenef. English translation.

Russian portraits. Eugene M. M. de Vogüé. English translation.

Russian priest. ...I. N. Potapenko. English translation.

Russian proprietor and other stories. ...Lyeff Nikolaievich Tolstoi. English translation.

Saveli's expiation. ...Alice M. Durand. (Pseud., Henry Gréville.)

Sowers. ...H. Seton Merriam.

Strange stories of a Nihilist. ...W. Le Quex. English translation.

Terrible Czar. ...Lyeff Nikolaievich Tolstoi.

Through Russian snows: a tale of Napoleon's retreat from Moscow. ...George Alfred Henty.

Two-legged wolf.· ...N. N. Karazin.

Under the black eagle. ...Andrew Hilliard.

Vera ; or, Russian princess and English earl. ...Charlotte L. H. Dempster.

Vera Vorontzoff. ...Sonia K. Kovalevsky.

War and peace. Lyeff Nikolaievich Tolstoi. English translation.

With fire and sword. ...Henryk Sienkiewicz. English translation.

Yanko, the musician. ...Henryk Sienkiewicz. English translation.

SAINT AUGUSTINE, AURELIUS. *Lived 354-430.*

Alypius of Tagaste: a tale of the early church. ...Mrs. J. B. Peploe (Webb).

SAINT BARTHOLOMEW'S DAY. 1572.

Astrologer's daughter : an historical novel. ...Rose E. Temple. (Pseud., R. E. Hendricks.)

Catharine de Médicis; or, Queen mother. ...Louisa S. Costello.

Chaplet of pearls; or, White and black Ribaumont. ...Charlotte Mary Yonge. (Historical reference not strictly correct.)

Chronique du temps de Charles IX., 1572. ...Prosper Mérimée.

La comtesse de Tende. ...Comtesse de Lafayette.

La dame d'entremont, récit du temps de Charles IX. ...Ernest d'Hervilly.

La fille d'honneur. ...Alexandrine S. C. de C. baronne da Bawr.

François de Guise. ...Joseph Mathurin Brisset.

SAINT BARTHOLOMEW'S DAY. 1572 — *Continued.*

François le Balafré.   ...Charles Buet.

Hamilton of Bothwellhaugh : a dark scene in Paris.   ...Henry W. Herbert.   (Pseud., Frank Forester.)

Henry of Guise ; or, States of Blois.   ...George Payne Rainsford James.

History of Nicholas Muss.   ...Charles DuBois-Melly.

House of the wolf : a romance.   ...Stanley J. Weyman.

In the vulture's nest ; or, Huguenots at the court of France in 1572.   ...Mildred Fairfax.

Margaret de Valois : a historical romance.   ...Alexandre Dumas. English translation.

Men at arms ; or, Henry de Cerons.   ...George Payne Rainsford James.

La reine Margot.   ...Alexandre Dumas.

Le roi Charlot : scènes de la Saint-Barthélemy.   ...Charles Buet.

Saint Bartholomew's eve : a tale of the Huguenot wars.   ...George Alfred Henty.

Sister Rose ; or, Eve of Saint Bartholomew.   ...Emily Sarah Holt.

See also **FRENCH HISTORY.**

SAINT CHRYSOSTOM, JOHN. *Lived 347–407.*

Gathering clouds.   ...Frederic William Farrar.

Quadratus : a tale of the world in the church.   ..Emma Leslie.

SAINT JEROME, EUSEBIUS HIERONYMUS. *Lived 340–420.*

Conquering and to conquer : a story of Rome in the days of Saint Jerome.   ...Elizabeth Charles.

Erling ; or, Days of Saint Olaf.   ...F. Scarlett Potter.

SAINT PAUL (SAUL). *Lived ?–67.*

Onesimus : memoirs of a disciple of Saint Paul.   ...Edwin A. Abbott.

Saint Paul in Greece.   ...Gerald S. Davies.

SALEM, MASSACHUSETTS.

Delusion ; or, Witch of New England.   ...Anon.

Fair Puritan : an historical romance of the days of witchcraft.   ...Henry W. Herbert.   (Pseud., Frank Forester.)

Lois the witch.   ...Elizabeth C. Gaskell.

Martha Corey : a tale of the Salem witchcraft.   ...Constance Goddard Du Bois.

Philip English's two cups.   ...Anon.

Rachel Dyer.   ...John Neal.

Salem : a tale of the seventeenth century.   ...D. R. Castleton.

Salem belle : a tale of 1692.   ...Anon.

Salem witchcraft : an Eastern tale.   ...R. C. Sands.

SALEM, MASSACHUSETTS — *Continued.*
Salem witchcraft; or, Adventures of Parson Handy from Punkapog. ...Anon.
Silent struggles. ...Ann S. W. Stephens.
South meadows. ...E. T. Disosway.
Spectre of the forest; or, Annals of the Housatonic. ...J. Mac-Henry.
Witch and the deacon. ...Cornelius Mathews.
Witch Hill: a history of Salem witchcraft. ...Z. A. Mudge.
Witch of Salem; or, Credulity run mad. ...John R. Musick.

SAN DIEGO, CALIFORNIA.
Monica, the Mesa maiden. ...Evelyn Raymond.

SAN FRANCISCO, CALIFORNIA.
Cat and the cherub, and other stories. ...Stephen Crane.
Mysteries and miseries of San Francisco. ...Anon.
Prodigal in love. ...Emma Wolf.

SARACENS. See **MOHAMMEDANS.**

SARATOGA, NEW YORK.
Grace Dudley; or, Arnold at Saratoga. ...Charles J. Peterson.
Samantha at Saratoga. ...Marietta Holley. (Pseud., Josiah Allen's wife.)
Saratoga: an Indian tale of frontier life: a true story of 1787. ...D. Shepherd.
Saratoga: a tale of 1787. ...Anon.

SAVONAROLA, GIROLAMO. *Lived 1452-1498.*
Agnes of Sorrento. ...Harriet Beecher Stowe.
House of Fiesole. ...Catherine Shaw.
Romola. ...Marian Evans Lewes Cross. (Pseud., George Eliot.)

SAXONS.
Die Bären von Augustusburg: eine Erzählung aus der sächsischen Geschichte des 18 Jahrhunderts. ...Carl Gustav Nieritz.
Bears of Augustusburg. ...Carl Gustav Nieritz. English translation.
Caedwalla; or, Saxons in the Isle of Wight. ...Frank Cowper.
Early dawn; or, Sketches of Christian life in England in the olden time. ...Elizabeth Charles.
Eldrick, the Saxon. ...A. S. Bride.
Ethelwold. ...Amélie Rives Chanler.
Fair Saxon. ...Justin McCarthy.
Gytha's message: a tale of Saxon England. ...Emma Leslie.
Harold, the last of the Saxon kings. ...Edward George Earle Lytton Bulwer-Lytton.

SAXONS — *Continued.*

Harold the Saxon. ...Alan Muir.
Hereward the Saxon. ...Charles Knight.
Hereward the Wake. ...Charles Kingsley.
Johann Georg I. von Sachsen. ...Franz Lubojatzky.
John George I. of Saxony. ...Franz Lubojatzky. English translation.
Die Kaiserlichen in Sachsen. ...Wilhelm R. Heller.
Lady Godiva: a tale of Saxon England. ...John B. Marsh.
Maurice, the elector of Saxony. ...Katharine Colquhoun.
Northumberland abbots. ...R. B. Werborton.
Pförtner Jugend. ...Friedrich Kenner.
Saxon serf. ...Henry W. Herbert. (Pseud., Frank Forester.)
Story of the stolen princes. ...Anon.
Tales of the Saxons. ...Emily Taylor.
Wager of battle. ...William Henry Herbert.
Wulf the Saxon. ...George Alfred Henty.
Wulf the Saxon. ...Ralph Peacock.
See also **ENGLISH HISTORY**.

SCANDINAVIA.

Description, History, Manners, and Customs.

By northern seas. ...Mary Bell.
Champion of Odin; or, Viking life in the days of old. ...J. Frederick Hodgetts.
Der Eisenkopf: eine historische Erzählung. ...Franz Hoffmann.
Erling; or, Days of Saint Olaf. ...F. Scarlett Potter.
Erling the Bold: a tale of the Norse sea-kings. ...Robert Michael Ballantyne.
Fältskärns berättelser. ...Zacharias Topelius.
Feats on the fiord. ...Harriet Martineau.
Gustaf Adolf och 30 åriga kript. ...Zacharias Topelius.
Gustavus Adolphus. ...Zacharias Topelius. English translation.
Gustavus III. ...Zacharias Topelius.
Heroes of the North. ...F. Scarlett Potter.
Hexan. ...Zacharias Topelius.
In God's way. ...Bjørnstjerne Bjørnson. English translation.
Iron head. ...Franz Hoffmann. English translation.
Ivar the viking. ...Paul Du Chaillu. English translation.
Ivo and Verena. ...Anon.
King Eric V. ...B. S. Ingemann.
Last of the freebooters. ...P. Sparre.
Liberty. ...Zacharias Topelius. English translation.
Maid of Stralsund. ...John B. de Liefde.

SCANDINAVIA — *Continued.*
Majiemi slott. ...Zacharias Topelius.
Modern vikings. ...Hjalmar Hjörth Boyesen. English translation.
Norseland tales. ...Hjalmar Hjörth Boyesen. English translation.
Norse stories retold from the Eddas. ...Hamilton W. Mabie.
Olaf the Glorious. ...Robert Leighton.
Princess Vasa. ...Zacharias Topelius. English translation.
Prinsessan af Vasa. ...Zacharias Topelius.
Rebell mot sin bycka. ...Zacharias Topelius.
Surgeon's stories. ...Zacharias Topelius. English translation.
Times of the battle and of rest. ...Zacharias Topelius. English
    translation.
Ulla. ...Edwin Lester Arnold.
Under Gustaf's första regeringsår. ...Zacharias Topelius.
Vikings of the Baltic: a tale of the North in the tenth century.
    ...George W. Dasent.
Waldemar. ...Bernhard Severin Ingemann.
Wonderful stories from northern lands. ...Julia Goddard.
Wood spirit. ...Ernst Jones
Yule-tide stories. ...Benjamin Thorpe.

SCEPTICISM.
Autobiography of Mark Rutherford. ...William Hale White.
    (Pseud., Mark Rutherford.)
Blessed Saint Certainty. ...William Mumford Baker.
Damnation of Theron Ware. ...Harold Frederic.
Le docteur Pascal. ...Émile Zola.
Doctor Pascal. ...Émile Zola. English translation.
Heresy of Mehetabel Clark. ...Annie T. Slosson.
History of David Grieve. ...Mary Augusta Ward (Mrs. Humphry
    Ward).
John Inglesant. ...Joseph Henry Shorthouse.
Legend of Thomas Didymus, the Jewish sceptic. ...James Freeman
    Clarke.
Mark Rutherford, dissenting minister. ...William Hale White.
    (Pseud., Mark Rutherford.)
Mark Rutherford's deliverance. ...William Hale White. (Pseud.,
    Mark Rutherford.)
Modern instance. ...William Dean Howells.
New minister. ...Kenneth Paul.
Phyllis of Philistia. ...Frederic Frankfort Moore.
Robert Elsmere. ...Mary Augusta Ward (Mrs. Humphry Ward).
Robert Falconer. ...George MacDonald.
Rome. ...Émile Zola.
Rome. ...Émile Zola. English translation.

SCEPTICISM — *Continued.*
Singular life.   ...Elizabeth Stuart Phelps Ward (Mrs. Herbert D. Ward).
See also **DOGMAS.**

SCHILLER, JOHANN CHRISTOPH FRIEDRICH. *Lived 1759–1805.*
Goethe and Schiller.   ...Clara M. Mundt.   (Pseud., Louise Mühlbach.) English translation.
Göthe und Schiller.   ...Clara M. Mundt.   (Pseud., Louise Mühlbach.)
Schiller. Kultur-historischer Roman.   ...Johannes Scherr.

SCHOOL-LIFE. *A Selection.*
Alwyn Morton, his school and his school fellows.   ...Anon.
Antony Brade.   ...Robert T. S. Lowell.
Boston Latin School.   ...Willis B. Allen.
Changing base.   ...William Everett.
Cuore.   ...Edmondo De Amicis.
Cuore.   ...Edmondo De Amicis.   English translation.
Day of my life; or, Every-day experiences at Eton.   ...George N. Banks.
Dick Rodney; or, Adventures of an Eton-boy.   ...James Grant.
District school as it was.   ...Warren Burton.
His masters : a story of school life forty years ago.   ...S. S. Pugh.
Luke Ashleigh; or, School life in Holland.   ...Alfred Elwes.
Master and pupil ; or, School life at the Old Baldwin.   ...Estelle D. Kendall.
My school life in Paris.   ...Margaret S. Jeune.
New senior at Andover.   ...Herbert D. Ward.
Northern cross ; or, Randolph's last year at the Boston Latin school.   ...Willis B. Allen.
Pförtner Jugend.   ...Friedrich Kenner. (School life in Saxony.)
Red-shanty boys.   ...Theron Brown.
Rocky Fork.   ...Mary Hartwell Catherwood.
Schoolboy days in Japan.   ...André Laurie.
Schoolboys' honour : a tale of Halminster college.   ...Henry C. Adams.
School days at Kingscourt.   ...Henry C. Adams.
School days at Rugby.   ...Thomas Hughes.
School-girls.   ...Annie Carey.
Tales of Charlton school.   ...Henry C. Adams.
Tim.   ...Edward Harrington
Tom Brown at Oxford.   ...Thomas Hughes.
Tom Brown's school days.   .. Thomas Hughes.
Unlessoned girl : a story of school life.   ...Elizabeth K. Tompkins.
Who was Philip ? a tale of public school life.   ...Henry C. Adams.
See also **COLLEGE AND UNIVERSITY LIFE.**

SCHRÖDER, FRIEDRICH LUDWIG. *Lived 1744–1816.*
Charlotte Ackermann: ein Hamburger Roman aus dem vorigen
Jahrhundert. ...Otto Müller.
Charlotte Ackermann: a theatrical romance. ...Otto Müller.
English translation.

SCHUBERT, FRANZ PETER. *Lived 1797–1838.*
Schubert und seine Zeitgenossen. ...Albert E. Brachvogel.

SCOTLAND.
Description, History, Manners, and Customs.

Abbot. ...Sir Walter Scott.
Adventures of Rob Roy. ...James Grant.
Antiquary. ...Sir Walter Scott.
Auld licht idylls. ...James Matthew Barrie.
Auld licht manse and other sketches. ...James Matthew Barrie.
Balmoral. ...Alexander Allardyce.
Barncraig: episodes in the life of a Scottish village. ...Gabriel
Setoun.
Beside the bonnie brier bush. ...John Maclaren Watson. (Pseud.,
Ian Maclaren.)
Black dwarf. ...Sir Walter Scott.
Border shepherdess. ...Amelia Edith Barr.
Bothwell; or, Days of Mary Queen of Scots. ...James Grant.
Braes of Yarrow. ...H. Gibbon.
Bride of Lammermoor. ...Sir Walter Scott.
Brownie of Bodsbeck. ...James Hogg.
Caged lion. ...Charlotte Mary Yonge.
Captain Macdonald's daughter. ...Archibald Campbell.
Captain of the guard. ...James Grant.
Castle Dangerous. ...Sir Walter Scott.
Catriona. ...Robert Louis Stevenson.
Chronicles of Glenbuckie. ...Henry Johnston.
Courting of Tnow Head's bell. ...James Matthew Barrie.
Crichton. ...William Harrison Ainsworth.
Dark year of Dundee. ...Deborah Alcock.
David Balfour in the year 1751. ...Robert Louis Stevenson.
David Elginbrod. ...George MacDonald.
David Rizzio. ...William H. Ireland.
Dawsons of Glenara. ...Henry Johnston.
Days of auld lang syne. ...John Maclaren Watson. (Pseud., Ian
Maclaren.)
Days of Bruce. ...Grace Aguilar.
Days of yore. ...Henrietta Keddie. (Pseud., Sarah Tytler.)
Donald Ross of Heimra. ...William Black.
Duke of Albany's own Highlanders. ...James Grant.

SCOTLAND — *Continued.*

Two penniless princesses. ...Charlotte Mary Yonge.

Waverley. ...Sir Walter Scott.

White cockade. ...James Grant.

White wings. ...William Black.

Who is the man ? a tale of the Scottish border. ...James Selwin Tait.

Who was lost and is found. ...Margaret O. W. Oliphant.

Window in Thrums. ...John Matthew Barrie.

X jewel : a Scottish romance. ...Frederick Moncrieff.

Yellow frigate. ...James Grant.

SEA STORIES.

Adrift on the sea. ...E. M. Norris.

Adventures of Captain Mago ; or, Phœnician expedition B. C. 1000. ...Léon Cahun. English translation.

Adventures of Harry Marline ; or, Notes from an American midshipman's lucky bag. ...David Dixon Porter.

Adventures of Reuben Davidger. ...James Greenwood.

Afloat and ashore. ...James Fenimore Cooper.

Anchor-watch yarns. ...E. Downey.

Antony Weymouth : or, Gentlemen adventurers. ...William Henry Giles Kingston.

Around the world in eighty days. ...Jules Verne. English translation.

Australian crusoe. ...Charles Rowcroft.

Autobiography of a man-o'-war's bell. ...C. R. Low.

Les aventures de Capitaine Magon ; ou, Une exploration phénicienne mille ans avant l'ère chrétienne. ...Léon Cahun.

Bad penny. ...J. T. Wheelwright.

Battles with the sea. ...Robert Michael Ballantyne.

Beautiful white devil. ...Guy Boothby.

Ben Brace, the last of Nelson's Agamemnons. ...Frederick Chamier.

Biscuits and grog. ...James Hannay.

Blockade runners. ...Jules Verne. English translation.

Born to command : a tale of the sea and of sailors. ...William Gordon Stables.

Boy life in the United States navy. ...H. H. Clark.

Boy tar. ...Mayne Reid.

By reef and palm. ...Louis Becke.

By sea and land. ...William Gordon Stables.

Campaigns and cruises in Venezuela and New Granada and the Pacific Ocean, 1817–1830. ...Anon.

Cape Cod and along the shore. ...Charles Nordhoff.

Le capitaine Paul. ...Alexandre Dumas.

SEA STORIES — *Continued.*

Captain of the Kittiewink. ...Herbert D. Ward.
Captain of the Rajah. ...H. Patterson.
Captain Paul. ...Alexandre Dumas. English translation.
Captain's room. ...Walter Besant.
Cast away in the cold. ...Isaac Israel Hayes.
Cast up by the sea. ...Samuel White Baker.
Charlie Laurel. ...William Henry Giles Kingston.
Chase of the Leviathan. ...Mayne Reid.
Chase of the Meteor. ...E. L. Bynner.
Children's Robinson Crusoe. ...F. Eliza Farrar.
Children's voyage. ...A. J. Cupples.
Clare Avery. ...Emily Sarah Holt.
Confederate flag on the ocean. ...William H. Peck.
Coralie. ...C. H. Eden.
Crew of the Dolphin. ...Hannah Smith. (Pseud., Hesba Stretton.)
Cruise of the Acorn. ...A. Smith.
Cruise of the Midge. ...Michael Scott.
Cruise of the Wasp. ...Henry Frith.
Cruising in the last war. ...Charles J. Peterson.
Cyril Hamilton: his adventures by land and sea. ...C. R. Low.
Dark colleen. ...Harriet Jay.
Death ship. ...William Clark Russell.
Derval Hampton: a story of the sea. ...James Grant.
Le désert de glace. ...Jules Verne.
Deserted ship. ...George Cupples.
Diccon the Bold. ...John Russell Coryell.
Dick Sands, the boy captain. ...Jules Verne. English translation.
Diego Pinzon and the fearful voyage he took into the unknown
    ocean. ...John Russell Coryell.
Dog Crusoe. ...Robert Michael Ballantyne.
Dog-fiend; or, Snarleyyow. ...Frederick Marryat.
Driven to sea. ...A. J. Cupples.
Ebbing of the tide: South Sea stories. ...Louis Becke.
Emigrant ship. ...William Clark Russell.
English family Robinson. ...Mayne Reid.
Erling the Bold; or, Days of Saint Olaf. ...Robert Michael Ballan-
    tyne.
Eustace Conyers. ...James Hannay.
Fast in the ice. ...Robert Michael Ballantyne.
Field of ice. ...Jules Verne. English translation.
Fighting the sea. ...E. A. Rand.
Fighting the whales. ...Robert Michael Ballantyne.
Fisher-maiden: a Norwegian tale. ...Bjørnstjerne Bjørnson. English
    translation.

SEA STORIES — *Continued.*

Die sogenannte insel Felsenburg. ...Johann G. Schnabel.

Solitary of Juan Fernandez. ...Joseph Xavier Boniface. (Pseud., Saintine.) English translation.

South Sea yarns. ...Basil Thomson.

Stories of the sea. ...James Fenimore Cooper.

Story of a shell. ...J. R. MacDuff.

Stranded ship. ...L. C. Davis.

Strange elopement. ...William Clark Russell.

Strange voyage. ...William Clark Russell.

Sunk at sea. ...Robert Michael Ballantyne.

Sunken rock. ...George Cupples.

Swiss family Robinson. ...Johann Rudolf Wyss. English translation.

Tales of a tar. ...W. N. Glascock.

Tales of a voyager to the Arctic ocean. ...R. P. Gillies.

Tales of old ocean. ...C. R. Low.

Tales of the marines. ...H. A. Wise.

Tales of the shore and ocean. ...William Henry Giles Kingston.

Terrapin Island. ...A. J. Cupples.

Three commanders; or, Active service afloat in modern days. ...William Henry Giles Kingston.

Three lieutenants. ...William Henry Giles Kingston.

Three midshipmen. ...William Henry Giles Kingston.

Toilers of the sea. ...Victor Hugo. English translation.

Tom Bowling. ...Frederick Chamier.

Tom Cringle's log. ...Michael Scott.

Le tour du monde en 80 jours. ...Jules Verne.

Les travailleurs de la mer. ...Victor Hugo.

Treasure Island. ...Robert Louis Stevenson.

True blue; or, Life and adventures of a British seaman of the old school. ...William Henry Giles Kingston.

'T was in Trafalgar's bay. ...Walter Besant and James Rice.

Twenty thousand leagues under the sea. ...Jules Verne. English translation.

Twice lost: a story of shipwreck. ...William Henry Giles Kingston.

Two admirals. ...James Fenimore Cooper.

Two frigates. ...George Cupples.

Two supercargoes. ...William Henry Giles Kingston.

Typee: a real romance of the Southern sea. ...Herman Melville.

Uncle John's first shipwreck; or, Loss of the "Nellie." ...C. Brace.

Under Drake's flag: a tale of the Spanish main. ...George Alfred Henty.

SEA STORIES — *Continued.*
Under the red ensign ; or, Going to sea. ...T. Gray.
Upper berth. ...Francis Marion Crawford.
Vingt mille lieues sous les mers. ...Jules Verne.
Voyage around the world. ...William Henry Giles Kingston.
Un voyage involontaire. ...Lucien Biart.
Voyage of the Steadfast. ...William Henry Giles Kingston.
Voyages imaginaires. ...C. G. T. Garnier. (35 volumes.)
Water-witch. ...James Fenimore Cooper.
Watchers on the long ships. ...J. F. Cobb.
Westward Ho! or, Voyages and adventures of Sir Amyas Leigh.
...Charles Kingsley.
White-jacket ; or, World in a man-of-war. ...Herman Melville.
White-wings : a yachting romance. ...William Black.
Wild adventures at the pole ; or, Cruise of the Snowbird in the
Arrandoon. ...William Gordon Stables.
William Allair. ...E. P. Wood.
Willis the pilot. ...Johann David Wyss.
Will Weatherhelm, the yarn of an old sailor. ...William Henry
Giles Kingston.
Wing-and-Wing ; or, Le Feu-follet. ...James Fenimore Cooper.
With the admiral of the Ocean Sea. ...C. P. MacKie.
Wonderful adventures by land and by sea of the seven queer travel-
lers who met at the inn. ...Josiah Barnes.
Wreck of the Grosvenor. ...William Clark Russell.
Wreck of the Red Bird. ...Edward C. Eggleston.
Wreckers. ...M. R. S. Kettle.
Yarns of an old mariner. ...Mary Cowden Clarke.
Yarns on the beach : a bundle of tales. ...George Alfred Henty.
Young Crusoe. ...Barbara Hofland.
Young middy. ...F. C. Armstrong.
Young yachtsman. ...A. Bowman.

SEMINOLE WAR. 1829-37.
Life and writings of Major Jack Downing of Downingville. ...Seba
Smith.
Light dragoon. ...Justin Jones. (Pseud., Harry Hazel.)
My thirty years out of the Senate. ...Seba Smith.
Osceola. ...Mayne Reid.
Osceola ; or, Fact and fiction : a tale of the Seminole war. ...Sey-
mour R. Duke.
Sergeant Atkins : a tale of adventure founded on fact. ...James L.
Donaldson.
She loved a sailor. ...Amelia Edith Barr.
Volunteer. ...Edward Z. C. Judson. (Pseud., Ned Buntline.)
Zachary Phips. ...Edwin Lassetter Bynner.

SEVEN WEEKS' WAR. 1866.

Die Czechin: Erzählung aus dem Kriege des Jahres 1866. ...Eugen H. von Dedenroth.

1866; oder in Böhmen und am Main.   ...Franz Lubojatzky.

Der Jäger von Königgrätz.   ...Eugen H. von Dedenroth.

Das Jahr 1866.   ...Johann Ferdinand Martin Oskar Meding. (Pseud., Gregor Samarow.)

Der Kaplan von Königgrätz.   ...Franz Lubojatzky.

Maid, wife, or widow?   ...Annie F. Hector. (Pseud., Mrs. Alexander.)

Die Rose von Sadowa.   ...Ludwig A. R. Kuerbis.

Scepter and crown.   ...Johann Ferdinand Martin Oskar Meding, (Pseud., Gregor Samarow.) English translation.

Um Scepter und Kronen.   ...Johann Ferdinand Martin Oskar Meding. (Pseud., Gregor Samarow.)

Unter Preussens Fahnen : historischer Roman aus dem Jahre 1866. ...Stanislaus S. A. Grabowski.

Vor dem Sturm.   ...Johann Ferdinand Martin Oskar Meding. (Pseud., Gregor Samarow.)

SEVEN YEARS' WAR. 1756–63.

Fallen star ; or, Scots of Frederick: a tale of the Seven Years' war.   ...Charles Lowe.

Die historischen Volkslieder des siebenjährigen Krieges.   ...Franz W. Ditfurth.

Die Kaiserlichen in Sachsen.   ...Wilhelm R. Heller.

Luck of Barry Lyndon : a romance of the last century.   ...William Makepeace Thackeray.

Prisoner's daughter: a story of 1758.   ...Esmé Stuart.

Rückwirkungen; oder wer regiert denn ?   ...Johann Heinrich D. Zschokke.

Verrath und Liebe.   ...Adolf Muetzelburg.

SHAKESPEARE, WILLIAM. *Lived 1564–1616.*

Anne Hathaway; or, Shakespeare in love.   ...Edward Severn.

Bailiff of Tewkesbury.   ...C. E. D. Phelps and Leigh North.

Citation and examination of William Shakespeare.   ...Walter Savage Landor.

Forest youth; or, Shakespere as he lived.   ...Henry Curling.

Geraldine Maynard; or, Abduction : a tale of the days of Shakespere.   ...Henry Curling.

How Shakespeare's skull was stolen and found by a Warwickshire man.   ...C. J. Langston.

Judith Shakespeare.   ...William Black.

Montchensey: the days of Shakespeare.   ...Nathan Drake.

Secret passion.   ...Robert Folk Williams.

SHAKESPEARE, WILLIAM — *Continued.*
Shakespeare and his friends; or, Golden age of merry England.
...Robert Folk Williams.
Shakespeare's funeral. ...Edward B. Hamley.
Shakespere the poet, the lover, the actor, the man. ...Henry
Curling.
Stolen mask ; or, Mysterious cash box. ...Wilkie Collins.
William Shakespeare. ...Heinrich J. Koenig. English translation.
William Shakespeare. Culturgeschichtlich-biographischer Roman.
...Heribert Rau.
Williams Dichter und Trachten. ...Heinrich J. Koenig.
Youth of Shakespeare. ...Robert Folk Williams.

SHELLEY, PERCY BYSSHE. *Lived 1792–1822.*
Shelley. Biographische Novelle. ...Wilhelm Ritter von Hamm.

SIBERIA.
Condemned as a Nihilist : a story of escape from Siberia. ...George
Alfred Henty.
Élisabeth ; ou, Les Exilés de Sibérie. ...Sophie Risteau Cottin.
Highway of sorrow. ...Hannah Smith. (Pseud., Hesba Stretton.)
In the dwellings of silence: a romance of Russia. ...Walker
Kennedy.
See also **RUSSIA.**·

SIDNEY, SIR PHILIP. *Lived 1554–1586.*
Penshurst Place in the time of Sir Philip Sidney. ...Emma Marshall.

SIEGES.
Alamo. Siege of 1836.
Inez : a tale of the Alamo. ...Augusta J. Evans (Mrs. Wilson).
Remember the Alamo. ...Amelia Edith Barr.

Alexandria. Siege of 1798.
Chapter of adventures ; or, Through the bombardment of Alexandria. ...George Alfred Henty.

Antwerp. Siege of 1585.
Alarum for London. ...Anon.
Shut in; or, Tale of the wonderful siege of Antwerp in 1585.
...Evelyn Everett Green.
Siege of Antwerp. ...William Kennedy.

Boston. Siege of 1775.
Lionel Lincoln ; or, Leaguers of Boston. ...James Fenimore Cooper.
Story of the siege of Boston. ...Horace E. Scudder. (In his
Stories and romances.)

SIEGES — *Continued.*

. **Bredá. Siege of 1625.**

El sitio de Bredá. ...Pedro Calderon de la Barca.

Breslau. Siege of 1474.

Matthias Corvinus. ...Franz A. Wentzel.

Bristol. Siege of 1645.

Prince and pedler. ...Ellen Pickering.

Bourges. Siege of 52 B. C.

La Avarchide. ...Luigi Alamanni.

Buda. Siege of 1541.

Zord Idő. ...Zsigmund Kemény.

Buda. Siege of 1686.

Der Pascha von Buda. ...Johann Heinrich D. Zschokke.
Siege of Buda. ...Anon.

Calais. Siege of 1347.

Die Belagerung von Calais. ...Carl Weichselbaumer.
Eustache de Saint Pierre. ...Henry W. Herbert. (Pseud., Frank
    Forester.)
Le siège de Calais. ...Pierre L. B. de Belloy.
Le siège de Calais. ...Claudine A. M. Tencin.

Carthagena. Siege of 1707.

Adventures of Roderick Random. ...Tobias George Smollett.

Charleston. Siege of 1865.

Out of a besieged city. ...Charles W. Hutson.

Chester. Siege of 1643-1646.

Lettice : a tale of the siege of Chester. ...Pauline Biddulph.

Colchester. Siege of 1640.

Siege of Colchester. ...G. F. Townsend.

Constantinople. Siege of 1453.

Die Eroberung von Constantinople. ...Alexander Petrick.
Prince of India; or, Why Constantinople fell. ...Lew Wallace.
Siege of Constantinople. ...C. R. Eaglestone.
Siege of Constantinople. ...Nicholas Michell.
Theodora Phranza; or, Fall of Constantinople. ...John Mason
    Neale. (Pseud., Aurelius Gratianus.)

SIEGES — *Continued.*
Lily of Leyden. ...William Henry Giles Kingston.
Rosalia Vanderiver. ...Anon.
Wind and wave fulfilling His word. ...Harriette E. Burch.

Lichfield. Siege of 1643.
Siege of Lichfield : a tale illustrative of the great rebellion.
...William Gresley.

Limerick. Siege of 1651.
Florence O'Neill. ...Agnes M. Stewart.

London. Siege of 994.
Siege of London. ...Henry James, Jr.

Londonderry. Siege of 1689.
Cousin Isabel : a tale of the siege of Londonderry. ...Marion
Andrews.
Derry : a tale of the revolution. ...Charlotte Elizabeth Tonna.
(Pseud., Charlotte Elizabeth.)
Londerias. ...Joseph Aickin.
Man's foes. ...E. H. Strain.

Louisburg. Siege of 1745.
Boys of 1745 at the capture of Louisburg. ...James Otis Kaler.
(Pseud., James Otis.)
Englishman's haven. ...W. J. Gordon.
William and Mary : a tale of the siege of Louisburg, 1745.
...Daniel Hickey.

Magdeburg. Siege of 1551.
Die Belagerung von Magdeburg. ...Friedrich L. Schmidt.

Malta. Siege of 1565.
Knight of Saint John. ...Anna Maria Porter.

Mexico. Siege of 1521.
Infidel ; or, Fall of Mexico. ...Robert Montgomery Bird.

New Orleans. Siege of 1812.
Creole. ...Joseph B. Cobb.

New York. Siege of 1776.
Burton. ...Joseph Holt Ingraham.

Nieuport. Siege of 1600.
Fisherman's daughter. ...Hendrik Conscience. English translation.
Siska van Roosemael, de ware Geschiedenis van eene juffer die nog
leeft. ...Hendrik Conscience.

SIEGES — *Continued.*

Norwich Castle. Siege of 1216.

Siege of Norwich Castle. ...M. M. Blake.

Orleans. Siege of 1428.

Jeanne d'Arc ; ou, Le siège d'Orléans. ...Martial Soullier.
Le Mistrere du siège d'Orléans publié pour la première fois. ...F.
Guessard et de Certain.

Paris. Siege of 1870–71.

American in Paris. ...Eugene Coleman Savidge.
Aulnay Tower. ...Blanche Willis Howard.
La bande rouge. ...Fortuné Abraham Du Boisgobey.
Comedy of Terrors. ...James De Mille.
Dreams and reality: a tale of the siege of Paris. ...Anon.
Geschichten aus der Pariser Belagerung. ...Paul d'Abrest.
Great mistake: a story of adventure. ...Thomas S. Millington.
Une idylle pendant le siège. ...François Coppée.
Kentucky's love; or, Roughing it about Paris. ...Edward King.
Leonie; or, Light out of darkness. ...Annie Lucas.
Mademoiselle. ...Frances M. Peard.
Les mauvais jours : Notes d'un bourru sur le siège de Paris.
...Zénaïde M. A. Fleuriot.
Parisians. ...Edward George Earle Lytton Bulwer-Lytton.
Une Parisienne sous la foudre. ...Zénaïde M. A. Fleuriot.
Red band; or, Adventures of a young girl during the siege of
Paris. ...Fortuné Abraham du Boisgobey. English translation.
Red republic. ...Robert W. Chambers.
Shut up in Paris. ...Nathan Sheppard.
Story of the siege of Paris. ...Henri and Louis.
Twins of St. Marcel: a tale of Paris incendié. ...Mrs. Alexander S.
Orr.
Die Verschwörung der Republikaner oder die Geheimnisse der
Belagerung von Paris. ...Ernst Kaiser.
Within iron walls: a tale of the siege of Paris. ...Annie Lucas.
Workman and soldier : a tale of Paris life during the siege and rule
of the Commune. ...James F. Cobb.

Pavia. Siege of 774.

Le siège de Pavie ; ou, La gloire de Charlemagne. ...Maria J. A.
Boieldieu.

Phalsbourg. Siege of 1870.

Blockade: an episode of the fall of the First Empire. ...Émile
Erckmann and Alexandre Chatrian. English translation.
Le blocus. ...Émile Erckmann et Alexandre Chatrian.

SIEGES — *Continued.*

Quebec. Siege of 1759.

Burton.   ...Joseph Holt Ingraham.
Le dernier boulet.   ...Joseph Marmette.
Seats of the mighty.   ...Gilbert Parker.

Rhodes. Siege of 1521.

Bertrand de la Croix.   ...George Payne Rainsford James.
Knight of Rhodes.   ...James B. Burges.
Knight of the white cross : a tale of the siege of Rhodes.   ...George
    Alfred Henty.
Siege of Rhodes.   ...William Davenant.

La Rochelle. Siege of 1627.

La malheur et la conscience.   ...Stéphanie Félicité, Comtesse de
    Genlis.
Le siège de La Rochelle.   ...Stéphanie Félicité, Comtesse de
    Genlis.
Siege of La Rochelle.   ...R. C. Dallas.
Siege of La Rochelle ; or, Christian hero.   ...Stéphanie Félicité,
    Comtesse de Genlis.   English translation.
Three musketeers.   ...Alexandre Dumas.   English translation.
Les trois mousquetaires.   ...Alexandre Dumas.

Rouen.

Espérance.   ...Mary Bramston.
Story of the siege of Rouen.   ...Mary Bramston.

Saragosse. Siege of 1808–1809.

Édouard ; ou, Le Siège de Saragosse.   ...Léopold Méry.
Zaragoza.   ...Benito Pérez Galdós.

Stralsund. Siege of 1715.

Brave resolve.   ...John B. de Liefde.
Maid of Stralsund : a tale of the thirty years' war.   ...John B. de
    Liefde.

Strasburg. Siege of 1870.

Max Kromer : a story of the siege of Strasburg, 1870.   ...Hannah
    Smith.   (Pseud., Hesba Stretton.)

Szigeth. Siege of 1566.

Die Belagerung von Szigeth.   ...Friedrich A. Werthes.
Zriny.   ...Theodor Koener.

Vienna. Siege of 1683.

Die Belagerung Wiens im Jahre 1683.   ...Caroline Pichler.
Rüdiger von Starhemberg.   ...Franz X. Huber.

SIEGES — *Continued.*

Siege of Vienna. ...J. Latchmore, Jr. Translated from the Turkish.

Story of the Turkish war of 1583. ...J. Latchmore, Jr. Translated from the Turkish.

For incidental accounts of sieges see the special country wanted.

For sieges not founded on fact see **IMAGINARY LANDS, CITIES, AND INSTITUTIONS**.

SILESIA.

Description, History, Manners, and Customs.

Debit and credit. ...Gustav Freytag. English translation.

Durch Nacht zum Licht. ...Friedrich Spielhagen.

Soll und Haben. ...Gustav Freytag.

Through night to light. ...Friedrich Spielhagen. English translation.

SIN.

Book of strange sins. ...Coulson Kernahan.

Can wrong be right? ...Anna Maria Hall.

Capillary crime. ...Francis D. Millet.

Confessions of a bank officer. ...William Taylor Adams.

Conscience. ...Hector Malot.

Conscience. ...Hector Malot. English translation.

Crime and punishment. ...Fedor M. Dostoevsky. English translation.

Damen auf Reisen. Jodocus Donatus Hubertus Temme.

Erema; or, My father's sin. ...Richard Doddridge Blackmore.

For fifteen years. ...Louis Ulbach. English translation.

Die Freiherren von Falkenbury. ...Jodocus Donatus Hubertus Temme.

Gervase Skinner. ...Theodore E. Hook.

Good people of Pawlooz. ...Coleman Mikszath.

Guilt and innocence. ...Marie Sophie.

Gustavus Lindorm. ...Emilie Flygare Carlen.

Harvest of wild oats. ...Florence Marryat.

Human document. ...George Meredith.

In prison and out. ...Hannah Smith. (Pseud., Hesba Stretton.)

John Guilderstring's sin. ...C. French Richards.

Lady of quality. ...Frances Hodgson Burnett.

Making haste to be rich. ...Timothy Shay Arthur.

Marble faun; or, Romance of Monte Beni. ...Nathaniel Hawthorne.

Memoirs of Jane Cameron, female convict. ...Frederick William Robinson.

Le mesonge de Sabine. ...Olga Cantacuzène.

Modern Hagar. ...Charlotte Moore Clark.

Moral dilemma. ...Annie Thompson.

SIN — *Continued.*

Murder considered as one of the fine arts. ...Thomas DeQuincey.
Murderer's last night. ...Thomas Doubleday.
Mysteries and miseries of San Francisco. ...Anon.
Mystery of Paul Chadwick. ...J. W. Postgate.
Night in a workhouse. ...James Greenwood.
Out of the past. ...E. Anson More.
Der Pfeifenhannes. ...Jodocus Donatus Hubertus Temme.
Quinze ans de bagne. ...Louis Ulbach.
Regicide's daughter. ...William H. Carpenter.
Romola. ...Marian Evans Lewes Cross. (Pseud., George Eliot.)
Scarlet letter. ...Nathaniel Hawthorne.
Silent witness. ...Mrs. J. H. Walworth.
Sinner's comedy. ...John Oliver Hobbes, pseud.
Sinners twain. ...J. Mackie.
Six sinners. ...H. Campbell.
So very human. ...Alfred B. Richards.
Story of Bessie Costrell. ...Mary Augusta Ward (Mrs. Humphry
    Ward).
Study in scarlet. ...Arthur Conan Doyle.
Study in temptation. ...John Oliver Hobbes, pseud.
Tale of sin. ...Ellen P. Wood.
Temptation of Katharine Gray. ...Mary Lowe Dickinson.
Terrible temptation. ...Charles Read.
Transgression of Terence Clancy. ...Harold Vallings.
Two marriages. ...Dinah Maria Mulock Craik.
Unforgiven sin. ...Caldwell Lipsett. (In his Where the Atlantic
    meets the land.)
Wages of sin. ...Mrs. W. Harrison. (Pseud., Malet Lucas.)
Wages of sin. ...Edward H. Yates.
Was he guilty? ...Eliza Ann Dupuy.
Wayward woman. ...Arthur Griffiths.
What will he do with it? ...Edward George Earle Lytton Bulwer-
    Lytton.
Which: right or wrong? ...M. L. Moreland.
Who is guilty? ...Philip Woolf.
Why Paul Ferroll killed his wife. ...Caroline Clive.
See also **CRIMINOLOGY.**

SLAVERY.

Alamontade, der Galeerensklave. ...Johann Heinrich D. Zschokke.
Alamontade, the galley slave. ...Johann Heinrich D. Zschokke.
    English translation.
Amanda. ...W. H. Brisbane.
Au bord du lac. ...Émile Souvestre.

.

SLAVERY — *Continued.*

Life of Captain Jack.   ...Daniel Defoe.

Little captive maid.   ...Florence L. Henderson.

Lord of himself.   ...Francis H. Underwood.

Lutchmee and Dilloo.   ...E. Jenkins.

Memoirs of Archy Moore.   ...Richard Hildreth.

Miss Ravenal's conversion from secession to loyalty.   ...John W. De Forest.

Modern Hagar.   ...Charlotte M. Clark.   (Pseud., Charles M. Clay.)

El Mulato Sab.   ...Gertrúdes Gomez de Avellaneda.

My Kalulu, prince, king, and slave.   ...Henry Moreland Stanley.

Negro slave.   ...Washington Nash.

North and the South.   ...Caroline E. Rush.

Old plantation.   ...James Hungerford.

Oroonoko ; or, Royal slave.   ...Anon.

Partisan leader.   ...Beverley.

Planter's daughter.   ...Eliza Ann Dupuy.

Poor Paddy's cabin ; or, Irish slavery.   ...By an Irishman.

Prince of Kashna.   ...Anon.

Quadroona ; or, Slave mother.   ...Percy B. Saint John.

Rachel Stanwood.   ...Lucy Gibbons Morse.

Recaptured negro.   ...Mary M. Sherwood.

Red rover.   ...James Fenimore Cooper.

Salome Muller, the white slave.   ...George Washington Cable.

Samantha on the race problem.   ...Marietta Holley.   (Pseud., Josiah Allen's wife.)

Der Scheik von Alessandria und seine Sklaven.   ...Wilhelm Hauff.

Sclaverei in Amerika.   ...F. A. Stubberg.

Sheik of Alexandria and his slaves.   ...Wilhelm Hauff.   English translation.

She loved a sailor.   ...Amelia Edith Barr.

Slave girl of Pompeii.   ...Emily Sarah Holt.

Slave, the serf and the freeman.   ...Anon.

Slaves and cruisers.   ...Samuel Whitchurch Sadler.

Slaves of Sabinus.   ...Charlotte Mary Yonge.

Slaves of the ring.   ...Frederick William Robinson.

Sunny South ; or, Southerner at home.   ...Joseph Holt Ingraham.

Toinette.   ...Albion Winegar Tourgée.

Trooper Peter Halket of Mashonaland. ...Olive Schreiner. (Pseud., Ralph Iron.)

Two runaways and other stories.   ...Harry S. Edwards.

Uncle Tom's cabin ; or, Life among the lowly.   ...Harriet Beecher Stowe.

Virginian plantation story.   ...Margaret J. Preston.

SLAVERY — *Continued.*
Wager of battle. ...William Henry Herbert.
What answer? ...Anna E. Dickinson.
White slave. ...Richard Hildreth.
Wife or slave? ...Mrs. Albert S. Bradshaw.
Yankee slave dealer. ...Anon.
Youma: the story of a West Indian slave. ...Lafcadio Chita Hearn.
See also **SOUTH, U. S. A.**

SLUM STORIES. *A Selection.*
Adrift in a great city. ...M. E. Winchester.
All sorts and conditions of men. ...Walter Besant.
Artie : a story of the streets and town. ...George Ade.
Boy's revolt. ...James Otis Kaler. (Pseud, James Otis.)
Child of the Jago. ...Arthur Morrison.
Chimmie Fadden. ...Edward W. Townsend.
Chuck Purdy. ...William Oliver Stoddard.
Cleg Kelley, arab of the city, his progress and adventures. ...Samuel Rutherford Crockett.
Daughter of the tenements. ...Edward W. Townsend.
Evil that men do. ...Edgar Fawcett.
Flotsam. ...H. S. Scott. (Pseud., H. Seton.)
George's mother. ...Stephen Crane.
History of Tom Jones, a foundling. ...Henry Fielding.
Iron rule, and Story of the West End. ...Mrs. Newton Crosland. (Pseud., Camilla Toulmin.)
Jerry's family. ...James Otis Kaler. (Pseud., James Otis.)
Lucky number : a book of stories of the Chicago slums. ...I. K. Friedman.
Maggie, a girl of the streets. ...Stephen Crane.
Marm 'Lisa. ...Kate Douglas Wiggin (Mrs. George Riggs).
Princess of the gutter. ...Mrs. E. T. Smith.
Slum stories of London. ...Henry W. Nevinson.
Sybaris and other storied homes. ...Edward Everett Hale.
Tales of mean streets. ...Arthur Morrison.
Tenement tales of New York. ...J. W. Sullivan.
Timothy's quest. ...Kate Douglas Wiggin (Mrs. George Riggs).
Toil and trouble : a story of London life. ...Mrs. Newton Crosland. (Pseud., Camilla Toulmin.)
Uncle Tom's tenement. ...Alice Wellington Rollins.
Valentine McClutchy. ...William Carleton.
Your little brother James. ...James Otis Kaler. (Pseud., James Otis.)
See also **SOCIAL PURITY.**

SOCIALISM.

Alton Locke.   ...Charles Kingsley.
City of refuge.   ...Walter Besant.
Demos : a story of English socialism.   ...George R. Gissing.
Felix Holt, the radical.   ...Marian Evans Lewes Cross.   (Pseud.,
    George Eliot.)
Hard times.   ...Charles Dickens.
King Mammon and the heir apparent.   ...George A. Richardson.
Looking backwards, 2000–1887.   ...Edward Bellamy.
Man of mark.   ...Anthony Hope Hawkins.   (Pseud., Anthony Hope.)
Marcella.   ...Mary Augusta Ward (Mrs. Humphry Ward).
Marzio's crucifix.   ...Francis Marion Crawford.
Master Craftsman.   ...Walter Besant.
Mr. East's experiences in Mr. Bellamy's world.   ...Conrad Wilbrant.
Murvale Eastman, Christian socialist.   ...Albion Winegar Tourgée.
1900 : a forecast and a story.   ...Marianne Farningham.
Pictures of the socialistic future.   ...Eugene Richter.   English
    translation.
Sir George Tressady.   ...Mary Augusta Ward (Mrs. Humphry
    Ward).
Sophos; or, Kidnapping the kings.   ...Albert Alberg.
Sozialdemokratische Zukunftbilder.   ...Eugene Richter.
Stephen Remarx.   ...James Adderley.
Sunrise.   ...William Black.
This son of Vulcan.   ...Walter Besant and James Rice.
Tragic comedians.   ...George Meredith.
Traveller from Altruria.   ...William Dean Howells.
Vera Verontzoff.   ...Sonia K. Kovalevsky.
Yeast.   ...Charles Kingsley.
See also **LABOR AND CAPITAL, CONFLICT OF.**

SOCIAL PURITY.

La baraonda.   ...Girolamo Rovetta.
Clarissa Harlowe.   ...Samuel Richardson.
Claude Beauclerc: a story of modern morality.   ...Ambofilius,
    pseud.
Courage of her convictions.   ...Caroline A. Huling and Therese
    Stewart, M. D.
De profundis : a tale of the social deposits.   ...William Gilbert.
Le docteur Pascal.   ...Émile Zola.
Doctor Pascal.   ...Émile Zola.   English translation.
His perpetual adoration.   ...Joseph F. Flint.
History of Tom Jones, a foundling.   ...Henry Fielding.
Human document.   ...George Meredith.
Is this your son, my Lord?   ...Helen A. Gardener.

SOCIAL PURITY — *Continued.*

Jude, the obscure. ...Thomas Hardy.
Lady of quality. ...Frances Hodgson Burnett.
Love is a spirit. ...Julian Hawthorne.
Manhattaners. ...E. S. Van Zile.
Manxman. ...Thomas Henry Hall Caine.
Maremma. ...Louise de la Ramé. (Pseud., Ouida.)
Miss Stuart's legacy. ...Flora Annie Steel.
Murder of Delicia. ...Marie Corelli.
Nobody's fault. ...Netta Syrett.
Passages from the journal of a social wreck. ...Margaret Floyd.
Pitiless passion. ...Ella MacMahon.
Platonic affections. ...John Smith.
Poppæa. ...Julia S. Cruger. (Pseud., Julien Gordon.)
Pray you, Sir, whose daughter? ...Helen H. Gardener.
Les roches blanches. ...Édouard Rod.
Rome. ...Émile Zola.
Rome. ...Émile Zola. English translation.
Rose of Dutcher's Coolly. ...Hamlin Garland.
Scarlet letter. ...Nathaniel Hawthorne.
Shock to society. ...Florence A. James.
Social experiment. ...Annie E. Searing.
Story of the Latin quarter. ...Frances Hodgson Burnett.
Tess of the D'Urbervilles, a pure woman faithfully presented.
    ...Thomas Hardy.
Tower Hill. ...William Harrison Ainsworth.
Transgression of Terence Clancy. ...Harold Vallings.
Trilby. ...George Du Maurier.
Trilby. ...George Du Maurier. English edition.
Trionfo della morte. ...Gabriele D'Annunzio.
Triumph of death. ...Gabriele D'Annunzio. English translation.
Unclassed. ...George Gissing.
Vawder's understudy: a study in platonic affections. ...James
    Knapp Reeves.
White rocks. ...Édouard Rod. English translation.
Wish. ...Hermann Sudermann. English translation.
Der Wunsch. ...Hermann Sudermann.
See also **RELATION OF THE SEXES** and **SIN**.

SOCIETY. *A Selection.*

African.

Gentleman digger. ...Anna de Bremont.

American.

Debutante in New York society. ...Rachel Buchanan.
Fortune hunter. ...Anna C. Ritchie.

24

SOCIETY — *Continued.*

Gramercy Park. ...John Seymour Wood.
In sight of the goddess. ...Harriet Riddle Davis.
Last assembly ball. ...Mary H. Foote.
Little men. ...Louisa May Alcott.
Little women. ...Louisa May Alcott.
Metropolitan. ...Jeanie Drake.
Miss Curtis. ...Kate Gannett Wells.
Modern man. ...Ella MacMahon.
Next door. ...Clara Louise Burnham. (Pseud., Edith Douglas.)
Old-fashioned girl. ...Louisa May Alcott.
Old plantation. ...James Hungerford.
On both sides. ...Frances Courtenay Baylor.
One fair daughter. ...Frankfort Moore.
Perils of fast living. ...Charles Burdett.
Planter's daughter. ...Eliza Ann Dupuy.
Professor Conant. ...Lucius Seth Huntington.
Shock to society. ...Florence A. James.
Social experiment. ...Annie E. Searing.
Story of Helen Troy. ...Constance Cary Harrison (Mrs. Burton Harrison).
Sweet bells out of tune. ...Constance Cary Harrison (Mrs. Burton Harrison).
Through one administration. ...Frances Hodgson Burnett.
Through winding ways. ...Ellen Olney Kirk.
Transplanted rose. ...Mary E. W. Sherwood.
Sunny South; or, Southerner at home. ...Joseph Holt Ingraham.
Upper ten thousand. ...Charles Astor Bristed.
Voyage of discovery. ...Aïdé Hamilton.
See also **NEW ENGLAND, AMERICA,** and the **SOUTH.**

English.

Alice Lorraine. ...Richard Doddridge Blackmore.
American peeress. ...H. C. Chatfield.
As in a looking-glass. ...F. C. Philips.
Barnaby Rudge. ...Charles Dickens.
Beau Nash. ...William Harrison Ainsworth.
Bread and butter Miss. ...George Paston.
Coningsby. ...Benjamin Disraeli.
Courtier of the days of Charles II. ...Catherine G. F. Gore.
Cranford and other tales. ...Elizabeth C. Gaskell.
Days of Lamb and Coleridge. ...Alice E. Lord.
Dean and his daughters. ...F. C Philips.
De Vere. ...Robert P. Ward.
Dictator. ...Justin McCarthy.

SOCIETY — *Continued.*

Dodo. ...E. F. Benson.
Endymion. ...Benjamin Disraeli.
Expedition of Humphrey Clinker. ...Tobias George Smollett.
Fool of quality. ...Henry Brooke.
Frank Hilton; or, Queen's own. ...James Grant.
Gipsy. ...George Payne Rainsford James.
Henrietta Temple. ...Benjamin Disraeli.
Hetty Hyde's lovers; or, Household brigade. ...James Grant.
King of Andaman: a saviour of society. ...J. Maclaren Cobban.
King of Bath. ...Mary C. Ware (Mrs. Hibbert Ware).
King's own son. ...Frederick Marryat.
Lady Bell. ...Henrietta Keddie. (Pseud., Sarah Tytler.)
Lady Grizel. ...Lewis Wingfield.
Lady of quality. ...Frances Hodgson Burnett.
Lakewood: a story of to-day. ...Mary Harriott Norris.
Linwoods. ...Catherine Maria Sedgwick.
Little Mrs. Murray. ...F. C. Philips.
Lord Harry Bellair. ...Anne Manning.
Lord Mayor of London. ...William Harrison Ainsworth.
Mainstone's housekeeper. ...Eliza Meteyard.
Major and minor. ...William E. Norris.
Manchester strike. ...Harriet Martineau.
Mansfield park. ...Jane Austen.
Mary Burton. ...E. C. Gaskell.
Memoirs of a lady-in-waiting. ...J. D. Fenton.
Minister's wooing. ...Harriet Beecher Stowe.
Miser's daughter. ...William Harrison Ainsworth.
Mr. Bailey Martin. ...Percy White.
Mustard leaves. ...E. Balch.
Nobody's fault. ...Netta Syrett.
North and South. ...Elizabeth C. Gaskell.
On both sides. ...Frances Courtenay Baylor.
Professor Conant. ...Lucius Seth Huntington.
Ralf Skirlaugh the Lincolnshire squire. ...Edward Peacock.
Right Honorable: a romance of society and politics. ...Justin
    McCarthy and Mrs. Campbell Praed.
Rival apprentices. ...Anon.
Rubicon. ...E. F. Benson.
Sense and sensibility. ...Jane Austen.
Sidonia the sorceress. ...Johann Wilhelm Meinhold.
Spendthrifts and other social photographs. ...Grenville Murray.
Storied holidays: a cycle of historic red-letter days. ...Elbridge S
    Brooks.
Strange adventures of Lucy Smith. ...F. C. Philips.

SOCIETY — *Continued.*

Surgeon's daughter.   ...Sir Walter Scott.
Tapestried chamber.   ...Sir Walter Scott.
Ten thousand a year.   ...Samuel Warren.
Tremaine.   ...Robert P. Ward.
Vicar of Wakefield.   ...Oliver Goldsmith.
Virginians.   ...William Makepeace Thackeray.
When George III. was king.   ...Anon.
White rose of Langley : a story of the court of England in the olden
    time.   ...Emily Sarah Holt.
Wild times.   ...Cecilia M. Caddell.
World went very well then.   ...Walter Besant.
Young Mr. Ainslie's courtship.   ...F. C. Philips.
Youth and manhood of Cyril Thornton.   . .Thomas Hamilton.
See also **ENGLAND** and **ENGLISH HISTORY.**

French.

De l'Orco.   ...George Payne Rainsford James.
Gaston de Blondville ; or, Court of Henry III. resting at Ardennes.
    ...Anna Radcliffe.
King of a day.   ...Florence Wilford.
Parisians.   ...Edward George Earle Lytton Bulwer-Lytton.
Rose d'Albert ; or, Troublous times.   ...George Payne Rainsford
    James.
Scènes de la vie Parisienne.   ...Honoré de Balzac.
Scenes from Parisian life.   ...Honoré de Balzac.   English translation.
See also **FRANCE, FRENCH HISTORY,** and **PARIS.**

German.

At odds.   ...Jemima M., Baroness von Tautphœus.
Aus der Höhe.   ...Berthold Auerbach.
Aus drei Kaiserzeiten.   ...Johann Georg L. Hesekiel.
Buchholz family.   ...Julius E. W. Stinde.   English translation.
Die Familie Buchholz : aus dem Leben der Haupstadt.   ...Julius E.
    W. Stinde.
For the right.   ...Karl Emil Franzos.   English translation.
Frau Wilhelmine : aus dem Leben der Haupstadt.   ...Julius E. W.
    Stinde.
Frau Wilhelmine : sketches of Berlin life.   ...Julius E. W. Stinde.
    English translation.
In hot haste.   ...Mary E. Hullah.
Ingo.   ...Gustav Freytag.
Ingraban.   ...Gustav Freytag.
Ein Kampf um's Recht.   ...Karl Emil Franzos.
Lost manuscript.   ...Gustav Freytag.   English translation.

SOCIETY — *Continued.*
Lutaniste of Saint Jacobi's. ...Catharine Drew.
Mit und ohne Vokation. ...E. von Grotthuss.
On the heights. ...Berthold Auerbach. English translation.
Die unsichtbare Loge. ...Jean Paul F. Richter.
Verlorene Handschrift. ...Gustav Freytag.
Verrath und Liebe. ...Adolf Muetzelburg.
See also **GERMAN HISTORY** and **GERMANY**.

Irish.

Adventures of Mick Callighin. ...William R. Anckettill.
Anglo-Irish in the nineteenth century. ...John Banim. (Pseud., The O'Hara family.)
Archie Mason, an Irish story. ...Anon.
Art Maguire ; or, Broken pledge. ...William Carleton.
At the rising of the moon. ...Frank Mathew.
Ballyblunder : an Irish story. ...Anon.
Barny O'Reirdon, the navigator. ...Samuel Lover.
Bit o' writin'. ...Michael Banim. (Pseud., The O'Hara family.)
Black baronet ; or, Chronicles of Ballytrain. ...William Carleton.
Bunch of shamrocks, being a collection of Irish tales and sketches. ...Elizabeth Casey. (Pseud., E. Owens Blackburne.)
Castle Rackrent. ...Maria Edgewood.
Crohoore of the Bill-hook. ...Michael Banim. (Pseud., The O'Hara family.)
Emigrants. ...William Carleton.
Eveline Wellwood : a story of modern Irish life. ...Norris Paul.
Evil eye ; or, Black spectre. ...William Carleton.
Fardorougha, the miser ; or, Convicts of Lisnamona. ...William Carleton.
Father Connell. ...Michael Banim. (Pseud., The O'Hara family.)
For her sake : a tale of life in Ireland. ...Gordon Roy.
Frank O'Donnell : a tale of Irish life. ...Allen H. Clington.
Grania. . .Emily Lawless.
Handy Andy : a tale of Irish life. ...Samuel Lover.
Hermit of Glenconella. ...Eneas MacDonnell.
Ierne. ...William Stewart Trench.
Irish idylls. ...Jane Barlow.
Ismay's children. ...May Laffan.
Jane Sinclair. ...William Carleton.
Kinkora : an Irish tale. ...Albert S. G. Canning.
Knicknagow ; or, Homes of Tipperary. ...Charles J. Kickham.
Knight of Gwynne. ...Charles James Lever.
Lawrence Bloomfield in Ireland ; or, New landlord. ...W. Allingham.
Legends and stories of Ireland. ...Samuel Lover.

SOCIETY — *Continued.*

Lights and shades of Irish life.   ...Anna Maria Hall.
Lord Kilgobbin : a tale of Ireland in our own time.   ...Charles
    James Lever.
Maid of Sker.   ...Richard Doddridge Blackmore.
Martins of Cro' Martin.   ...Charles James Lever.   (Pseud., Cor-
    nelius O'Dowd.)
Mary Lee ; or, Yankee in Ireland.   ...John Boyce.
Rory of the hills : an Irish tale.   ...Robert Curtis.
Shandon bells.   ...William Black.
Strangers at Lisconnet.   ...Jane Barlow.
Tales of my neighborhood.   ...Gerald Griffin.
Tales of the Munster festivals.   ...Gerald Griffin.
Tales of the O'Hara family.   ...John Banim.   (Pseud., The O'Hara
    family.)
Tupper Derg; or, Red well.   ...William Carleton.
Value of Fostertown : a tale of Irish life.   ...A. M. Donelan.
Wild Ireland : days and nights with Father Michael. ...B. Bonbavand.
See **IRELAND** and **IRISH HISTORY.**

SOCIOLOGY.   *A Selection.*

Ab o' th' Yate's soup kitchen.   ...Benjamin Brierley.   (Pseud., Ab
    o' th' Yate.)
Adrift in a great city.   ...M. E. Winchester.
All sorts and conditions of men.   ...Walter Besant.
Alton Locke.   ...Charles Kingsley.
Artie : a story of the streets and town.   ...George Ade.
Austin Elliot.   ...Henry Kingsley.
Bread-winners : a social study.   ...John Hay.
Child of the Jago.   ...Arthur Morrison.
Children of the Ghetto.   ...Isaac Zangwill.
City of refuge.   ...Walter Besant.
City side ; or, Passages from a pastor's portfolio.   ...Clara Belmont.
Cleg Kelley, arab of the city, his progress, and adventures.
    ...Samuel Rutherford Crockett.
Colloquies of Edward Osborne.   ...Anne Manning.
Daughter of humanity.   ...Edgar Maurice Smith.
Daughter of the tenements.   ...Edward W. Townsend.
Felix Holt, the radical.   ...Marion Evans Lewes Cross.   (Pseud.,
    George Eliot.)
Golden bottle.   ...Ignatius Donnelly.
Human document.   ...George Meredith.
Illustrations of political economy.   ...Harriet Martineau.   (19 vols.)
Iron crown.   ...Anon.
Jerry's family.   ...James Otis Kaler.   (Pseud., James Otis.)

SOCIOLOGY — *Continued.*
John Halsey, the anti-monopolist. ...R. U. Collins.
Jude, the obscure. ...Thomas Hardy.
King Mammon and the heir apparent. ...George A. Richardson.
Kitty's conquest. ...Charles King.
Kreutzer sonata. ...Lyeff Nikolaievich Tolstoi. English transla-
tion from the Russian.
Last days of the Carnival. ...J. Kostromiten.
Life's little ironies. ...Thomas Hardy.
Looking backwards, 2000–1887. ...Edward Bellamy.
Lucky number : a book of stories of the Chicago slums. ...I. K.
Friedman.
Marcella. ...Mary Augusta Ward (Mrs. Humphry Ward).
Master Craftsman. ...Walter Besant.
Mr. East's experiences in Mr. Bellamy's world. ...Conrad Wilbrant.
Murvale Eastman, Christian socialist. ...Albion Winegar Tourgée.
Nicholas Minturn. ...Josiah Gilbert Holland.
Our great West. ...Julian Ralph.
Outcast of the islands. ...Joseph Conrad.
Put yourself in his place. ...Charles Reade.
Redemption of Edward Strahan. ...W. J. Dawson.
Rival apprentices. ...Anon.
Ruined race. ...Hester Sigerson.
Slum stories of London. ...Henry W. Nevinson.
Social crime. ...Minnie L. Armstrong and G. N. Sceets.
Social study. ...Helen Brent.
Sybil Knox ; or, Home again. ...Edward Everett Hale.
Tales of mean streets. ...Arthur Morrison.
Tenement tales of New York. ...J. W. Sullivan.
Traveller from Altruria. ...William Dean Howells.
White satin and home-spun. ...Katrina Trask.
Wonderful visit. ...Henry G. Wells.
Yeast. ...Charles Kingsley.
Your little brother James. ...Caroline H. Pemberton.
See also **LABOR AND CAPITAL, CONFICT OF; IMAGINARY
LANDS, CITIES, AND INSTITUTIONS;** and **WORKING
CLASSES.**

SOLDIERS. *A Selection.*
Un ami de la Reine. ...Paul Gaulot.
L'argent des autres. ...Émile Gaboriau.
Ben Brace, the last of Nelson's Agamemnons. ...Frederic Chamier.
Bivouac ; or, Stories of the Peninsular war. ...William H. Maxwell.
Blockade : an episode of the fall of the first empire. ...Émile
Erckmann and Alexandre Chatrian. English translation.

SOLDIERS — *Continued.*

Le blocus.   ...Émile Erckmann et Alexandre Chatrian.

Bombardier H. and Corporal Dose ; or, Military life in Prussia.
...Friedrich W., Ritter von Hacklaender.   English translation.

Boy soldiers of 1812.   ...Everett T. Tomlinson.

Cadet days : a story of West Point.   ...Charles King.

Captain Blake.   ...Charles King.

Colonel's daughter ; or, Winning his spurs.   ...Charles King.

Courting of Dinah Shadd.   ...Rudyard Kipling.

Deserter.   ...Charles King.

Drums of the fore and aft.   ...Rudyard Kipling.

Every inch a soldier.   ...Henrietta E. V. Stannard.   (Pseud., John
Strange Winter.)

Foes in ambush.   ...Charles King.

Friend of the queen.   ...Paul Gaulot.   English translation.

From the ranks.   ...Charles King.

La grande famille.   Roman militaire.   ...Jean Grave.

Histoire du plébiscite.   ...Émile Erckmann et Alexandre Chatrian.

Historical tales ; or, Romance of reality.   ...Charles Morris.

Incarnation of Krishna Mulvaney.   ...Rudyard Kipling.

John Inglesant.   ...Joseph Henry Shorthouse.

Kenneth ; or, Rear-guard of the grand army.   ...Charlotte Mary
Yonge.

King's own borderers : a military romance.   ...James Grant.

Kitty's conquest.   ...Charles King.

Der letzte Bombardier.   ...Friedrich W., Ritter von Hacklaender.

Madness of private Ortheris.   ...Rudyard Kipling.

Man who was.   ...Rudyard Kipling.

Marion's faith.   ...Charles King.

Maurice Tiernay, the soldier of fortune.   ...Charles James Lever.
(Pseud., Cornelius O'Dowd.)

Micah Clarke : his statement as made to his three grandchildren.
...Arthur Conan Doyle.

Monsieur Jack : a tale of the old war-time.   ...Alfred H. Engelbach.

Old revolutionary soldier.   ...Joseph Alden.

Otterbourne.   ...Edward Duras.

Phantom regiment ; or, Stories of "Ours."   ...James Grant.

Quaker soldier ; or, British in Philadelphia.   ...J. Richter Jones.

Red badge of courage.   ...Stephen Crane.

Scottish soldiers of fortune, their adventures and achievements in
the armies of Europe.   ...James Grant.

Shadow of the sword.   ...Robert W. Buchanan.

Das Soldatenleben im Frieden.   ...Friedrich W., Ritter von Hacklaender.

Soldier stories.   ...Rudyard Kipling.

SOLDIERS — *Continued.*

Soldiers three. ...Rudyard Kipling.

Starlight ranch and other stories of army life on the frontier. ...Charles King.

Story of the plébiscite. ...Émile Erckmann and Alexandre Chatrian. English translation.

Taking of Lungtungpen. ...Rudyard Kipling.

Two-legged wolf. ...N. N. Karazin.

Two soldiers and a politician. ...Clinton Ross.

Two soldiers and Dunraven ranch. ...Charles King.

Under fire. ...Charles King.

When Greek meets Greek. ...Joseph Hatton.

With the main guard. ...Rudyard Kipling.

Workman and soldier: a tale of Paris life during the siege and the rule of the commune. ...James F. Cobb.

See also **ARMY AND NAVY, BATTLES, CIVIL WAR, REBELLIONS, REVOLUTIONS,** and **WAR.**

SOUTH, U. S. A. *A Selection.*

Adventures of Huckleberry Finn. ...Samuel Langhorne Clemens. (Pseud., Mark Twain.)

Among the pines. ...James R Gilmore. (Pseud., Edmund Kirke.)

Ante bellum: Southern life as it was. ...Mary L. Cook.

Anti-fanaticism: a tale of the South. ...Martha H. Butt.

Balaam and his master and other sketches and stories. ...Joel Chandler Harris.

Beatrice of Bayou Teche. ...Alice Ilgenfritz Jones.

Beech-bluff: a tale of the South. ...Fanny Warner.

Behind the Blue Ridge. ...Frances C. Baylor.

Black ice. ...Albion Winegar Tourgée.

Bricks without straw. ...Albion Winegar Tourgée.

Bonaventure: a prose pastoral of Acadian Louisiana. ...George Washington Cable.

Border foes: a romance of early Kentucky. ...E. Willett.

Bright days in the old plantation time. ...Mary R. Banks.

Captain Smith and Princess Pocahontas: an Indian tale. ...J. Davis.

Colonel Dunwoddie, millionaire. ...William Mumford Baker.

Colonial cavalier; or, Southern life before the revolution ...Maud Wilder Goodwin.

Cora; oder, die Sklavin. ...Urich Baudissin.

Creole orphans. ...James S. Peacock.

Dangerfield's rest. ...Henry Sedley.

Daughter of the South. ...Constance Cary Harrison (Mrs. Burton Harrison).

SOUTH, U. S. A. — *Continued.*

Doctor Sevier.   ...George Washington Cable.

Doctor Vandyke.   ...John Esten Cooke.

Dred: a tale of the Great Dismal Swamp.   ...Harriet Beecher Stowe.

Earthly paragon.   ...Eva Wilder McGlasson.

Eleanor Mirton; or, Life in Dixie.   ...Anon.

Etna Vandemir.   ...Sallie J. Hancock.

Eustis: a novel of Southern life.   ...Robert A. Boit.

Fairfax: a chronicle of the valley of the Shenandoah.   ...John Esten Cooke.

First settlers of Virginia: an historical novel.   ...J. Davis.

Fool's errand by one of the fools.   ...Albion Winegar Tourgée.

Grandissimes.   ...George Washington Cable.

Hatchie, the guardian slave.   ...William T. Adams.

House of bondage.   ...Octavia V. Rogers Albert.

How I found it: North and South.   ...John H. Woodbury.

In war times.   ...M. E. M. Davis.

Iron furnace.   ...John H. Aughey.

John March, Southerner.   ...George Washington Cable.

Justice in the by-ways.   ...Francis C. Adams.

Justin Harley: a romance of old Virginia.   ...John Esten Cooke.

Kenneth Cameron.   ...L. Q. C. Brown.

Late Mrs. Null.   ...Frank Richard Stockton.

Lights and shadows of life.   ...Madeleine V. Dahlgren.

Manuel Pereira.   ...Francis C. Adams.

Marston Hall.   ...L. Ella Byrd.

Master's house.   ...Logan, pseud.

Meh Lady.   ...Thomas Nelson Page.

Mr. Absalom Billingslea and other Georgia folks.   ...Richard M. Johnston.

Mose Evans.   ...William Mumford Baker.

My Southern friends.   ...James R. Gilmore. (Pseud., Edmund Kirke.)

New Timothy.   ...William Mumford Baker.

Nick of the woods.   ...Robert Montgomery Bird.

Old Creole days.   ...George Washington Cable.

Old plantation.   ...James Hungerford.

On both sides.   ...Frances C. Baylor.

On the border: a story of Southern life and character.   ...James R. Gilmore. (Pseud., Edmund Kirke.)

Philip Randolph.   ...Mary Gertrude.

Pleasant waters.   ...Graham Claytor.

Rodman the keeper: Southern sketches.   ...Constance Fenimore Woolson.

SOUTH, U. S. A. — *Continued.*

Sable cloud : a Southern tale with Northern comments. ..." Nehemiah," pseud.

She loved a sailor. ...Amelia Edith Barr.

Sister Jane. ...Joel Chandler Harris.

Sons of Ham : a tale of the New South. ...Louis Pendleton.

Summer stories of the South. ...Thomas Addison Richards.

Sunny South ; or, Southerner at home. ...Joseph Holt Ingraham.

Tales of a time and place. ...Grace King.

Thirty-four years. ...John Marchmont.

Uncle Tom's cabin; or, Life among the lowly. ...Harriet Beecher Stowe.

Under the magnolias. ...L. W. Denton.

Under the water oaks. ...Marian Brewster.

Union : a story of the great rebellion. ...John R. Musick.

Valerie Aylmer. ...Frances C. Tiernan. (Pseud., Christian Reid.)

Voodoo tales as told among the negroes of the Southwest. ...Mary Alicia Owen.

White and black : a story of the Southern States. ...A. Biggs.

Yankee school-teacher in Virginia. ...Lydia W. Baldwin.

See also **SLAVERY.**

SOUTH AMERICA. See **AMERICA,** South.

SOUTH CAROLINA.

Cassique of Kiawah. ...William Gilmore Simms.

For the major. ...Constance Fenimore Woolson.

Forayers. ...William Gilmore Simms.

In old St. Stephens. ...Jeanie Drake.

Katharine Walton. ...William Gilmore Simms.

Mellichampe : a legend of the Santel. ...William Gilmore Simms.

Milrose ; or, Cotton planter's daughter. ...John H. Robinson.

Rombert : a tale of Carolina. ...Anon.

Scout. ...William Gilmore Simms.

Sea-island romance. ...William P. Brown.

Woodcraft. ...William Gilmore Simms.

SPAIN.

Description, Manners, and Customs.

Alhambra. ...Washington Irving.

Alice Lorraine. ...Richard Doddridge Blackmore.

Beyond the seas, being the surprising adventures and ingenious opinions of Ralph Lord Saint Keyne. ...Oswald Crawfurd.

Buccaneer chief : a romance of the Spanish main. ...Gustave Aimard. English translation.

Cardinal Pole. ...William Harrison Ainsworth.

SPAIN — *Continued.*

Christian prince, Wolfgang, prince of Anhalt.   ...Franz Hoffmann. English translation.

Clara Avery: a story of the Spanish Armada.   ...Emily Sarah Holt.

Conquest of Granada.   ...Washington Irving.

Constancia's household: a story of the Spanish reformation. ...Emma Leslie.

Edict: a tale of 1492.   .. Grace Aguilar.

Escape: a tale of 1755.   ...Grace Aguilar.

Estevan: a story of the Spanish conquests.   ...John R. Musick.

Fawn of Sertorius.   ...Robert Landor.

Felix Alvarez; or, Manners in Spain.   ...Alexander R. C. Dallas.

For the right.   ...Karl Emil Franzos.   English translation.

Fürst Wolfgang: historische Erzählung.   ...Franz Hoffmann.

Gilded man.   ...Adolf Francis Bandelier.

Gonzalve de Cordoue.   ...Jean Pierre Claris de Florian.

Gonsalvo of Cordova; or, Conquest of Granada.   ...Jean Pierre Claris de Florian.   English translation.

La grande filibuste.   ...Gustave Aimard.

Great captain.   ...Ulrick R. Burke.

Heiress of Burges.   ...Thomas C. Grattan.

Jews of Barnow.   ...Karl Emil Franzos.   English translation.

Die Juden von Barnow.   ...Karl Emil Franzos.

Kaiser Karl des Fünften erste Jugendliebe.   ...Ludwig A. Arnim.

Ein Kampf um's Recht.   ...Karl Emil Franzos.

Karl von Spanien. Roman.   ...Ludwig Storch.

Leila; or, Siege of Granada.   ...Edward George Earle Lytton Bulwer-Lytton.

Leon Roch.   ...Benito Pérez Galdós.

Lichtenstein.   ...Wilhelm Hauff.

Lichtenstein; or, Swabian league.   ...Wilhelm Hauff.   English translation.

Martyrs of Spain and liberators of Holland.   ...Elizabeth Charles.

Mercedes of Castile; or, Voyage to Cathay.   ...James Fenimore Cooper.

Moritz von Sachsen.   ...Friedrich C. Schlenkert.

Der neue Hiob.   ...Leopold Sacher-Masoch.

New Job.   ...Leopold Sacher-Masoch.   English translation.

Pastor's fireside.   ...Jane Porter.

Phantom regiment; or, Stories of Ours.   ...James Grant.

Romance of war; or, Highlanders in Spain.   ...James Grant.

Saint Leon: a tale of the sixteenth century.   ...Caleb Williams Godwin.

Saragossa; or, Houses of Castello and De Arno.   ...E. A. Archer.

SPAIN — *Continued.*

Die Schlacht bei Drakenburg. ...Werner Bergman.

Sibylle von Cleve. ...Julius Bacher.

Sister Saint Sulpice. ...Armando Palacio Valdés.

Spanish barber. ...Anne Manning.

Spanish brothers : a tale of the sixteenth century. ...Deborah Alcock.

Spanish cavalier: a story of Seville. ...Charlotte Tucker. (Pseud., A. L. O. E.)

Spanish match ; or, Charles Stuart at Madrid. ...William Harrison Ainsworth.

Stories of Torres Vedras. ...John G. Millingen.

Subalterns. ...George R. Gleig.

Tabithe von Geyersberg. ...Amalie Schoppe.

Tales from Spanish history. ...Elizabeth J. Brabazon.

Vale of cedars ; or, Martyrs. ...Grace Aguilar.

Wayside cross; or, Raid of Gomez: a tale of the Carlist war. ...Edward A. Milman.

Young buglers: a tale of the Peninsular war. ...George Alfred Henty.

See also **NETHERLANDS** and **SPANISH HISTORY.**

SPANISH HISTORY. *A Selection.*

Early and General.

Abderahman : the founder of the dynasty of the Ommiades in Spain. ...Washington Irving. (In his Spanish papers.)

Alhambra : a series of tales and sketches of the Moors and Span- iards. ...Washington Irving.

Almahide; ou, L'esclave reine. ...Madeleine de Scudéry.

El bastardo de Castilla. ...Jorge W. Montgomery.

Bernardo del Carpio. ...Jorge W. Montgomery.

Castilian. ...Joaquin Telesforo de Trueba y Cosio.

Chronicle of Fernan Gonzalez. ...Washington Irving. (In his Spanish papers.)

Chronicle of the conquest of Granada. ...Washington Irving.

Cid : a short chronicle founded on the early poetry of Spain. ...George Dennis.

El conde Fernan Gonzalez. ...Narciso B. Selva.

Count Julian ; or, Last days of the Goth. ...William Gilmore Simms.

La cruz del Moro. ...Juan P. Criado y Dominguez.

Doña Isabel de Solis, reina de Granada. ...Francisco Martinez de la Rosa.

Dove of Tabenna. ...John Mason Neale. (In his Tales illustrating church history.)

SPANISH HISTORY — *Continued.*

Fawn of Sertorius.   ...Robert Landor.

Foreign tales and traditions.   ...George G. Cunningham.

Historia de los vandos de los Zegries, y Abenzerrages, Cavalleros Moros de Granada.   ...Gines Perez de Hita.

Historical tales of the Muslims in Spain.   ...Elizabeth J. Brabazon.

In the shadow of the Alhambra; or, Last of the Moorish kings. ...William MacC. Greenlee.

King's page: a legend of the Moorish wars in Spain and other stories.   ...Anna T. Sadlier.

Legend of Count Julian and his family.   ..Washington Irving. (In his Spanish papers.)

Legend of Don Roderick.   ...Washington Irving.   (In his Spanish papers.)

Legend of Palayo.   ...Washington Irving.   (In his Spanish papers.)

Leila; or, Siege of Granada.   ...Edward George Earle Lytton Bulwer-Lytton.

Moors of Granada.   ...Henri Guenot.   English translation.

Le More de Grenade.   ...Henri Guenot.

Old court life in Spain.   ...Frances Minto Elliot.

Pelayo: a story of the Goth.   ...William Gilmore Simms.

Piquillo Alliaga; ou, Les Maures sous Philippe III.   ...Eugene Scribe.

Rescue: a tale of the Moorish conquest of Spain.   ...John Mason Neale.   (In his Tales illustrating church history.)

Romance of history. Spain.   ...Joaquin Telesforo de Trueba y Cosio.

Spanish cavalier: a story of Seville.   ...Charlotte Tucker.   (Pseud., A. L. O. E.)

Story of the Christians and Moors in Spain.   ...Charlotte Mary Yonge.

Die Tochter der Alhambra.   ...Valeska Voigtel.   (Pseud., Arthur Stahl.)

Fourteenth Century.

Le bâtard de Mauléon.   ...Alexandre Dumas.

Half brothers; or, Head and the hand.   ...Alexandre Dumas. English translation.

Romances históricos.   ...Angel de Saavedra (Duque de Rivas).

Fifteenth Century.

Alhambra.   ...Washington Irving.

Cardinal Ximenes.   ...Jean Bertheroy.   English translation.

Chronicle of the conquest of Granada.   ...Washington Irving.

Dolores.   ...Gertrúdes Gomez de Avellaneda.

SPANISH HISTORY — *Continued.*
Doña Blanca de Navarra, crónica del siglo XV.   ...Francisco Navarro Villoslada.
Doña Blanca of Navarra.   ...Francisco Navarro Villoslada.   English translation.
Edict : a tale of 1492.   ...Grace Aguilar.
Ferdinand und Isabella.   ...Amalie Schoppe.
Gonzalve de Cordoua.   ...Jean Pierre Claris de Florian.
Gonsalvo of Cordova ; or, Fall of Granada.   ...Charles Hood.
Great captain.   ...Ulrick R. Burke.
Leila ; or, Siege of Granada.   ...Edward George Earle Lytton Bulwer-Lytton.
Mercedes of Castile ; or, Voyage to Cathay.   ...James Fenimore Cooper.
Romances históricos.   ...Angel de Saavedra (Duque de Rivas).
Tales from Spanish history.   ...Elizabeth J. Brabazon.
Vale of cedars ; or, Martyrs.   ...Grace Aguilar.
Ximénès.   ...Jean Bertheroy.

Sixteenth Century.

By little and little.   ...Emma Leslie.
Clare Avery: a story of the Spanish armada.   ...Emily Sarah Holt.
Constancia's household : a story of the Spanish reformation. ...Emma Leslie.
Don Carlos.   ...César Vichard, Abbé de Saint Réal.
Don Carlos.   ...César Vichard, Abbé de Saint Réal.   English translation.
Drake and the Dons.   ...R. Lovett.
Fatal revenge ; or, Family of Montorio.   ...Charles Robert Maturin.
  \ (Pseud., Dennis Jasper Murphy.)
For love and liberty : a tale of the sixteenth century in Spain. ...A. Harcourt.
Last look : a tale of the Spanish inquisition.   ...William Henry Giles Kingston.
Martyrs of Spain and the liberators of Holland.   ...Elizabeth Charles.
Ni rey ni roque.   ...Patricio de la Escosura.
Princess of Viarna ; or, Spanish inquisition in the reign of the Emperor Charles V.   ...T. Picton.
Romances históricos.   ...Angel de Saavedra (Duque de Rivas).
Rosa : a tale of the Spanish inquisition.   ...Derwent Tremorne.
Saint Leon : a tale of the sixteenth century.   ...Caleb Williams Godwin.
Sitio de Mons por el Duque de Alba.   ...Alonso de Ramon.
La sombra de Felipe II.   ...Romon Ortega y Frias.

SPANISH HISTORY — *Continued.*

Spanish brothers: a tale of the sixteenth century. ...Deborah Alcock.

Spanish cavalier. ...Charlotte Tucker. (Pseud., A. L. O. E.)

Word, only a word. ...Georg Moritz Ebers. English translation.

Ein Wort. ...Georg Moritz Ebers.

See also **CHARLES V. OF GERMANY AND I. OF SPAIN.**

Seventeenth Century.

Berber; or, Mountaineer of the Atlas. ...William S. Mayo.

Fatal revenge; or, Family of Montorio. ...Charles Robert Maturin. (Pseud., Dennis Jasper Murphy.)

Fortunate fool. ...Alonso Jerónimo Salas Barbadillo. English translation.

History and life of Squire Marcos de Obregon. ...Vicente Espinel. English translation.

Leila; or, Siege of Granada. ...Edward George Earle Lytton Bulwer-Lytton.

Monk's pardon. ...Marie de Navery.

El necio bien afortunado. ...Alonso Jerónimo Salas Barbadillo.

Relaciones de la vida del escudero Marcos de Obregon. ...Vincente Espinel.

Ruy Blas; or, King's rival. ...H. L. Williams.

Spanish match; or, Charles Stuart at Madrid. ...William Harrison Ainsworth.

Eighteenth Century.

Angel of the world, an Arabian tale; and Sebastian, a Spanish tale. ...George Croly.

Bravest of the brave; or, With Peterborough in Spain. ...George Alfred Henty.

Escape: a tale of 1755. ...Grace Aguilar.

Held fast for England: a tale of the siege of Gibraltar. ...George Alfred Henty.

Historia del famoso predicador Fray Gerundio de Campazas aliàs Zotes. ...José Francisco de Isla.

Pastor's fireside. ...Jane Porter.

Nineteenth Century.

Adventures of an aid-de-camp; or, Campaign in Calabria. ...James Grant.

Alice Lorraine. ...Richard Doddridge Blackmore.

Los apostólicos. ...Benito Pérez Galdós.

Los Arcos: a Spanish carlist romaunt. ...George Ryder.

Bailén. ...Benito Pérez Galdós.

La batalla de los Arapiles. ...Benito Pérez Galdós.

SPANISH HISTORY — *Continued.*

Bivouac; or, Stories of the Peninsular war. ...William Hamilton Maxwell.

Cádiz. ...Benito Pérez Galdós.

Charles O'Malley, the Irish dragoon. ...Charles James Lever. (Pseud., Cornelius O'Dowd.)

Los cien mil hijos de San Luis. ...Benito Pérez Galdós.

La corte de Cárlos IV. ...Benito Pérez Galdós.

Don Alonso; ou, L'Espagne, histoire contemporaine. ...Narcisse A. Salvandy.

Édouard; ou, Le siège de Saragosse. ...Léopold Méry.

El equipage del rey José. ...Benito Pérez Galdós.

Felix Alvarez; or, Manners in Spain. ...Alexander R. C. Dallas.

Gerona. ...Benito Pérez Galdós.

El grande oriente. ...Benito Pérez Galdós.

Her heart was true: a story of the Peninsular war. ...E. M. Cuttim.

Juan Martin el empecinado. ...Benito Pérez Galdós.

King's own borderers: a military romance. ...James Grant.

Memorias de un cortesano de 1815. ...Benito Pérez Galdós.

Napoléon en Khamartin. ...Benito Pérez Galdós.

El 19 de Marzo y el 2 de Mayo. ...Benito Pérez Galdós.

Phantom regiment; or, Stories of Ours. ...James Grant.

Romance of war; or, Highlanders in Spain. ...James Grant.

Rosaura: a tale of Madrid. (In Tales from Blackwood, vol. 9.)

Saragossa; or, Houses of Castello and De Arno. ...E. A. Archer.

La segunda casaca. ...Benito Pérez Galdós.

Spanish barber. ...Anne Manning.

Spanish cavalier: a story of Seville. ...Charlotte Tucker. (Pseud., A. L. O. E.)

Stories of Torres Vedras. ...John G. Millingen.

Subaltern. ...George Robert Gleig.

El terror de 1824. ...Benito Pérez Galdós.

Trafalgar. ...Benito Pérez Galdós.

Trafalgar. ...Benito Pérez Galdós. English translation.

Un voluntario realista. ...Benito Pérez Galdós.

Wayside cross; or, Raid of Gomez: a tale of the Carlist war. ...Edward A. Milman.

Young buglers: a tale of the Peninsular war. ...George Alfred Henty.

Zaragoza. ...Benito Pérez Galdós.

SPIRITUALISM.

Dream child. ...Florence Huntley.

Fearless investigator. ...Anon.

Fools of nature. ...Alice Brown.

SPIRITUALISM — *Continued.*

Great Amherst mystery. ...Walter Hubbell.
Herr Paulus, his rise, his greatness, and his fall. ...Walter Besant.
Island of Doctor Moreau. ...Henry G. Wells.
Passing of Alix. ...Marjorie Paul.
Six cent Sam's. ...Julian Hawthorne.
Undiscovered country. ...William Dean Howells.

SPORTING STORIES.

All adrift. ...W. T. Adams.
Annals of a sportsman. ...Ivan Sergyevich Turgenef. English translation.
Beaten on the post; or, Joe Morton's mercy. ...J. P. Wheldon.
Blue ribbon of the turf: a chronicle of the race of the Derby. ...Louis Henry Curzon.
Brise-de-Mai; ou, Les trappeurs de l'Hudson. ...V. Lamy.
Camp-fires of the Everglades; or, Wild sports in the South. ...Charles E. Whitehead.
Cinder-path tales. ...William Lindsey.
Fairport nine. ...Noah Brooks.
False start. ...Hawley Smart.
Forest, the jungle, and the prairie; or, Scenes with the trapper and hunter in many lands. ...William H. D. Adams.
Gaut Gurley; or, Trappers of Umbagog. ...D. P. Thompson.
Hendricks the hunter: a tale of Zululand. ...William Henry Giles Kingston.
Home of the wolverene and the beaver. ...Charles H. Eden.
Hunting girl. ...Mrs. Edward Kennard.
Long odds. ...Hawley Smart.
Maggie Jacket: a tale of the turf. ...Nathaniel Gould.
Marjorie's Canadian winter. ...Agnes Maule Machar.
Our base-ball club. ...Noah Brooks.
Poker stories as told by statesmen, soldiers, and lawyers. ...J. F. B. Lillard.
Pound of cure: a story of Monte Carlo. ...William Henry Bishop.
Rifle, rod, and gun in California: a sporting romance. ...Theodore Strong Van Dyke.
Rodney Stone. ...Arthur Conan Doyle.
Sea spray; or, Facts and fancies of a yachtsman. ...Samuel G. W. Benjamin.
Sporting tales and characters. ...Henry W. Herbert. (Pseud. Frank Forester.)
That hated Saxon. ...Beatrice V. Greville.
That pretty little horse-breaker. ...Mrs. Edward Kennard.
Tom Crackenthorpe. ...Charles Clarke.

SPORTING STORIES — *Continued.*

Trappers of New York. ...Jeptha R. Simms.

Up the north branch. ...C. A. J. Farrar.

Wedded to sport. ...Mrs. Edward Kennard.

Wild woods life ; or, Trip to Parmachenee. ...Charles A. J. Farrar.

With pack and rifle in the far West. ...Achilles Daunt.

Without love or license. ...Hawley Smart.

Young moose-hunters. ...Charles A. Stephens.

STAËL-HOLSTEIN, ANNE LOUISE GERMAINE NECKER. *Lived 1766–1817.*

Frau von Staël : biographische Roman. ...Amalie C. E. M. Boelte.

Madame de Staël : an historical novel. ...Amalie C. E. M. Boelte. English translation.

STEPHEN OF ENGLAND. *Reigned 1135–1154.*

Brian Fritz-count : a story of Wallingford Castle and Dorchester Abbey. ...Augustine D. Crake.

King Stephen ; or, Battle of Lincoln. ...David W. Paynter.

SWEDEN.

Description, History, Manners, and Customs.

Adventures of Gustavus Vasa ; or, Dawning of light in Sweden. ...L. S. Griffith.

Un ami de la reine. ...Paul Gaulot.

Arwed Gyllenstierna : eine Erzählung aus dem Anfange des achtzehnten Jahrhunderts. ...Carl F. Van der Velde.

Arwed Gyllenstierna : a tale of the early part of the eighteenth century. ...Carl F. Van der Velde. English translation.

Bilder ur verklighten. ...August T. Blanche.

Blameless knights ; or, Lützen and La Vendée. ...Alice H. F. Byng, Viscountess Enfield.

Brave resolve ; or, Siege of Stralsund. ...John B. de Liefde.

Camilla. ...Richard von Kock.

Carine. ...Louis Énault.

Carine ; or, Story of Sweden. ...Louis Énault. English translation.

En dagbok. ...Frederika Bremer.

Deluge. ...Henryk Sienkiewicz. English translation.

Diary and other tales. ...Frederika Bremer. English translation.

Fältskarns berättelser. ...Zacharias Topelius.

Familjen H***. ...Frederika Bremer.

Fifteen years. ...Thérèse Albertine Luise von Jakob Robinson. (Pseud., Talvj.) English translation.

Der Freibeuter. ...Ludwig Storch.

Friend of the queen. ...Paul Gaulot. English translation.

SWEDEN — *Continued.*

Fünfzehn Jahre. ...Thérèse Albertine Luise von Jakob Robinson. (Pseud., Talvj.)

Grannarne. ...Frederika Bremer.

Gustaf Lindorm. ...Emilia Flygare-Carlen.

Gustav Adolph: historischer Roman. ...Joseph E. C. Bischoff.

Gustav Adolf och 30 åriga kriget. .. Zacharias Topelius.

Gustav den Tredje och hans hof. ...Carl A. Kullberg.

Gustavus Adolphus. ...Zacharias Topelius. English translation.

Gustavus Lindorm ; or, Lead us not into temptation. ...Emilia Flygare-Carlen. English translation.

Gustavus III. ...Zacharias Topelius. English translation.

Gustav IIIs testamente eller 1792 och 1815. ...Henrik af Tolle.

Gustav Wasa. ...Louisa M. Gräfin von Robiano.

Gustavus Vasa and his stirring times. ...Albert Alberg.

H— family. ...Frederika Bremer. English translation.

Hemmet, eller familje-sorger och frojder. ...Frederika Bremer.

Home ; or, Life in Sweden. ...Frederika Bremer. English translation.

L'homme de neige. ...Amantine L. A. D. Dudevant. (Pseud., George Sand.)

Iron head. ...Franz Hoffmann. English translation.

Jacobite exile : adventures in the service of Charles XII. of Sweden. ...George Alfred Henty.

Karl X. Gustav: historischer Roman. ...Carl G. von Berneck.

Last of the free buuters. ...P. Sparre.

Lion of the North : a tale of the times of Gustavus Adolphus and the wars of religion. ...George Alfred Henty.

Maid of Stralsund : a story of the thirty years' war. ...John B. de Liefde.

Memoirs of a cavalier ; or, Military journal of the wars in Germany and England. ...Daniel Defoe.

My Lady Rotha. ...Stanley J. Weyman.

Neighbors : a story of everyday life. ...Frederika Bremer. English translation.

Nono ; or, Golden house : a tale of Swedish life. ...Sarah S. Baker.

Ogniem i mieczem. ...Henryk Sienkiewicz.

Presidentens döttrar. ...Frederika Bremer.

President's daughter. ...Frederika Bremer. English translation.

Princess Vasa. ...Zacharias Topelius. English translation.

Prinsennan af Vasa. ...Zacharias Topelius.

Die Schweden in Prag (1648). ...Caroline Pichler.

Seton. ...Carl S. F. von Zeipel.

Snow man. ...Amantine L. A. D. Dudevant. (Pseud., George Sand.)

SWEDEN — *Continued.*

Die Söhne Gustav Wasas.   ...Carl Berkow.

Surgeon's stories.   ...Zacharias Topelius.   English translation.

Times of Charles XII.   ...Zacharias Topelius.   English translation.

Times of Gustav Adolf.   ...Zacharias Topelius.   English translation.

Under Gustaf IIIs första regeringsår.   ...Zacharias Topelius.

Ur Karl XIIs ungdom.   ...Johann Borjesson.

With fire and sword.   ...Henryk Sienkiewicz.   English translation.

SWITZERLAND.

Description, History, Manners, and Customs.

Addrich im Moos ; historische Erzählung.   ...Johann Heinrich D. Zschokke.

Anne of Geierstein ; or, Maiden of the mist.   ...Sir Walter Scott.

At the Red-Glove.   ...Katharine Macquoid.

Aus der Schweizergeschichte.   ...Gustav A. von Heeringen.

Calvin : culturhistorischer Roman.   ...Theodor Koenig.

Chatelaine of La Trinité.   ...Henry B. Fuller.

Chillon ; or, Protestants of the sixteenth century.   ...Jane L. Willyams.

City and the castle : a story of the reformation in Switzerland. ...Annie Lucas.

Claudine ; or, Humility the basis of all the virtues.   ...Anon.

Der Freihof von Aarau.   ...Johann Heinrich D. Zschokke.

Geneva's shield : a story of the Swiss reformation.   ...William M. Blackburn.

Good old times : a tale of Auvergne.   ...Anne Manning.

Guillaume Tell ; ou, La Suisse libre. ...Jean Pierre Claris de Florian.

Hermit of Livry : a tale of the days of Calvin.   ...Emma Leslie.

Hour will come : a tale of an Alpine cloister.   ...Wilhelmine von Hillern.   English translation.

Idyl of the Alps.   ...Anne Manning.

Jean Roubaix : a tale of the Swiss mountains.   ...M. Montgomery Campbell.

Der Kaufmann von Luzern.   ...Gustav A. von Heeringen.

Der Knabe von Luzern : historischer Roman aus der Schweizer-geschichte.   ...Gustav A. von Heeringen.

Leonard and Gertrude.   ...Johann H. Pestalozzi. English translation.

Lienhard und Gertrud : ein Buch für das Volk.   ...Johann H. Pestalozzi.

Louis Belat : a tale of the reformation in Savoy.   ...Mrs. Alexander S. Orr.

Mountain patriots : a tale of the reformation in Savoy.   ...Mrs. Alexander S. Orr.

SWITZERLAND — *Continued.*
Der Müller von Sempach. Kulturgeschichtliche Erzählung aus der
Zeit des Sempacher Krieges. ...J. Bucher.
Nikolaus Manuel. Roman aus der Zeit der schweizerischen Glau-
benskämpfe. ...Ludwig Eckardt.
Pleasant life. ...Mary Howitt.
Realmah. ...Sir Arthur Helps.
Rivers of ice : a tale illustrative of Alpine adventure and glacial
action. ...Robert Michael Ballantyne.
Les roches blanches. ...Edouard Rod.
Rose of Disentis. ...Johann Heinrich D. Zschokke. English
translation.
Die Rose von Disentis. ...Johann Heinrich D. Zschokke.
Saly's Revolutionstage. ...Ulrich Hegner.
Secret quest. ...George Manville Fenn.
Story of a noble life ; or, Zurich and its reformer, Ulrich Zwingli.
...Janet Hardy.
Tales and traditions of Switzerland. ...William Westall.
Und sie kommt doch! Erzählung aus einem Alpenkloster. ...Wil-
helmine von Hillern.
White rocks. ...Edouard Rod. English translation.
William Tell. ...Jean Pierre Claris de Florian. English translation.
William Tell. ...Agnes Strickland. (In Stories from history.)
Züricher novellen. ...Gottfried Keller.
See also **REFORMATION, Switzerland**.

SYRIA.
   Description, History, Manners, and Customs.
Mirage. ...Julia C. Fletcher. (Pseud., George Fleming.)
Peculiar people. ...William S. Balch.

TELL, WILLIAM. *Lived ? – 1350.*
Guillaume Tell ; ou, La Suisse libre. ...Jean Pierre Claris de Florian.
William Tell. ...Jean Pierre Claris de Florian. English translation.
William Tell. ...Agnes Strickland. (In her Stories from history.)

TEMPERANCE. See **INTEMPERANCE**.

TENNESSEE.
Down in Tennessee. ...James R. Gilmore. (Pseud., Edmund Kirke.)
In the " stranger people's " country. ...Mary Noailles Murfree.
(Pseud., Charles Egbert Craddock.)
In the Tennessee mountains. ...Mary Noailles Murfree. (Pseud.,
Charles Egbert Craddock.)
Prophet of the Great Smoky Mountains. ...Mary Noailles Murfree.
(Pseud., Charles Egbert Craddock.)

TENNESSEE — *Continued.*
Sergeant Slasher. ...Herrick Johnson.
Where the battle was fought. ...Mary Noailles Murfree. (Pseud.,
Charles Egbert Craddock.)
See also **UNITED STATES HISTORY**.

TEXAS.
Archbishop. ...O. S. Bellisle.
Aurifodina. ...G. W. Peck.
Baby Rue: her friends and her enemies. ...Charlotte M. Clark.
(Pseud., Charles M. Clay.)
Bernard Lisle: an historical romance, embracing the periods of
the Texas revolution and the Mexican war. ...Jeremiah
Clemens.
Biglow papers. 1st series. ...James Russell Lowell.
Cabin book; or, Scenes and sketches of the late American and
Mexican war. ...Charles Sealsfield (formerly Karl Postl).
Das Cajütenbuch; oder, Nationale Charakteristiken. ...Charles
Sealsfield (formerly Karl Postl).
House of Yorke. ...Mary A. Tincker.
Inez: a tale of the Alamo. ...Augusta J. Evans (Mrs. Wilson).
Life and death of Sam in Virginia. ...Anon.
Lone star of Texas. ...J. W. Dallam.
Mary Rock. ...Percy B. Saint John.
Mercedes: a tale of the Mexican war. ...Frederick C. Wraxall.
Michael Bonham; or, Fall of Bexar.
Mustang Gray. ...Jeremiah Clemens.
Nathan, der Squatter-Regulator; oder, Der erste Regulator in Texas.
...Charles Sealsfield (formerly Karl Postl).
Overland. ...John W. DeForest.
Piney woods tavern. ...Samuel A. Hammett.
Power of the "S. F.": a tale developing the secret action of parties
during the presidential campaign of 1844. ...Thomas D.
English.
Rangers and regulators of the Tanaha: a tale of the republic of
Texas. ...Alfred W. Arrington.
Remember the Alamo. ...Amelia Edith Barr.
Les scalpeurs blancs. ...Gustave Aimard.
Simple heart. ...S. B. Elliot.
Stanhope Burleigh. ...Charles Edward Lester.
Stray Yankee in Texas. ...Samuel A. Hammett.
Talbot and Vernon. ...McConnell.
Texas ranger. ...Captain Flack.
Vidette: a tale of the Mexican war. ...N. M. Curtis.
Virginians in Texas. ...William Mumford Baker.

TEXAS — *Continued.*
White scalper: a story of the Texan war. ...Gustave Aimard
English translation.
Wild life; or, Adventures of the frontier: a tale of the early days
of the Texan republic. ...Thomas M. Reid.

THANKSGIVING DAY.
Chanticleer. ...Cornelius Mathews.
John Norton's Thanksgiving. ...William H. H. Murray.
Old fashioned Thanksgiving. ...Louisa May Alcott. (In Aunt Jo's
scrap-bag.)

THEATRICAL LIFE.
Actor's wooing. ...Louise Jordon Miln.
Born player. ...Mary West.
Le capitaine Fracasse. ...Théophile Gautier.
Captain Fracasse. ...Théophile Gautier. English translation.
Charlotte Ackermann: a theatrical romance. ...Otto Müller.
English translation.
Charlotte Ackermann, ein Hamburger Roman aus dem Vorigen
Jahrhundert. ...Otto Müller.
Cross currents. ...Mary Angela Dickens.
Diana's hunting. ...Robert Buchanan.
Dolly Dillenbeck. ...James L. Ford.
Dorothy Wallis. ...Walter Besant.
King Zub. ...Walter Herries Pollock.
Lost in a great city. ...Amanda M. Douglas.
Miser Farebrother. ...Benjamin Leopold Farjeon.
Peg Woffington. ...Charles Reade.
Play actress. ...Samuel Rutherford Crockett.
Rising star: a story of the stage. ...David Christie Murray.
Strolling players. ...Charlotte Mary Yonge and Christabel R.
Coleridge.
Der Theaterartzt und andere Humoresken. ...Arthur Bornstein.

THEODORIC THE GREAT. See **DIETRICH OF BERN.**

THEOSOPHY.
Atman: the documents in a strange case. ...Francis Howard
Williams.
Brethren of Mount Atlas. ...Hugh E. M. Shutfield.
Dr. Mirabel's theory. ...Ross George Dering.
Dreams of the dead. ...E. Stanton, pseud.
Etidorhpa. ...Anon.
For God and humanity. ...Haskett Smith.
Mahatma's pupil. ...R. Marsh.

THEOSOPHY — *Continued.*
Morial the Mahatma. ...Mabel Collins.
On the height of Himalay. ...H. van der Naillen.
Six cent Sam's. ...Julian Hawthorne.
Stone dragon. ...Murray Gilchrist. •
Strange friend. ...Julian Hawthorne.
That fiddler fellow. ...Horace Hutchinson.
Veiled beyond: a romance of the adepts. ...Sigmund B. Alexander.
Witch of Prague. ...Francis Marion Crawford.

THIRTY YEARS' WAR. 1618–1648.
Der abenteuerliche Simplicissimus. ...Hans J. C. von Grimmelshausen.
Allwin: ein Roman. ...Friedrich H. C. La Motte Fouqué.
Axel. ...Carl F. van der Velde.
Axel: a tale of the thirty years' war. ...Carl F. van der Velde. English translation.
Baron and squire: a story of the thirty years' war. ...Wilhelm Noeldechen.
Bernhard, Herzog zu Sachsen-Weimar. ...Friedrich G. Schlenkert.
Bílá hora, aneb; Tři léta z třiceti. ...Ludwig Rellstab.
Blameless knights ; or, Lützen and La Vendée. ...Alice H. F. Byng, viscountess Enfield.
Brave resolve ; or, Siege of Stralsund. ...John B. de Liefde.
Der Bürgermeister von Neisse. ...Georg Hartwig.
Der Christian Königlichen, Fürsten Herculiscus und Herculadisla Wundergeschichte. ...Andreas H. Buchholtz. ·
La conspiration de Walstein. ...Jean François Sarrasin.
Der deutsche Krieg: historischer Roman in drei Büchern. ...Heinrich Laube.
Der deutsche Michel. ...Johann Georg L. Hesekiel.
Ein deutsches Schneiderlein. ...Franz Isidor Proschko.
Der dreissigjährige Krieg. ...Clara M. Mundt. (Pseud., Louise Mühlbach.)
Duellists : a tale of the thirty years' war. ...Anon. (In Tales from Blackwood, vol. 10.)
Fair Else, Duke Ulrich, and other tales. ...Margaret Roberts.
Fältskärns Berättelser. ...Zacharias Topelius.
Fides: eine Erzählung aus der Zeit des dreissigjährigen Krieges. ...F. Jordan.
Der Findling. ...Carl A. F. von Witzleben.
Frau Schatz Regine. ...Johann Georg L. Hesekiel.
Gabriel: a story of the Jews in Prague. ...Salomon Kohn. English translation.

THIRTY YEARS' WAR — *Continued.*

Gawriel : historische Erzählung aus dem dreissigjährigen Kriege. ...Salomon Kohn.

Geglänzt und erloschen.   ...Ferdinand Pflug.

Geschichte der Gräfin Thekla von Thurn.   ...Christiane Benedicte Eugenie Naubert.

Gustaf Adolf och 30 åriga kriget.   ...Zacharias Topelius.

Gustave Adolf and the thirty years' war.   ...Zacharias Topelius. English translation.

Gustavus III.   ...Zacharias Topelius.

Heidelberg.   .....George Payne Rainsford James.

Herzog Bernhard.   ...Hans Blum.

Herzog Bernhard.   ...Heinrich Laube.

Die historisch-politischen Volkslieder des dreissigjährigen Krieges. ...Franz W. Ditfurth.

Im Schloss zu Heidelberg.   ...Eva Hartner.

Ingerstein Hall and Chadwick Rise.   ...James Routledge.

Johann Georg I. von Sachsen.   ...Franz Lubojatzky.

King's service : a story of the thirty years' war.   ...Anon.

Klosterheim.   ...Thomas De Quincey.

Das Kräuterweible von Wimpfen.   ...Conrad Fron.

Kruitzner ; or, German's tale.   ...Harriet Lee.   (In her Canterbury tales.)

Lichtenstein.   ...Wilhelm Hauff.

Lichtenstein ; or, Swabian league.   ...Wilhelm Hauff.   English translation.

Die Lichtensteiner.   ...Carl F. van der Velde.

Lichtensteins : a tale of the thirty years' war.   ...Carl F. van der Velde.   English translation.

Die Lieder des dreissigjährigen Krieges.   ...Edward Weller.

Lion of the North : a tale of the times of Gustavus Adolphus and the wars of religion.   ...George Alfred Henty.

Maid of Stralsund : a story of the thirty years' war.   ...John B. de Liefde

Memoirs of a cavalier ; or, Military journal of the wars in Germany and England from 1632–'48.   ...Daniel Defoe.

Memoirs of the honourable Col. Andrew Newport.   ...Daniel Defoe.

Mönch und Gräfin.   ...Anton J. Gross-Hoffinger.

Die Mörder Wallensteins.   ...George C. R. Herlossohn.

Odolan Pěptipeský.   ...Beneš-Trebizský.

Philip Rollo ; or, Scottish musketeers.   ...James Grant.

Princess Vasa.   ...Zacharias Topelius.   English translation.

Prinsessan af Vasa.   ...Zacharias Topelius.

Schoolmaster and his son.   ...Carl H. Caspari.   English translation.

THIRTY YEARS' WAR — *Continued.*
Der Schulmeister und sein Sohn: eine Erzählung aus dem dreissigjährigen Kriege. .. Carl H. Caspari.
Die Schweden in Prag. ...Caroline Pichler.
Thorn fortress: a tale of the thirty years' war. ...Mary Bramston.
Times of Gustaf Adolf. ...Zacharias Topelius. English translation.
Die Tochter des Piccolomini. ...G. C. R. Herlosssohn.
Um den Kaiserstuhl. ...Wilhelm Jensen.
Under Gustaf III.'s första regeringsår. ...Zacharias Topelius.
Waldemar. ...W. H. Harrison.
Waldner von Wildenstein. ...Jos. Schreyvogel.
Waldstein. ...Heinrich Laube.
Wallenstein. ...Ernst Wilkomm.
Wallensteins erste Liebe. ...George C. R. Herlosssohn.
Wallenstein's letzte Tage. ...Franz Lubojatzky.
Young carpenters of Freiberg: a tale of the thirty years' war. ...Anon.
Young deserter. ...Anon.

TURKEY.
Description, History, Manners, and Customs.
Adventures of Hajji Baba in Turkey, Persia, and Russia. ...James Morier.
Among the Turks. ...Verney L. Cameron.
Armenians : a tale of Constantinople. ...Charles MacFarlane.
Barabbas. ...Marie Corelli.
Die Belagerung Wiens im Jahre 1683. ...Caroline Pichler.
Bertrand de la Croix. ...George Payne Rainsford James.
Boatman of the Bosphorus: a tale of Turkey. ...Anon.
Captain of the Janizaries : a story of the times of Sanderbeg and the fall of Constantinople. ...James M. Ludlow.
Chrisna. ...Xavier B. Saintine.
Chrisna, Queen of the Danube : a story of Montenegro. ...Xavier B. Saintine. English translation.
Cross above the crescent : a romance of Constantinople. ...Horatio Southgate.
Czar and Sultan : the adventures of a British lad in the Russo-Turkish war of 1877-'78. ...Archibald Forbes.
Djambek the Georgian : a tale of Modern Turkey. ...A. Gundaccar, Freiherr von Suttner.
Lazar-house of Leros : a tale of the Eastern church. ...Anon. (In Tales illustrating church history.)
Little blue lady. ...Mitchell.
Mahmound, the life of a Turkish apostate. ...Anon.
Prince of India; or, Why Constantinople fell. ...Lew Wallace.
Romance of the harem. ...Julia Pardoe.

TURKEY — *Continued*.

Siege of Rhodes.   ...William Davenant.

Siege of Vienna : a story of the Turkish war of 1683.   ...J. Latch-more.   Translated from the Turkish.

Story of Jewad.   ...E. J. W. Gibb.   Translated from the Turkish.

Story of the Turkish war.   ...Anon.

Tame Turk.   ...Olive Harper.

Tcherkiss and his victim : sketches illustrative of life in Constanti-nople.   ...Anon.

Die Türken vor Wiens.   ...Carl Mueller.

Turkish tales.   ...Henry W. Weber.

Um den Halbmond.   ...Johann Ferdinand Martin Oskar Meding. (Pseud., Gregor Samarow.)

Zord Idő.   ...Zsigmund Kemény.

See also **CONSTANTINOPLE** and **RUSSIA**.

TYROLESE.

Andreas Hofer.   ...Clara M. Mundt.   (Pseud., Louise Mühlbach.)

Andreas Hofer.   ...Clara M. Mundt.   (Pseud., Louise Mühlbach.) English translation.

Andreas Hofer.   ...Maximilian von Schenkendorf.

At odds.   ...Jemima Montgomery, Baroness von Tautphœus.

1809 : historischer Roman.   ...Eduard Breier.

Die Geier-Wally : eine Geschichte aus den Tyroler Alpen.   ...Wil-helmine von Hillern.

Geier-Wally : a tale of the Tyrol.   ...Wilhelmine von Hillern. English translation.

Hofer.   ...Bramley-Moore.

Peter Mayr : der Wart und der Mahr.   ...Peter K. Rosegger.

Year nine : a tale of the Tyrol.   ...Anne Manning.

UNITED STATES HISTORY. *A Selection.*

Early History.   See **AMERICA, North.**

Seventeenth Century.

Bay-path : a tale of New England colonial life.   ...Josiah Gilbert Holland.

Braddock : a story of the French and Indian wars.   ...John R. Musick.

Captain Kyd ; or, Wizard of the sea.   ...Joseph Holt Ingraham.

Cassique of Kiawah.   ...William Gilmore Simms.

Cavaliers of Virginia.   ...W. A. Caruthers.

Doctor Le Baron and his daughters.   ...Jane Goodwin Austin.

First of the Knickerbockers.   ...P. H. Meyers.

First settlers in Virginia.   ...J. Davis.

First settlers of New England.   ...Lydia Maria Child.

Gaut Gurley ; or, Trappers of Umbagog.   ...Daniel P. Thompson.

UNITED STATES HISTORY — *Continued.*

Golden hair : a tale of the Pilgrim Fathers. ...F. C. L. Wraxall.
Green Mountain boys. ...Daniel P. Thompson.
Hansford. ...Saint George Tucker.
Hobomok. ...Lydia Maria Child.
Indian princess. ...J. N. Barker.
King Noanett. ...Anon.
Knights of the Horseshoe. ...W. A. Caruthers.
Lily and the totem. ...William Gilmore Simms.
Merry-Mount : a romance of the Massachusetts colony. ...John Lothrop Motley.
Mount Hope. ...George H. Hollister.
Nameless nobleman. ...Jane Goodwin Austin.
Naomi ; or, Boston two hundred years ago. ...Eliza Buckminster Lee.
Narragansett chief. ...B. J. Peirce.
Nix's mate. ...R. Dawes.
Peep at the Pilgrims in 1636. ...Harriet V. Cheney.
Pilgrims of New England : a tale of the early American settlers. ...Mrs. J. B. Peploe (Webb).
Pocahontas. ...Mary W. Mosby.
Rebels ; or, Boston before the revolution. ...Lydia Maria Child.
Rejected wife. ...Ann S. W. Stephens.
Salem : a tale of the seventeenth century. ...D. R. Castleton.
Seeking a home ; or, Home of the Pilgrims. ...Edward N. Hoare.
True hero ; or, Story of William Penn. ...William Henry Giles Kingston.
Vasconselos. ...William Gilmore Simms.
Wept of Wish-ton-wish. ...James Fenimore Cooper.
White chief among the red men ; or, Knight of the golden Melice. ...J. T. Adams.
Woman of Shawmut : a romance of colonial times. ...E. J. Carpenter.
Young patroon. ...P. H. Meyers.
Youth of the old Dominion. ...S. Hopkins.
See also **NEW ENGLAND, PLYMOUTH, PILGRIMS, PURITANS, QUAKERS**, and **WITCHCRAFT**.

Eighteenth Century.

Alamance ; or, Great and final experiment. ...Calvin Wiley.
Amelia ; or, Faithless Briton. ...Anon.
American hunter : a tale from incidents which happened during the war with America. ...Anon.
American spy ; or, Freedom's early sacrifice : a tale of the revolution. ...Jeptha R. Simms.

UNITED STATES HISTORY— *Continued.*

Clayton's rangers ; or, Quaker partisans.   ...E. H. Williamson.
Conspirator.   ...Eliza Ann Dupuy.
Creole.   ...J. B. Cobb.
Daughters of the revolution and their times.   ...Kate Chopin.
Doctor Vandyke.   ...John Esten Cooke.
Doom of the tory's guard.   ...Newton M. Curtis.
Double masquerade : a romance of the revolution.   ...Charles R. Talbot.
Dutch dominie of the Catskills ; or, Times of the " Bloody Brandt " : a tale of the revolution.   ...David Murdoch.
Eagle of Washington.   ...Burkitt J. Newman.
Ellen Grafton, the lily of Lexington ; or, Bride of liberty.   ...Benjamin Barker.
Emma Corbett ; or, Miseries of the civil war.   ...Samuel J. Pratt.
Ethan Allen ; or, King's men.   ...Melville, pseud.
Ethel Hamilton ; or, Lights and shadows of the war of independence.   ...Anna T. Sadlier.
Eutaw.   ...William Gilmore Simms.
Fairfax : a chronicle of the valley of Shenandoah.   ...John Esten Cooke.
Fair maid of Wyoming.   ...Gabriel Alexander.
Female spy ; or, Treason in the camp : a story of the revolution.   ...Emerson Bennett.
Forayers.   ...William Gilmore Simms.
Forest and shore ; or, Legends of the Pine tree State.   ...Charles P. Ilsley.
Forest tragedy.   ...Grace Greenwood.
Foresters : an American tale.   ...Jeremy Belknap.
Forsaken.   ...Richard P. Smith.
Foundling of the Mohawk.   ...Newton M. Curtis.
Franklin's oath : a tale of Wyoming one hundred years ago.   ...C. I. A. Chapman.
Frederick de Algeroy, the hero of Camden plains : a revolutionary tale.   ...Giles Gazer.
From colony to commonwealth : stories of the revolutionary days in Boston.   ...Nina M. Tiffany.
Fugitives ; or, Quaker scout of Wyoming : a tale of the massacre of 1778.   ...Edward S. Ellis.
Gideon Godbold : a tale of Arnold's treason.   ...N. C. Iron.
Golden eagle ; or, Privateer of 1776.   ...Sylvanus Cobb, Jr.
Grace Dudley ; or, Arnold at Saratoga.   ...Charles J. Peterson.
Great treason : a story of the war of independence.   ...Mary A. M. Hoppus.
Green mountain boys.   ...Daniel P. Thompson.

UNITED STATES HISTORY — *Continued.*

Old Fort Duquesne: a tale of the early toils, struggles, and adventures of the first settlers at the forks of the Ohio in 1754. ...Anon.

Old Fort Duquesne; or, Captain Jack the scout. ...Charles Mc-Knight.

Old Put. ...Justin Jones. (Pseud., Harry Hazel.)

Old revolutionary soldier. ...Joseph Alden.

Old stone house. ...Joseph Alden.

Out of a besieged city. ...Charles W. Hutson.

Overland. ...J. W. De Forest.

Partisan. ...William Gilmore Simms.

Patriot and tory, one hundred years ago. ...Julia M. Wright.

Paul and Persis; or, Revolutionary struggle in the Mohawk valley. ...Mary E. Bunce.

Paul Ardenheim, the monk of Wissahikon. ...George Lippard.

Paul Jones: a romance. ...Allen Cunningham.

Paul Jones. ...Theodor Muegge.

Paul Redding: a tale of the Brandywine. ...T. Buchanan Read.

Pemberton; or, One hundred years ago. ...Henry Peterson.

Peter and Polly; or, Home life in New England a hundred years ago. ...Annie D. Greene.

Pilot: a tale of the sea. ...James Fenimore Cooper.

Pride of Lexington. ...W. Seton.

Quaker soldier; or, British in Philadelphia: a romance of the revolution. ...J. Richter Jones.

Ralph the drummer boy: a story of the days of Washington. ...Théophile L. Rousselet.

Ralphton; or, Young Carolinian of 1776. ...A. H. Brisbane.

Rangers and regulators of the Tanaha: a tale of the republic of Texas. ...Alfred W. Arrington.

Rangers; or, Tory's daughter. ...Daniel P. Thompson.

Rebels and the tories; or, Blood of the Mohawk. ...Lawrence Labree.

Rejected wife. ...Ann S. Stephens.

Revolutionary times: sketches of our country one hundred years ago. ...Edward Abbott.

Rosalie du Pont; or, Treason in the camp. ...Emerson Bennett.

Russell and Sidney. ...Eliza Leslie.

Saratoga: an Indian tale of frontier life: a true story of 1787. ...D. Shepherd.

Saratoga: a tale of 1787. ...Anon.

Satanstoe. ...James Fenimore Cooper.

Les scalpeurs blancs. ...Gustave Aimard.

Scout. ...William Gilmore Simms.

Septimius Felton. ...Nathaniel Hawthorne.

26

UNITED STATES HISTORY — *Continued.*

'76. ...John Neal.

Sketches of Western adventure. ...John A. McClung.

Spinning-wheel stories. ...Louisa May Alcott.

Spy: a tale of the neutral ground. ...James Fenimore Cooper.

Standish the Puritan: a tale of the American revolution. ...Eldred Grayson, pseud.

Stories of the old Dominion from the settlement to the end of the revolution. ...John Esten Cooke.

Stories of the revolution. ...Josiah Priest.

Storm mountain. ...Edward S. Ellis.

Story of a Hessian: a tale of the revolution in New Jersey. ...Lucy Ellen Guernsey.

Thankful blossom: a story of the Jerseys. ...Francis Bret Harte.

Three girls of the revolution. ...Lucy Ellen Guernsey.

Ticonderoga; or, Black eagle. ...George Payne Rainsford James.

Tory spy; or, Britisher "done brown." ...J. Colfort Clifton.

True stories of the American fathers. ...Rebecca MacConkey.

True to the old flag: a tale of the American war of independence. ...George Alfred Henty.

Twice taken: an historical romance of the maritime British provinces. ...Charles W. Hall.

Walter Thornley; or, Peep at the past. ...Susan R. Sedgwick.

Washington and his generals; or, Legends of the revolution. ...George Lippard.

Washington and his men. ...George Lippard.

Washington and seventy-six. ...Lucy Ellen Guernsey.

Water waif: a story of the revolution. ...Elizabeth S. Bladen.

Wau-nan-gee; or, Massacre at Chicago: a romance of the American revolution. ...John Richardson.

White scalper: a story of the Texan war. ...Gustave Aimard. English translation.

White slave; or, Memoirs of Archy Moore. ...Richard Hildreth.

Woodcraft. ...William Gilmore Simms.

Yankee champion; or, Tory and his league. ...Sylvanus Cobb, Jr.

Yemasee. ...William Gilmore Simms.

Yorktown: an historical romance. ...Anon.

Young rebels: a story of the battle of Lexington. ...Robert Hope Moncrieff. (Pseud., Ascot R. Hope.)

See also **FRENCH AND INDIAN WARS, 1756–1775,** and **INDIANS OF NORTH AMERICA.**

Nineteenth Century.

Bernard Lile: an historical romance embracing the periods of the Texas revolution and the Mexican war. ...Jeremiah Clemens.

UNITED STATES HISTORY — *Continued.*
Sustained honor: a story of the war of 1812.   ...John R. Musick.
Talbot Vernon.   ...McConnell.
Three scouts.   ...John T. Trowbridge.
Tiger-lilies.   ...Sidney Lanier.
Union : a story of the great rebellion.   ...John R. Musick.
See also **WAR OF 1812, CIVIL WAR OF 1861, REVOLUTIONS,**
America, 1775–1789, and **MEXICAN WAR OF 1845.**

VEHMGERICHT.   See **FEMGERICHTE.**

VENEZUELA.
Description, History, Manners, and Customs.

Die Blauen und Gelben.   ...F. Gerstaecker.
Campaigns and cruises in Venezuela, New Granada, and in the
Pacific Ocean.   ...Anon.
Frank Redcliffe : a story of travel and adventure in the forests of
Venezuela.   ...Achilles Daunt.
Young Llanero.   ...William Henry Giles Kingston.

VENICE, ITALY.
Bravo.   ...James Fenimore Cooper.
Le fils du Titien.   ...Alfred de Musset.
Giacomo.   ...W. C. Bamburgh.
Haunted hotel.   ...Wilkie Collins.
Lion of Saint Mark : a tale of Venice.   ...George Alfred Henty.
Venetian June.   ...Anna Fuller.
Winged lion.   ...James De Mille.

VERMONT.
Eagle's plume.   ...Sarah E. Heald.
Green Mountain boys : a historical tale of the early settlement of
Vermont.   ...Daniel P. Thompson.

VICTORIA OF ENGLAND. *Reigned 1837–*
Alton Locke.   ...Charles Kingsley.
Beatrice and Benedick : a romance of the Crimea.   ...Hawley Smart.
Coningsby.   ...Benjamin Disraeli.
Convict.   ...George Payne Rainsford James.
Corporal Bruce of the Balaklava "six hundred."   ...George B. Perry.
Endymion.   ...Benjamin Disraeli.
Fall of Sebastopol; or, Jack Archer in the Crimea.   ...George
Alfred Henty.
Frederick Gordon ; or, Storming of the Sedan.   ...Anon.
Gaythorne Hall.   ...John M. Fothergill.
Henri de la Tour ; or, Comrades in arms.   ...J. Frederick Smith.

VICTORIA OF ENGLAND — *Continued.*
Interpreter: a tale of the Crimean war. ...George John Whyte
  Melville.
Jack Archer: a tale of the Crimea. ...George Alfred Henty.
James Woodford, carpenter and chartist. ...Henry Solly.
Lady Wedderburn's wish: a tale of the Crimean war. ...James
  Grant.
Laura Everingham; or, Highlanders of Glen Ora. ...James Grant.
Lord Hermitage. ...J .ies Grant.
Lothair. ...Benjamin Disraeli.
Mainstone's housekeeper. ...Eliza Meteyard. (Pseud., Silver pen.)
Manchester strike. ...Harriet Martineau.
Martyrs to circumstances. ...Maria Thérèse Longworth Yelverton.
Mary Barton. ...Elizabeth C. Gaskell.
North and South. ...Elizabeth C. Gaskell.
One of the " six hundred." ...James Grant.
Ravenshoe. ...Henry Kingsley.
Soldier born : or, Adventures of a subaltern of the Ninety-fifth in
  the Crimea, and Indian mutiny. ...John P. Groves.
Sybil; or, Two nations. ...Benjamin Disraeli.
Under the Red Dragon. ...James Grant.
Vera; or, Russian princess and the English earl. ...Charlotte
  Dempster.
Wenderholme : a story of Lancashire and Yorkshire. ..Philip
  Gilbert Hamerton.
Yeast: a problem. ...Charles Kingsley.
See also **CRIMEAN WAR** and **INDIA**.

VIENNA, AUSTRIA.
Auf dem Wiener Kongress. ...Julius Bacher.
Die Belagerung Wiens im Jahre 1683. ...Caroline Pichler.
Das Buch von den Wiencrn.., ...Michael Beheim.
Der Congress zu Wien. ...Eduard Breier.
1805; oder, Die Franzosen zum erstenmal in Wien. ...Eduard
  Breier.
Die Geheimnisse des Waldschlosses. ...Reinhard Edmund Haln.
Johann Sobieski, der grösse Polenkönig und Befreier Wiens.
  ...Georg F. Born.
Moderne Grössen. ...B. Aba.
Pastor's fireside. ...Jane Porter.
Die revolution der Wiener im 15 Jahrhundert. ...Eduard Breier.
Romance of Vienna. ...Frances Trollope.
Die Rosenkreuzer in Wien. ...Eduard Breier.
Rüdiger von Starhemberg. ...Franz X. Huber.
Siege of Vienna. ...J. Latchmore, Jr. Translated from the Turkish.

VIENNA, AUSTRIA — *Continued.*
Thekla.   ...William Armstrong.
Die Türken vor Wiens.   ...Carl Mueller.
Wien und Rom.   ..Eduard Breier.
Wien vor vierhundert Jahren.   ...Eduard Breier.

VIKINGS.   *A Selection.*
Champion of Odin.   ...J. Frederick Hodgetts.
Gunnar : a tale of Norse heroism.   ...Hjalmar H. Boyesen.   English translation.
Ivar, the Viking.   ...Paul Du Chaillu.
Modern Vikings : stories of life and sport in the Norseland.   ...Hjalmar H. Boyesen.
Olaf the Glorious : a historical story of the Viking age.   ...Robert Leighton.
Sentimental Vikings.   ...R. V. Risley.
Sons of the Vikings : an Orkney story.   ...John Gunn.
Ulla.   ...Edwin Lester Arnold.
Viking tales of the North.   ...Rasmus B. Anderson.
Vikings of the Baltic.   ...George W. Dasent.
See also **NORWAY.**

VIOLINS AND VIOLIN MUSIC.
Den Bergtekne.   ...Kristofer Nagel Janson.
Cord from a violin.   ...Winifred Agnes Haldane.
Faience violin.   ...Jules F. F. H. Fleury. (Campfleury.) English translation.
Fiddler of Lugau.   ...Margaret Roberts.
First violin.   ...Jessie Fothergill.
Flute and violin.   ...James Lane Allen.
Kun en Spillemand.   ...Hans Christian Andersen.
Lay of the Scottish fiddle.   ...James K. Paulding.
Lost Stradivarius.   ...J. Meade Falkner.
Marie.   ...Laura E. Richards.
Markof, the Russian violinist.   ...Alice M. Durand.
Only a fiddler.   ...Hans Christian Andersen. English translation.
Otto's inspiration.   ...Mary H. Ford.
Spell-bound fiddler : a Norse romance.   ...Kristofer Nagel Janson. English translation.
Teacher of the violin and other tales.   ...John Henry Shorthouse.
That fiddler fellow.   ...Horace Hutchinson.
Le violon de faïence. ...Jules F. F. H. Fleury. (Pseud., Champfleury.)

VIRGINIA.
Aunt Dorothy : an old Virginia plantation story.   ...Margaret J. Preston.

VIRGINIA — *Continued.*

Bonnybel Vane. ...John Esten Cooke.

Captain Macdonald's daughter. ...Archibald Campbell.

Captain Smith and the Princess Pocahontas. ...J. Davis.

Chief's daughter. ...Anon.

Christian Indian. ...Anon.

Christmas week at Bigler's mill. ...Dora E. W. Spratt.

Doctor Vandyke : a story of Virginia in the last century. ...John Esten Cooke.

Fairfax : a chronicle of the valley of the Shenandoah. ...John Esten Cooke.

First settlers in Virginia. ...J. Davis.

Flower de hundred : the story of a Virginia plantation. ...Constance Cary Harrison (Mrs. Burton Harrison).

From the school-room to the bar. ...W. H. W. Moran.

Girl's life in Virginia before the war. ...Letitia M. Burwell.

Hansford : a tale of Bacon's rebellion. ...Saint George Tucker.

Henry Saint John : a tale of 1774–5. ...John Esten Cooke.

Holcombes. ...Mary T. Magill.

Homoselle. ...Mary S. Tiernan.

Indian princess. ...J. N. Barker.

In ole Virginia ; or, Marse Chan and other stories. ...Thomas Nelson Page.

Judith : a chronicle of old Virginia. ...Mary Virginia Terhune. (Pseud., Marion Harland.)

Justin Harley : a romance of old Virginia. ...John Esten Cooke.

King Noanett : a story of old Virginia and Massachusetts Bay. ...F. J. Stimpson.

Knights of the Horseshoe. ...W. A. Caruthers.

Last of the foresters. ...John Esten Cooke.

Leather stockings and silk ; or, Hunter John Myers and his times : a story of the Valley of Virginia. ...John Esten Cooke.

Life and death of Sam in Virginia. ...Gardiner.

Life of Captain Jack. ...Daniel Defoe.

Man of honor. ...George C. Eggleston.

Miss Lou. ...Edward Payson Roe.

My Lady Pokahontas : a true relation of Virginia. ...John Esten Cooke.

Paths in the great waters. ...Edward N. Hoare.

Philip Randolph : a tale of Virginia. ...Mary Gertrude.

Pleasant waters : a story of Southern life and character. ...Graham Claytor.

Pocahontas : a story of Virginia. ...John R. Musick.

Roland Blake. ...S. Weir Mitchell.

Rose-hill. ...Anon.

VIRGINIA — *Continued.*
Ruth Emsley, the betrothed maiden : a tale of the Virginia massacre.   ...William H. Carpenter.
Stories of the old Dominion.   ...John Esten Cooke.
Storm mountain.   ...Edward S. Ellis.
Story of Don Miff.   ...Virginius Dabney.
Tanis the sand-digger.   ...Amélie Rives Chanler.
Throckmorton.   ...Molly E. Seawell.
Two little confederates.   ...Thomas Nelson Page.
Virginia Bohemians.   ...John Esten Cooke.
Virginia comedians.   ...John Esten Cooke.
Virginia inheritance.   ...Edward Pendleton.
Virginia of Virginia.   ...Amélie Rives Chanler.
Virginia plantation story.   ...Margaret J. Preston.
Virginians.   ...William Makepeace Thackeray.
White aprons.   ...Maud Wilder Goodwin.
Witch of Jamestown.   ...J. T. Bowyer.
Yankee school-teacher in Virginia.   ...Lydia W. Baldwin.
Youth of the old Dominion.   ...S. Hopkins.
See also **POCAHONTAS.**

WALDENSES AND ALBIGENSES.
Adelaïde de Monfort ; ou, La guerre des Albigeois.   ...Justin É. M. Cénac-Moncaut.
Albigenses : a romance.   ...Charles R. Maturin.
Casella ; or, Children of the valley.   ...Martha Finley.
Christine ; or, Persecuted family.   ...J. Dillon.
Le comte de Toulouse. Roman.   ...Melchior Frédéric Soulié.
De Foix.   ...Anna Eliza Bray.
Exiles of Lucerna.   ...John R. Macduff.
Idyl of the Alps.   ...Anne Manning.
In His name : a story of the Waldenses seven hundred years ago.   ...Edward Everett Hale.
In letters of fame : a tale of the Waldenses.   ...Clara L. Matéaux.
Julio : a tale of the Vaudois.   ...Mrs. J. B. Peploe (Webb).
Pierre and his family : a story of the Waldenses.   ...Miss Grierson.
Six sisters of the valley : an historical romance.   ...William Bramley Moore.
Le vicomte de Béziers.   ...Melchior Frédéric Soulié.

WALLENSTEIN, ALBRECHT WENCESLAS EUSEBIUS.   *Lived 1583–1634.*
La conspiration de Walstein.   ...Jean François Sarrasin.
Mönch und Gräfin.   ...Anton J. Gross-Hoffinger.
Die Mörder Wallensteins.   ...George C. R. Herlosssohn.
Die Schweden in Prag.   ...Caroline Pichler.

WALLENSTEIN, ALBRECHT WENCESLAS EUSEBIUS — *Continued.*
Die Tochter des Piccolomini. ...George C. R. Herlosssohn.
Waldstein. ...Heinrich Laube.
Wallenstein. ...Ernest Willkomm.
Wallensteins erste Liebe. ...George C. R. Herlosssohn.
Wallensteins letzte Tage. ...Franz Lubojatzky.

WALPOLE, HORACE. *Lived 1717–1797.*
Strawberry Hill. ...Robert Folk Williams.

WANDERING JEW.
Ahasvérus. ...Edgar Quinet.
Le Juif errant. ...Marie Joseph Eugène Sue.
Prince of India ; or, Why Constantinople fell. ...Lew Wallace.
Salathiel, the immortal. ...George Croly.
Wandering Jew. ...Marie Joseph Eugène Sue. English translation.

WAR. *A Selection.*
Agincourt. ...George Payne Rainsford James.
Alice Bridge of Norwich : a tale of the time of Charles the First. ...Andrew Reed.
Alice Lorraine. ...Richard Doddridge Blackmore.
Un ami de la reine. ...Paul Gaulot.
L'argent des autres. ...Émile Gaboriau.
Attack on the mill. ...Émile Zola. English translation.
At the sign of the guillotine. ...Harold Spender.
Belfry of Saint Jude. ...Esmé Stuart.
Ben Brace : the last of Nelson's Agamemnons. ...Frederick Chamier.
Bivouac ; or, Stories of the Peninsular war. ...William H. Maxwell.
Black and gold. ...Patten Saunders.
Blockade : an episode of the fall of the first French empire. ...Émile Erckmann and Alexandre Chatrian. English translation.
Le blocus. ...Émile Erckmann et Alexandre Chatrian.
Castle Dangerous. ...Sir Walter Scott.
Child of the Ganges. ...Robert N. Barrett.
Comedy of terrors. ...James De Mille.
Cornet of horse : a tale of Marlborough's wars. ...George Alfred Henty.
Creole. ...J. B. Cobb.
Cressy and Poictiers ; or, Story of the Black Prince's page. ...John George Edgar.
Days of Bruce. ...Grace Aguilar.

WAR — *Continued.*
Delaplaine.   ...Mansfield Tracy Walworth.
Diana's crescent.   ...Anne Manning.
Edmond Dantes.   ...Alexandre Dumas.
Edmond Dantes.   ...Alexandre Dumas.   English translation.
Eighty years ago.   ...Harriet Morton.
Elba und Waterloo.   ...F. Stolle.
Enemy's friendship.   ...S. M. S. Clarke.
Felix Alvarez; or, Manners in Spain.   ...Alexander R. C. Dallas.
Fighting the Saracens; or, Boy knight.   ...George Alfred Henty.
For king or country.   ...James Barnes.
Frank Hilton.   ...James Grant.
Friend of the queen.   ...Paul Gaulot.   English translation.
Gate and the glory beyond it.   ...Onyx, pseud.
La grande famille: Roman militaire.   ...Jean Grave.
Ground arms.   ...Bertha von Suttner.
Gun-bearer.   ...Edward A. Robinson.
Helen Treveryan.   ...John Roy.
Henry Masterton; or, Adventures of a young cavalier.   ...George Payne Rainsford James.
Histoire d'un homme du peuple.   ...Émile Erckmann et Alexandre Chatrian.
Histoire du plébiscite.   ...Émile Erckmann et Alexandre Chatrian.
Humbled pride: a story of the Mexican war.   ...John R. Musick.
In blue uniform.   ...George Israel Putnam.
Invasion of France in 1814: comprising the night march of the Russian army past Phalsburg.   ...Émile Erckmann and Alexandre Chatrian.   English translation.
L'invasion; ou, Le fou Yégof.   ...Émile Erckmann et Alexandre Chatrian.
Kenneth; or, Rear-guard of the grand army.   ...Charlotte Mary Yonge.
Kentucky's love; or, Roughing it about Paris.   ...Edward King.
King's own borderers: a military romance.   ...James Grant.
Lady Betty's governess; or, Corbet chronicles.   ...Lucy Ellen Guernsey.
Leonie; or, Light out of darkness.   ...Annie Lucas.
Little blue lady.   ...Mitchell.
Maid of Stralsund: a story of the thirty years' war.   ...John B. de Liefde.
Margaret Muller: a story of the late war in France.   ...Eugénie Bersier.
Max Kromer: a story of the siege of Strasburg.   ...Hannah Smith. (Pseud., Hesba Stretton.)
Memoirs of troublous times.   ...Emma Marshall.

WAR — *Continued.*

Micah Clarke : his statement as made to his three grandchildren. ...Arthur Conan Doyle.

Monsieur Jack : a tale of the old war-time. ...Alfred H. Engelbach.

New centurion : a tale of automatic war. ...James Eastwick.

Other people's money. ...Émile Gaboriau. English translation.

Parisians. ...Edward George Earle Lytton Bulwer-Lytton.

Phantom regiment ; or, Stories of Ours. ...James Grant.

Physiology of war. ...Lyeff Nikolaievich Tolstoi. English translation.

Powder and gold : a story of the Franco-Prussian war. ...Levin Schuecking. English translation.

Pulver und gold. ...Levin Schuecking.

Red badge of courage. ...Stephen Crane.

Red cap and blue jacket. ...G. Dunn.

Red cockade. ...Stanley J. Weyman.

Le roman d'un brave homme. ...Edmond About.

Romance of war. ...L. Bellstab.

Romance of war ; or, Highlanders in Spain. ...James Grant.

Saint Katharine's by the tower. ...Walter Besant.

Six years ago. ...James Grant.

Stanfield Hall. ...J. Frederick Smith.

Story of an honest man. ...Edmond About. English translation.

Story of the plébiscite. ...Émile Erckmann and Alexandre Chatrian. English translation.

Stories of the wars. ...J. Tillotson.

Stories of Torres Vedras. ...John G. Milligen.

Subaltern. ...George Robert Gleig.

Sustained honor : a story of the war of 1812. ...John R. Musick.

Tales of the wars of Montrose. ...J. Hogg.

Union : a story of the great rebellion. ...John R. Musick.

Valentin : a French boy's story of Sedan. ...Henry Kingsley.

Washingtons : a tale of a country parish in the seventeenth century. ...John N. Simpkinson.

Waterloo : suite du conscrit de 1813. ...Émile Erckmann et Alexandre Chatrian.

Waterloo. ...Émile Erckmann and Alexandre Chatrian. English translation.

Wayside cross ; or, Raid of Gomez : a tale of the Carlist war. ...Edward A. Milman.

When Greek meets Greek. ...Joseph Hatton.

White aprons. ...Maud Wilder Goodwin.

White slave. ...Richard Hildreth.

Within iron walls : a tale of the siege of Paris. ...Annie Lucas.

Woodstock ; or, Cavalier : a tale of the year 1651. ...Sir Walter Scott.

WAR — *Continued.*

Workman and soldier : a tale of Paris life during the siege and the rule of the Commune. ...James F. Cobb.

Young buglers : a tale of the Peninsular war. ...George Alfred Henty.

Young Franc-Tireurs ; and their adventures in the Franco-Prussian war. ...George Alfred Henty.

Young Llanero. ...William Henry Giles Kingston.

Youth and manhood of Cyril Thornton. ...Thomas Hamilton.

See also **ARMY AND NAVY LIFE,** and special wars under their titles.

WAR OF 1812.

Alida ; or, Town and country. ...Susan R. Sedgwick.

American cruiser : a tale of the war. ...Anon.

Big brother : a story of Indian war. ...George C. Eggleston.

Blue jackets of 1812, with account of the French and Indian war of 1798. ...Willis J. Abbot.

Boy soldiers of 1812. ...Everett T. Tomlinson.

Canadian brothers ; or, Prophecy fulfilled : a tale of the American war. ...John Richardson.

Captain Sam ; or, Boy scouts of 1814. ...George C. Eggleston.

Champions of freedom ; or, Mysterious chief. ...Samuel Wood-worth.

Chilhowee boys in war time. ...Sarah E. Morrison.

Creole. ...J. B. Cobb.

Cruising in the last war. ...Charles J. Peterson.

Cub's apple ; or, "Next time." ...Lucy Ellen Guernsey.

Les fiancés de 1812. ...Joseph Doutre.

For king and country : a story of 1812. ...Anon.

Haunted hearts : a tale of New Jersey. ...Maria S. Cummins.

L'héroïne de Chateauguay. ...Heinrich Émile Chevalier.

Lafitte, the pirate of the Mexican Gulf. ...Joseph Holt Ingraham.

Laura Secor. ...Mrs. Curzon.

Lay of the Scottish fiddle : a tale of Havre de Grace. ...James K. Paulding.

Loyal traitor : a story of the naval war of 1812. ...James Barnes.

Mafilda Montgomerie ; or, Prophecy fulfilled. ...John Richardson.

Mount Benedict ; or, Violated tomb : a tale of the Charlestown convict. ...Peter McCorry.

Ocean rovers. ...W. H. Thomas.

Queenstown : a tale of the Niagara frontier. ...Anon.

Search for Andrew Field. ...Everett T. Tomlinson.

Signal boys ; or, Captain Sam's company. ..George C. Eggleston.

Sustained honor : a story of the war of 1812. ...John R. Musick.

WAR OF 1812 — *Continued.*
Tory's daughter : a romance of the North-west, 1812–1813. ...Albert G. Riddle.
Wau-nan-gee ; or, Massacre at Chicago. ...John Richardson.
Western captive ; or, Times of Tecumseh. ...Seba Smith.
White slave ; or, Memoirs of Archy Moore. ...Richard Hildreth.
Will of the wisp. ...R. C. Rogers.
Within the capes. ...Howard Pyle.
Zachary Phips. ...Edwin Lassetter Bynner.

WAR OF PHILIP OF POKANOKET. 1607.
Mount Hope, or Philip, King of Wampanoags. ...G. H. Hollister.
Narragansett chief. ...B. J. Peirce.
Romance of King Philip's war. ...Fanny B. Workman. (In New England Mag., May, 1887.)
Wept-of-wish-ton. ...James Fenimore Cooper.

WAR OF THE ROSES. 1455–1485.
At ye Greene Griffin ; or, Mrs. Treadwell's cook. ...Emily Sarah Holt.
Black arrow : a tale of the two roses. ...Robert Louis Stevenson.
Chantry Priest of Barnet : a tale of the two Roses. ...Alfred John Church.
Civile wars between the two houses of Lancaster and Yorke. ...Samuel Daniel.
Earl printer : times of Caxton. ...C. M. M.
Fortress : an historical tale of the fifteenth century. ...Anon.
Grisly Grisell ; or, Laidly lady of Whitburn : a tale of the war of the Roses. ...Charlotte Mary Yonge.
Historical tales of the Lancastrian times. ...Henry P. Dunster.
In the wars of the Roses. ...Evelyn Everett Green.
Judged by appearances : a tale of the civil wars. ...Eleanor Lloyd.
Last of the Barons. ...Edward George Earle Lytton Bulwer-Lytton.
Maid of Warsaw. ...Ernst Jones.
Malvern Chase : an episode of the wars of the Roses and the battle of Tewkesbury. ...W. S. Symonds.
Queen's badge. ...Frances M. Wilbraham.
Red and white : a tale of the war of the Roses. ...Emily Sarah Holt.
Richard of York ; or, White Rose of England. ...Anon.
Rival Roses : a romance of English history. ...Eliza S. Francis.
Rocking stone : chronicle of the times of the wars of the Roses. ...William Henry Giles Kingston.
Stormy life. ...Georgiana C. L. G. Fullerton.
War of the Roses. ...John George Edgar.
York and Lancaster rose. ...Annie Keary.

WASHINGTON, DISTRICT OF COLUMBIA.
Across the chasm.   ...Julia Magruder.
Alice Brand : a romance of the capital.   ...Albert G. Riddle.
Child of the century.   ...John T. Wheelwright.
Gilded age.   ...Samuel Langhorne Clemens (pseud., Mark Twain),
and Charles Dudley Warner.
Her Washington season.   ...Jeanie G. Lincoln.
Honest John Vane.   ...John W. DeForest
In sight of the goddess.   ...Harriet Riddle Davis.
Justine's lovers.   ...John W. DeForest.
Scenes at Washington : a story of the last generation.   ...Anon.
Through one administration.   ...Frances Hodgson Burnett.
Washington winter.   ...Madeleine V. Dahlgren.
Zachariah, the congressman.   ...Gilbert A. Pierce.
See also **POLITICS, American.**

WASHINGTON, GEORGE.   *Lived 1732-1799.*
Boys of Greenaway court.   ...Hezekiah Butterworth.
Braddock : a story of the French and Indian wars.   ...John R.
Musick.
Eagle of Washington.   ...Burkitt J. Newman.
Farmer boy and commander-in-chief.   ...W. M. Thayer.
John Littlejohn of J.   ...George Morgan.
Lionel Lincoln ; or, Leaguer of Boston.   ...James Fenimore Cooper.
Story of Washington.   ...Elizabeth Eggleston Seelye.
True stories of the days of Washington.   ...Anon.
Washington and his generals.   ...George Lippard.
Washington and his men.   ...George Lippard.
Washington and '76.   ...Lucy Ellen Guernsey.

WEST INDIES.

Description, History, Manners, and Customs.

At last : a Christmas in the West Indies.   ...Charles Kingsley.
Buccaneers : a romance of our own country in its ancient days.
...S. B. H. Judah.
Bug-Jargal.   ...Victor Hugo.
Conspiracy : a Cuban romance.   ...Adam Badeau.
Cruise of the Midge.   ...Michael Scott.
Daughter of Cuba.   ...Helen M. Bowen.
Doris and Theodora.   ...Margaret J. Janvier. (Pseud., Margaret
Vandegrift.)
Enriqueta Faber.   ...Andrès Clemente Vazquez.
Free flag of Cuba.   ...H. M. Hardimann.
Hispaniola plate.   ...John B. Burton.
Humbled pride : a story of the Mexican war.   ...John R. Musick.

WILLIAM I. OF GERMANY — *Continued.*
Held und Kaiser.   ...Johann Ferdinand Martin Oskar Meding.
(Pseud., Gregor Samarow.)
Kaiser Wilhelm und seine Zeitgenossen.   ...Clara M. Mundt.
(Pseud., Louise Mühlbach.)
Prince Bismarck, friend or foe.   ...Minny Bothmer.

WILLIAM II. OF GERMANY. *Reigned 1888-* ——
Bis zum Kaiserthron.   ...Bruno Garlepp.
Vanished emperor.   ...Percy Andreae.

WILLIAM THE CONQUEROR.   See **NORMAN CONQUEST.**

WITCHCRAFT.
Delusion; or, Witch of New England.   ...Anon.
Fair Puritan: an historical romance of the days of witchcraft.
...Henry W. Herbert.   (Pseud., Frank Forester.)
Jane Seton.   ...James Grant.   (Scottish.)
Lancashire witches.   ...William Harrison Ainsworth.
Langham revels.   ...Lucy Ellen Guernsey.
Lois the witch.   ...Elizabeth C. Gaskell.
Mar's white witch.   ...Gertrude Douglas.
Maria Schweidler die Bernsteinhexe.   ...Johann Wilhelm Mein-
hold.
Maria Schweidler the amber witch.   ...Johann Wilhelm Meinhold.
English translation.
Martha Corey: a tale of the Salem witchcraft.   ...Constance God-
dard DuBois.
Necromancer.   ...George W. M. Reynolds.
Philip English's two cups.   ...Anon.
Rachel Dyer.   ...John Neal.
Salem: a tale of the seventeenth century.   ...D. R. Castleton.
Salem belle: a tale of 1692.   ...Anon.
Salem witchcraft: an Eastern tale.   ...R. C. Sands.
Salem witchcraft; or, Adventures of Parson Handy, from Punka-
pog.   ...Anon.
Secret of Narcisse.   ...Edmund W. Gosse.
Sidonia the sorceress.   ...Johann Wilhelm Meinhold.
Silent struggles.   ...Ann S. W. Stephens.
South meadows.   ...Ella T. Disosway.
Spectre of the forest; or, Annals of the Housatonic.   ...J. Mac-
Henry.
White witch of Moher.   ...Frank Mathew.   (In his At the rising
of the moon.)
Witch and the deacon.   ...Cornelius Mathews.
Witch hill: a history of Salem witchcraft.   ...Z. A. Mudge.

WITCHCRAFT — *Continued.*
Witch of Jamestown. ...J. T. Bowyer.
Witch of Prague. ...Francis Marion Crawford.
Witch of Salem ; or, Credulity run mad. ...John R. Musick.
Witches. ...Edmund W. Gosse.

WOLSEY, THOMAS, CARDINAL OF ROME. *Lived 1471–1550.*
Agnes Martin ; or, Fall of Cardinal Wolsey. ...Anon.
Darnley ; or, Field of the cloth of gold. ...George Payne Rainsford
 James.
Freston tower : a tale of the times of Wolsey. ...Richard Cobbold.
Lettice Eden : a tale of the last days of Henry the Eighth. ...Emily
 Sarah Holt.

WORKING-CLASSES. *A Selection.*
All sorts and conditions of men. ...Walter Besant.
Alton Locke. ...Charles Kingston.
Austin Elliot. ...Henry Kingsley.
Barnaby Rudge. ...Charles Dickens.
Break o' day and other stories. ...G. Wharton Edwards.
Breton mills. ...Charles J. Bellamy.
Cæsar's column. ...Ignatius Donnelly. (Pseud., Edmund Bois-
 gilbert.)
Claribel, the sea maid. ...E. M. Stewart.
Colloquies of Edward Osborne. ...Anne Manning.
Complaining millions of men. ...Edward Fuller.
Daughter of humanity. ...Edgar Maurice Smith.
David's loom : a story of Rochdale life in the nineteenth century.
 ...John T. Clegg.
Devil's hat. ...Melville Philips.
Eustace Marchmont, a friend of the people. ...Evelyn Everett
 Green.
Felix Holt, the radical. ...George Eliot.
Foure prentises of London. ...Thomas Heywood.
Gun-maker of Moscow. ...Sylvanus Cobb, Jr.
Hard times. ...Charles Dickens.
John Halifax, gentleman. ...Dinah Maria Mulock Craik.
Lace makers of Ireland. ...Louise A. Meredith.
Lamplighter. ...Maria Susanna Cummins.
Lawton girl. ...Harold Frederic.
Life in the iron mills. ...Rebecca Harding Davis.
Madonna of the tubs. ...Elizabeth Stuart Phelps Ward (Mrs.
 Herbert D. Ward).
Mainstone's housekeeper. ...Eliza Meteyard. (Pseud., Silver Pen.)
Manchester strike. ...Harriet Martineau.

WORKING CLASSES — *Continued.*

Marcella. ...Mary Augusta Ward (Mrs. Humphry Ward).
Mark Dennis; or, Engine-driver. ...M. C. J. G.
Mary Barton: a tale of Manchester life. ...Elizabeth C. Gaskell.
Master of his fate. ...Amelia Edith Barr.
Memoirs of a working man. ...Thomas Carter.
Mills of Tuxbury. ...Virginia F. Townsend.
Nicholas Minturn. ...Josiah Gilbert Holland.
North and South. ...Elizabeth C. Gaskell.
Palissy, the Huguenot potter. ...Cecilia L. Brightwell.
Prince Dusty: a story of the oil regions. ...Kirk Munroe.
Probation. ...Jessie Fothergill.
Put yourself in his place. ...Charles Reade.
Queen Philippa and the hurrer's daughter. ...Elizabeth M. Stewart.
Reuben Foreman, the village blacksmith. ...Darley Dale.
Rival apprentices. ...Anon.
Rival houses of Hobbs and Dobbs. ...Crotchet Crayon, pseud.
Sabaris and other homes. ...Edward Everett Hale.
Seed she sowed: a tale of the great dock strike. ...Emma Leslie.
Seed time and harvest. ...Fritz Reuter. English translation.
Sir George Tressady. ...Mary Augusta Ward (Mrs. Humphry Ward).
Sketches of working women. ...Ellen Barlee.
Social crime. ...Minnie L. Armstrong and G. N. Sleets.
Stone pastures. ...Eleanor Stuart.
Susannah. ...Mary E. Mann.
Tale of Lowell. ...Argus, pseud.
That lass o' Lowrie's. ...Frances Hodgson Burnett.
Through the fray: a tale of the Ruddite riots. ...George Alfred Henty.
Ut mine Stromtid. ...Fritz Reuter.
Wenderholm: a story of Lancashire and Yorkshire. ...Philip Gilbert Hamerton.
Yeast: a problem. ...Charles Kingsley.
Yekl: a tale of the New York Ghetto. ...A. Cahan.
See also **LABOR AND CAPITAL, CONFLICT OF.**

WORLD'S COLUMBIAN EXPOSITION. *Held at Chicago, 1893.*

Against odds: a romance of the Midway Plaisance. ...E. Murdock Van Deventer. (Pseud., Laurence L. Lynch.)
Samantha at the World's Fair. ...Marietta Holley. (Pseud., Josiah Allen's wife.)
Sweet clover: a romance of the White City. ...Clara Louise Burnham. (Pseud., Edith Douglas.)

WYCLIFFE, JOHN (*lived 1324-1384*), **AND WYCLIFFITES.**
Before the dawn: a story of Paris and the Jacquerie. ...George
Dulac.
Conrad. ...Emma Leslie.
Coulying castle; or, Knight of the olden days. ...Agnes Giberne.
Dearer than life: a tale of the times of Wycliffe. ...Emma Leslie.
For or against? ...Francis M. Wilbraham.
Geoffrey the Lollard. ...Mrs. D. C. Knevels. (Pseud., Frances
Eastwood.)
Gilbert Wright, the gospeller. ...F. S. Merryweather.
Gladys of Harleck. ...Anon.
Hubert Ellerdale. ...W. O. Rhind.
In Wycliffe's days. ...G. Stebbing.
Jack of the mill. ...William Howitt.
John de Wycliffe, the first of the reformers. ...Emily Sarah Holt.
Knight of Dilham: a story of the Lollards. ...Arthur Brown.
Lollard. ...Minnie K. Davis.
Lollard priest. ...Henry C. Adams.
Lollards. ...Thomas Gaspey.
Margary's son: a fifteenth-century tale of the court of Scotland.
...Emily Sarah Holt.
Mistress Margery: a tale of the Lollards. ...Emily Sarah Holt.
Richard de Lacy: a tale of the later Lollards. ...Charles E.
Maurice.
Richard Hunne: a story of old London. ...George E. Sargent.
White rose of Langley: a story of the court of England in the olden
time. ...Emily Sarah Holt.
Wycliffites; or, England in the fifteenth century. ...Margaret
Mackay.

YELLOWSTONE PARK, U. S. A.
Pursued. ...William J. Gordon.
Three Tetons: a story of the Yellowstone. ...Alice Wellington
Rollins.

YOSEMITE VALLEY, CALIFORNIA.
Tišayac of the Yosemite. ...Mary B. M. Tolland.
Zanita: a tale of the Yo-semite. ...Maria Thérèse Longworth
Yelverton.

YOUNG PRETENDER'S REBELLION. See **REBELLION, Young
Pretender's.**

ZOÖLOGY.
Adventures of a young naturalist. ...Lucien Biart. English
translation.
Animals and their social powers. ...M. T. Andrews.

ZOÖLOGY — Continued.

Aventures d'un jeune naturaliste. ...Lucien Biart.

Beautiful Joe. ...Howard Pyle.

Big cypress. ...Kirk Munroe.

Black Beauty. ...Anna Sewall.

Bruno; or, Lessons of patience, fidelity, and self-control taught by a dog. ...Jacob Abbott.

Cat of Bubastes. ...George Alfred Henty.

Diamond lens. ...Fitz James O'Brien.

Dog of Constantinople. ...Izora C. Chandler.

Dog stories. ...From the Spectator, selected and arranged by J. Saint Loe Strachey.

From monkey to man; or, Society in the Tertiary age. ...Austin Bierbower.

Horned cat. ...J. Maclaren Cobban.

Hunter cats of Connorloa. ...Helen Hunt Jackson.

Jungle-book. ...Rudyard Kipling.

Kentucky cardinal. ...James Lane Allen.

Leila; or, Island. ...Anna Fraser Tytler.

Letters from a cat. ...Helen Hunt Jackson.

Live toys: anecdotes of our four-legged and other pets. ...Emma Davenport.

Mammy Tittleback and her family. ...Helen Hunt Jackson.

Old rough the miser. ...L. F. Wesselhoeft.

Our dumb companions. ...Thomas Jackson.

Our dumb neighbors. ...Thomas Jackson.

Paddy O'Leary and learned pig. ...Elizabeth W. Champney.

El pájaro verde. ...Juan Valera.

Peter: a cat o' one tail. ...Charles Morley.

Public and private life of animals. ...Honoré de Balzac. English translation.

Quickening of Caliban. ...J. Compton Rickett.

Rab and his friends and other dogs and men. ...John Brown.

Scènes de la vie des animaux. ...Honoré de Balzac.

Story of a shell. ...John R. MacDuff.

Toilers of the sea. ...Victor Hugo. English translation.

Travailleurs de la mer. ...Victor Hugo.

Under the lilacs. ...Louisa May Alcott.

Water babies. ...Charles Kingsley.

ZULULAND.

Description, History, Manners, and Customs.

Gun-runner. ...Bertram Mitford.

Hendricks, the hunter: a tale of Zululand. ...William Henry Giles Kingston.

ZULULAND — *Continued.*
Lost heiress. ...Ernst Glanville.
Luck of Gerard Ridgley. ...Bertram Mitford.
Nada, the lily. ...H. Rider Haggard.
Quickening of Caliban. ...J. Compton Rickett.
Ula in veldt and laeger. ...Charles H. Eden.
Virgil. ...Charles Montague.

FINIS.

www.ingramcontent.com/pod-product-compliance
Lightning Source LLC
Chambersburg PA
CBHW032305280326
41932CB00009B/706